Abominable Snowmen:
LEGEND COME TO LIFE

OTHER BOOKS BY IVAN T. SANDERSON

ANIMAL TREASURE
LIVING TREASURE
CARIBBEAN TREASURE
FOLLOW THE WHALE
THE CONTINENT WE LIVE ON
ANIMAL TALES, AN ANTHOLOGY

Reference
HOW TO KNOW THE NORTH AMERICAN MAMMALS
LIVING MAMMALS OF THE WORLD
THE MONKEY KINGDOM

Juveniles
ANIMALS NOBODY KNOWS
THE SILVER MINK
JOHN AND JUAN IN THE JUNGLE

Abominable Snowmen:
LEGEND COME TO LIFE

*The Story of Sub-Humans
on Five Continents
from the Early Ice Age
Until Today*

IVAN T. SANDERSON
F.L.S., F.R.G.S., F.Z.S.

Illustrated

CHILTON COMPANY · BOOK DIVISION
Publishers
PHILADELPHIA AND NEW YORK

Published in Philadelphia by Chilton Company,
and simultaneously in Toronto, Canada,
by Ambassador Books, Ltd.

Library of Congress Catalog Card Number 61-10907
Designed by William E. Lickfield
Manufactured in the United States of America
by Quinn & Boden Company, Inc., Rahway, N. J.

DEDICATION

To BERNARD AND MONIQUE HEUVELMANS
and
MY OWN ALMA
And also to the Following

Today finds a surprising host of assorted students in this odd field, but
also a few professional scientists whose labors I would like first to note,
at the same time thanking them for their long-standing encouragement,
constructive criticism, and many forms of direct help, not only in this
book but also in my other studies of similar matters. In addition to Dr.
BERNARD HEUVELMANS, who has become the doyen of the whole business,
these are most especially Professor W. C. OSMAN HILL, presently Pro-
sector of the Zoological Society of London; Professor GEORGE A. AGOGINO,
Assistant Professor of Anthropology at the University of Wyoming; Professor TEIZO OGAWA, Department of Anatomy, University of Tokyo; Professor B. F. PORSHNEV of the Academy of Sciences of the U.S.S.R.;
Professor CORRADO GINI, President of the Institut International de Sociologie, Rome, Italy; and Dr. JOHN NAPIER, of the Royal Free Hospital
School of Medicine at the University of London, England. Dr. WALDIMIR
TSCHERNEZKY, of Queen Mary's College, London, has lent me much invaluable advice; and Dr. JORGE IBARRA, Director of the National Museum
of Guatemala, has pursued more specific details for me in his country.

There is, then, another category of students not primarily engaged in
scientific pursuits but without whose labors little would be known about
this subject, and without whose generous help this book could not have
been written. This class is headed by TOM SLICK, of San Antonio, Texas,
whose work is more fully acknowledged in the course of my story. Next,
J. W. BURNS of San Francisco, who has spent over half a lifetime in pursuit of the *Sasquatches*, and JOHN GREEN, newspaper publisher of Agassiz, B.C., on whose shoulders Mr. Burns' mantle has fallen. Then, there
is my old school friend, W. M. (GERALD) RUSSELL, and PETER BYRNE,
who separately and together did so much to clarify ABSMery in the
Himalayan region. In the same class is my friend and associate, KENNETH
C. (CAL) BROWN.

In still another category is a devoted and more or less dedicated little
band of my immediate associates. Foremost is my wife, who has worked
with me for over a quarter of a century—in the field, in my researches,
and on all my books—doing much more than merely typing and collating
roomfuls of material.

Next, I would like to acknowledge two of the most remarkable young
men I have had the pleasure and honor of meeting in scholarship—Rabbi
YONAH N. IBN AHARON and UMBERTO ORSI. Yonah is the recipient of
degrees from the University of Yemen and a philologist of remarkable
knowledge and talents, accredited to the U.N., who obtained his M.A.
degree upon production of the first (and only) Basrai Aramaic Lexicon.

[v]

He is, as detailed later, conversant with all the basic dialects upon which the larger number of languages of eastern Eurasia are today founded.

Umberto Orsi has given me vast assistance via his specialty, bibliographical research. He is not just a literary sleuth, but a true bloodhound when it comes to rescuing rare items from the mazes of modern libraries. Without his invaluable assistance I would not have dared to issue this work. Then, there is JOHANNA LINCH, who somehow reproduced all my maps, outside of office hours, in just two weeks. Then, too, our good friend, RAIZEL HALPIN, who gave great help on the manuscript, merely out of kindness and her interest in the subject.

There come next three new friends who have given their own particular technical skills to immeasurably further this work, and I don't quite know how to thank them. They are, first, LJUBICA POPOVICH and BENJAMIN ROTHBERG, both of Philadelphia, who translated some hundred thousand words of technical material from Russian originals of hitherto unpublished publications of the Special Commission of the Soviet Academy of Sciences. Coming after these two stalwarts was ETHEL WAUGH, who transcribed their translations from tape recordings—including place names in goodness knows how many languages. To all of these, and particularly to Ben Rothberg upon whom the greatest onus devolved, I hereby give my sincerest thanks. Actually, these three together accomplished a work of considerable significance to anthropology, which will, I hope, soon see the light of day in complete and technical form.

I would like to say, also, that I have been the recipient of splendid guidance and encouragement from the Chilton Company—Book Division, both as a whole and from all its departments. They have kept a fine old publishing tradition in a bright new setting—a novel experience, and a most delightful one to a latter-day writer who has seldom enjoyed such co-operation in the past.

Finally, there is another army of good people, many named in the body of the story but many more are not named, who have furthered the cause of ABSMery generally by coming out with their own stories in face of ridicule and censure so extreme as sometimes to have resulted in loss of their jobs. These people are pioneers—if not, on occasion, actually martyrs —in their pursuit of truth and the disproof of "official" mendacity, prejudice, and stupidity. I can only pray that one day their fortitude will be rewarded with full popular and scientific recognition.

IVAN T. SANDERSON

Foreword

The possible existence of the Yeti, Sasquatch, and other Abominable Snowman forms has long been a point of conjecture among travelers, naturalists, and scientists. While most of this evidence is circumstantial and inconclusive as yet, it provides a tantalizing mystery filled with enough interest and promise to warrant the attention of both serious students and casual readers.

In this book, Ivan T. Sanderson summarizes current world evidence regarding ABSMs (abominable snowmen), drawing from records and reports that are world-wide in scope and cover a broad period of time. For completeness he discusses all prevailing views, both pro and con, ranging from highly plausible accounts to reports that border on the absurd. The result is as thorough an evaluation of all known ABSM sightings as could possibly be compiled at this time.

My own approach to the ABSM problem was one of extreme skepticism. Three years ago I dismissed all such evidence as either hoax or legend, and in hopes of a confirmation of this viewpoint served as coordinator of laboratory research for several "abominable snowman" expeditions into the Himalayas. Today my skepticism is somewhat shaken, and I accept as plausible, perhaps even probable, the existence of the Yeti in the Tibetan plateau and view with growing interest the "global" sightings of similar creatures.

Since my own research has been in connection with the Himalayan Yeti, I will restrict my comments to this area alone. If I accept the results of serological tests, analysis of faeces for content and parasites, examination of hair, hide, and tracks and evaluation of mummified Yeti shrine items, then I must support the existence of a large unknown animal, the Yeti, in the Himalayas. However, the following question once disturbed my acceptance of this conclusion. Is it possible for any large animal to be sought systematically for over a decade without a single specimen being captured or killed?

For an example bearing on this question, I return to the Tibetan plateau. Here in Western Szechwan, China, on the very edge of the Tibetan border, a large animal, the Giant Panda, was once hunted unsuccessfully for over seventy years before one was captured alive. This search proves that a large animal can exist yet elude the best efforts of professional collectors to secure one. The story behind this hunt is fascinating.

In 1869, Abbé Armand David, a noted French missionary, observed a strange bear-like skin in Szechwan province located on the edge of the Tibetan plateau. This skin, much like that of a modest-size black and white bear, was the first tangible proof that the Bei-Shung (white bear) of Szechwan did actually exist. Excitedly, Father David, a long-time naturalist and conservationist, traveled to this animal's reported habitat, a high mountain bamboo forest, and engaged local hunters to secure a living specimen. In twelve days they returned. The hunters had captured a living Giant Panda, but since the animal proved troublesome in traveling, it was dispatched to make transportation more convenient. Although Father David was disappointed that he had failed to secure a living animal, he shipped the remains to the Paris Museum, providing the first tangible evidence that the "legendary" Bei-Shung actually existed and could be caught in the Szechwan bamboo forests.

Captivated by such evidence, several scientific institutions supported field teams staffed by professional collectors. The world waited to see which of several well-equipped expeditions to Szechwan would capture the first living specimen. This was in 1869. By 1900 the world was still waiting. Scientific interest was great, for the once mythical Bei-Shung had been given the scientific name, *Ailuropoda melanoleucus,* and a separate family of its own. In spite of professional excitement, no new Giant Pandas were even seen until 1915, and no new remains were obtained until 1929 when two sons of President Roosevelt, Theodore, Jr., and Kermit, shot one out of a hollow pine tree. By this time most zoologists had decided that the Panda was extinct, so that the Roosevelt shot, while killing a Giant Panda, at the same time punctured several scientific egos.

Assured that the Giant Panda was not extinct, several new expeditions were outfitted. Each contributed to the threat of extinction by shooting Giant Pandas, but living animals still defied capture. In 1931 a specimen was shot for the Philadelphia Academy of Natural Sciences, and in 1934 another was killed for the American Museum of Natural History. Two other specimens were killed, one by Captain Brocklehurst in 1935 and the second by Quentin Young in 1936. In 1936 Floyd T. Smith managed to get a Giant Panda as far as Singapore before it died of natural causes. Finally, an inexperienced woman collector, Ruth Harkness, succeeded where the others had failed by capturing two live specimens, the first in 1937 and the second in 1938. Both animals survived the trans-Pacific trip and were sent to the Brookfield Zoo in Chicago. Within months the animals had captured the imagination of American youngsters, and stuffed Panda Bears are still considered a necessary part of college dormitory life.

In retrospect, the hunt for the Giant Panda serves as an important lesson in regard to animal collecting. From 1869 until 1929, a period of sixty years, a dozen well-staffed and well-equipped professional zoological collecting teams unsuccessfully sought an animal the size of a small bear in a restricted area. During this time not a single specimen living or dead was obtained. The lesson is clear. The Giant Panda lives in the same general area and at the same general elevation (6,000–12,000 feet) as the Yeti, yet this animal remained hidden for over sixty years. The Yeti can well be a similar case. At any rate, one can no longer dismiss the Yeti just because it has eluded moderate search for a single decade.

While admittedly no living Giant Panda was captured during an intensive seventy-year search, several animals were killed by gunfire during the last few years (1929–1936) of that period. Why don't we have similar reports of Yeti killings? The truth is we do, but for the most part these reports come from behind the Communist curtain and cannot be substantiated. Nepal is the only country in the Free World with the Yeti ABSM form, and here killing a Yeti is a criminal offense with severe penalties. As a result, violators remain secret and reports are all but impossible to trace.

I have been asked if it is possible for modern science, fortified by great improvements in world transportation and communication, to miss completely authentic reports on the Yeti, if indeed such reports exist. It can be understood how the Bei-Shung could be mentioned in a seventh-century A.D. Chinese manuscript yet not be seen by any outsider until some 1200 years later. This was a period of an isolated and mysterious Far East—the land of the dragon, Shangri-La, the Great Wall, and the unknown oriental mind. The period from 1869 to 1929 was only relatively more progressive. Look how transportation has reduced our world since the time of the Model A Ford and the *Spirit of Saint Louis*. Look how communication has improved since the megaphone of Rudy Vallee and the early "talking pictures." Today our world is much smaller and nothing seems isolated any more. Could we find a case similar to the search for the Giant Panda which has occurred in more recent times?

Such a case would be the discovery of living Coelacanths in the Indian Ocean. Fossil remains of Coelacanth fish forms have been found in rocks of the Devonian Period some three hundred million years ago and up to the end of the Cretaceous Period sixty million years ago. No fossilized remains have been found in more recent deposits, and it was assumed that the Coelacanth died out at this time. Fossil Coelacanths were a most unique form of life as they lived in several different aquatic environments. Their fossilized remains have been found under conditions that indicate that the living fish could be found in both salt and fresh water, including rivers, lakes, and even swamps. In addition to a diverse habitat, these fish had a world-wide distribution. It now seems indeed strange that no remains have been found of this fish in rocks of the past sixty million years, for there is no doubt that this fish never became extinct and in fact exists in fair numbers today.

In December, 1938, a specimen of the "long extinct" Coelacanth was found in the fishnet of a British trawler working off the coast of East London in South Africa. Caught alive, the huge fish rolled steel blue eyes and waddled about the ship deck on clumsy fins that were used like stubby legs. The fish bit the inquisitive captain and oozed oil from its heavy scales for three hours before dying. Identified only after decay had rendered the fleshy parts useless for scientific purposes, it proved to be a heavy disappointment for ichthyologist James Smith of Rhodes University, Grahamstown, S.A. Fossil remains show skeletal structure, and the importance of the recent "catch" lay in the chance to study the unknown fleshy parts of the fish. Now this was impossible. Professor Smith realized that, if one such fish existed, others similar to it must also exist, and he began a fifteen-year search for a second living Coelacanth. For the next decade and a half he visited islands and coral reefs in the West Indian Ocean, asking, looking, fishing. Finally, in December, 1952, a fishing trawler off the Anjouan and Comoro Islands between Madagascar and the mainland of Africa caught another Coelacanth. Prompt action by ichthyologist Smith allowed him to obtain and preserve this specimen in excellent shape. Then came the big shock. For fourteen years he had tracked down all leads, talked to countless fishermen, without avail. Now within the next two years, three more Coelacanths were obtained, and there were indications that the native population in this part of the world had fished for and eaten these "living fossils" for several generations. Although not a common item in native diets, there is no doubt that, while Professor Smith dreamed of finding a second Coelacanth, a dozen or more had probably been served and eaten.

Here was an example where science, with all its modern improvements in communication and transportation, was unaware that what was to be one of the great "discoveries" of the twentieth century had long been a simple item of diet for the native population. Even Professor Smith, active in the area and specifically after a Coelacanth, was caught unaware. But who would think of looking in a fish market for a "living fossil" like a Coelacanth?

For a final illustration, let me turn to my own field of archeology. Prior to 1926, the general belief was that the American Indian was post-glacial in age, and as a consequence glacial strata were rarely examined by professional archeologists. The few archeologists who claimed to find cultural evidence were criticized for their ineptitude and then quickly dismissed. In 1846 a human pelvis was found with several ground sloth skeletons in Mammoth ravine near Natchez, Mississippi. Before the century ended, positive association was demonstrated by fluorine tests, yet not only was the discovery disregarded, but the actual bones were lost and the incident forgotten. All other finds met with a similar fate until the discovery in 1926 of the unique Folsom projectile points with the extinct glacial *Bison antiquus* near Folsom, New Mexico. In three years' research, nineteen Folsom points were found in direct association with twenty-three extinct bison, and the antiquity of the Paleo-Indian was firmly established. Now the long-neglected glacial strata were examined. Archeologists looked for additional Folsom sites wherever man, wind, or weather had scarred the surface of the land, exposing the glacial earth levels to the human eye. Within a decade of the Folsom, New Mexico, discovery, Paleo-Indian sites were found from Alaska to Patagonia and from coast to coast. These sites had been exposed to the eye of man for decades, but they were only found AFTER man was convinced that Ice Age Indians actually existed. *Again it shows that man must believe before he looks, and must look before he finds anything.* Important things may be all around us, but we will never find them unless we look for them. Perhaps one reason why we haven't more definite information on ABSMs is because not enough people have actually looked for ABSMs long enough or with enough dedication.

GEORGE A. AGOGINO
Assistant Professor of Anthropology
University of Wyoming

Contents

[xiii]

List of Maps

List of Maps

Abominable Snowmen:
LEGEND COME TO LIFE

1. A Certain Unpleasantness

Upon the detection of an unpleasant odor most people move off, while everybody wishes that it would go away. Nobody wants it around, yet it is seldom that anybody tries to determine its origin.

In 1887, a major in the Medical Corps of the British Indian Army, Lawrence Austine Waddell, LL.D., C.B., C.I.E., F.L.S., F.A.I.—i.e. Doctor of Laws, Commander of the Bath, Commander of the Indian Empire, Fellow of the Linnean Society, Fellow of the Anthropological Institute—was meandering about in the eastern Himalayas doing what that rather remarkable breed of men were wont to do: that is, a bit of shooting, some subdued exploring, and a certain amount of "politicking." Like many others of his ilk, he wrote a somewhat uninspired and uninspiring book about it, uninspiringly named *Among the Himalayas*. The Major was a normal sort of chappie and a sportsman, but his hunting was not of the feverish ninety-one-gun-in-closet variety of today; quite the contrary, he would take a few birds of types he considered to be legitimate game for his pot or to keep his eye in for grouse-shoots on his next home-leave in Scotland, and he banged away at "tygarr" whenever the local natives could rustle one up. But he was not scrambling about the Himalayas primarily for what we nowadays call "sport." He was just puttering—that lost 19th-century British art—because he had some time off, and official sanction to make use of it as he would.

Despite the limited intelligence attributed to 19th-century British-Indian Army colonels, they were really a most remarkable breed—almost a mutation—for, from some hidden depths of their public-school educations, and the remoter recesses of their ancient family traditions, they dredged up a wealth of wisdom, and they often developed an extraordinarily keen interest in the world about them wherever they happened to land. Most of them were sort of mild philosophers; many turned out to be brilliant linguists and great scholars; and they were often both leaders of men and students of animal life. They have been grossly maligned by almost everybody, laughed at as super-Blimps, and neglected as historians. But if you will just read their maunderings carefully, you will garner therefrom a trove of both literary and factual gems.

Take this Major Waddell, for instance. While pounding over one of the unpleasanter bits of Sikkim, in vile weather, he came upon a set of tracks made by some creature walking on two legs and bare feet that, he says, went on and on, over the freezing snow, not only taking the line of least resistance at every turn but marking out a course in conformity with the easiest gradients that brought whoops of admiration even from the Major's mountain-born porters. He remarks almost casually upon this remarkable achievement and wonders vaguely not what manner of man, but *what sort of creature* could have made them, and why it should have decided to cross this awful pass in the first place. The Major did not realize when he penned this thought just what he was starting; though "starting" is perhaps not the exact word to describe his remarks, for what he recorded was already ancient history when Columbus sailed for the West Indies. It just so happens that, as far as popular recognition is concerned, his was one of the earliest mentions to appear in print in the English language, in what may be called modern times, of what has latterly become known as the "abominable snowman."*

* You will find that, by the time I have said all that I am able to say within the compass of this book, there remain only two sets of evidence for the existence of ABSMs. One is subjective—i.e. reports; the other objective—i.e. tracks. All the other evidence, and of all kinds—such as scalps, hairs, excrement, myths, legends, folklore and so forth—may be questioned and often seriously on one ground or another. The one item that both protagonists and skeptics have to explain is tracks. They happen also to be both the commonest items in ABSMery, and the ones most readily recorded and analyzed. The study of foot-tracks is called Ichnology and the principles of this, together with its particular reference to our subject, will be found in Appendix B.

At that time nobody in what we now call the Western World paid the slightest attention to this extraordinary report —at least as far as we know. It just went into the record as a statement; for one could hardly, in that day and age, call any pronouncement on the part of anybody with such notable honors a lie, or even a "traveler's tale." It was therefore assumed that some religious chap must have preceded the gallant Major over that particular route and somehow managed not to die of frostbite, sun-blindness, or starvation; and it was remarked that he had done a dashed good job of negotiating the pass. There the matter rested.

Major Waddell's book was one of many written about the end of the last century when the Western World was complacently sure that it knew more or less everything about all countries, with the possible exceptions of Tibet and the holy city of Mecca which, it was then considered, were rather unsporting in that they did not welcome civilized Englishmen. All sorts of sporting gentry went wandering about the fringes of "The Empire" with rod and gun and later wrote about their experiences. Their effusions were read by both the previous and the upcoming generations of colonial pioneers, but by few others. What they said was not taken too seriously by the general, nonempire-building public. However, many of these gentry also submitted official reports on certain less publicized aspects of their activities to their superiors; and these were taken very seriously.

Unfortunately the great body of such reports are not published and many of them are either lost in some archive or truly lost forever. There are others that are still top-secret and unavailable, so that their very existence is often conjectural. Yet every now and then one stumbles upon such a report that is extremely tantalizing. Tracking down the original is a frightful chore and one of the most time-consuming and frustrating experiences. One is balked at every turn but not, I would stress, by any deliberate or organized defense on the part of authority. Official archives are preserved for the benefit of all and are open to inspection by all, and even the topmost secrets are in time released as mere historical dejecta. The trouble is simply that the original reporters, and more so those reported to, did not lay any store by or place any specific value on esoterica, or anything other than the primary matter at hand, which was often of a diplomatic or political nature, so that the items that interest us most were never indexed or catalogued. You just have to plow through mountains of material quite extraneous to your particular quarry and hope to stumble upon casual asides that are pertinent to it. But one does occasionally so stumble.

Now I should state, without further ado and quite frankly, that I am prejudiced in favor of official as opposed to any other form of reports and for the following reasons. In this country we do not, let's face it, have much respect for the law or its potential until we have recourse to it or it requires our submission. Until we have been on a witness stand, almost all of us believe that perjury—which is simply a legal term for lying in the law's presence—should be the easiest thing in the world, but even those of us who say that laws are made only to be broken, soon find that it is not. Few think twice about telling a fish story in the corner bar, but there are very few, even congenital idiots, who won't think before telling it in a court of law. When, therefore, somebody voluntarily makes an official statement, when there is no profit motive involved, I have always felt it reasonable to assume that it is quite likely true. The British happen to have a particular respect for their law, and British officialdom, despite what has been said about its colonial policies, has always been remarkably altruistic. British consuls and other officials just did not report a lot of rubbish to their service headquarters. Even paper was scarce in minor British outposts and the field officers did not clutter up essential reports with bizarre trivia unless they considered them to be of real import. We approach, therefore, the following official report with a certain quota of awe.

It appears that in 1902 British Indian officialdom was concerned with the stringing of the first telegraph line from Lhasa, the capital of Tibet, to Kalimpong, Darjeeling in Bengal Province of India just south of the Sikkim border (see Map).

The job entailed, first, going into Tibet and then stringing the cable out. When the crew reached a pass named Chumbithang near a place called Jelep-La on the Tibet-Sikkim border, an incident occurred that prompted an official report. A dozen workers failed to return to camp one evening and a military posse was sent next day to search for them at the scene of their operations. No trace of the missing men was found, but the soldiers during their wide search for them found a remarkable creature asleep under a rock ledge—or so the report goes. The soldiers were Indians, not Ghurkhas or mountain folk, and this is of significance because had they been they would doubtless have acted differently. The Indians had no qualms about shooting this creature to death immediately. It proved to be human rather than animal in form, though covered with thick hairy fur. Up to this point the report is official. Then it becomes unofficial but for one minor aside to the effect that a full report, together with the beast, was shipped to the senior British political officer then resident in Sikkim, who is correctly named as one Sir Charles Bell.

The unofficial sequence I take from an extraordinary book only recently published by a Mr. John Keel entitled *Jadoo*. This is the more startling in that it even mentions an incident apparently lost and certainly forgotten over half a century before, yet states that the information therein given was obtained firsthand. The author states that he met in 1957 in Darjeeling a retired Indian soldier named Bombahadur Chetri, who claimed that he was among the party that killed this creature, and that he personally examined it. He is also alleged to have said that it was about 10 feet tall, covered with hair but for a naked face, and that it had "long yellow fangs." Further, Mr. Keel says that Bombahadur Chetri told him that the carcass had been packed in ice and shipped to this same Sir Charles Bell, but that he did not hear anything further of it. Nor, apparently, did Mr. Keel; and nor have I, though I have spent a lot more time and energy than the item might seem to warrant in a fruitless endeavor to trace further reports, official or otherwise. This is the more aggravating since it is the earliest report that I have found on the actual (or even the alleged) capture of any form of what we shall henceforth be calling an ABSM—i.e. "the abominable snowmen," by what we must, also for lack of any established over-all name, call the "Western World," in the Oriental Region.*

Nevertheless, it is by no means the only such report, nor actually the earliest on record, for as we shall presently see, it was preceded in two if not three other continents by just as definitive statements and in some cases official ones at that.

And this brings up another point that I should endeavor to clear up forthwith.

I would have preferred to start this story where all stories should begin, which is to say at the beginning. However, despite a chronology that I have compiled over the years, such a procedure would be open to at least two serious defects. First, it is almost daily, and now with increasing tempo, being added to almost all along the line, while its origins are regressing ever farther into the recorded past; second, it would be extraordinarily dry and overformal in the eyes of any but extreme specialists. I have felt, therefore, that the history of this whole ABSM business will be much better understood if it is unfolded upon the chronology of *its* discovery and progress: a sort of history of a history. This is, further, herein recorded deliberately from what we called above the "Western" point of view, in that it is a chronological record of how the matter was brought to the attention of the Western World. In this, it will soon be seen that a greater part of the discoveries made have come to light in reverse. For instance, it has only been within most recent years that the earliest accounts have come to light, and the further research workers probe into the whole matter, the farther back the origins of the whole ABSM affair recede, while the wider does their distribution

* The term "ABSM" is coined from the best-known name for one kind of those creatures of which we speak, namely the Abominable Snowman. As is explained later, this term is incorrect, inappropriate, and misleading even in the case in which it was first applied; while it cannot possibly be applied to at least 80 per cent of the apparently most varied and quite different creatures involved, and now reported from five continents. The term "Western World" in this case has a cultural rather than a regional sense; but by the Oriental Region is to be understood a very precise geographical unit, as is explained.

[6]

become both in fact and in report. Thus, in treating of the history of this matter, we must bear in mind that what appear to us to be discoveries are more nearly revelations, because the majority of the world—which is, of course, *non*-Western—has, to some degree or another, known all about the business for centuries, while we have remained completely oblivious of and to it.

For these reasons, I divide our chronology into five stages and call these as follows: (1) the ancient period, prior to the 15th-century expansion of Europe, (2) the dark ages, from 1500 to 1880, (3) that of the Explorers, from about 1880 to 1920, (4) that of the mountaineers, 1920 to 1950 and (5) that of the searchers, from 1950 to the present day. All of this, however, applies primarily and most essentially to the Himalayan area of the Oriental Region wherein this business was primarily unfolded for us. The same periods, of course, exist in time elsewhere, such as North America, but they cannot be founded on the same criteria or named after the same classes of entrepreneurs. Behind this chronology and everywhere lies an immense period of what I call native knowledge. This trails off into the dim mists of the extreme past and into folklore and myth; an area which is only just now being taken into account as serious history rather than mere make-believe. Thus, in other parts of the world our story has often jumped straight out of the "native" period into that of scientific study.

While ABSMs were not only reported but also reported upon, and even officially, in other parts of the world—vide: Canada—long before the travels of Major Waddell, and while specimens (as it now turns out) are alleged to have been captured or killed long before that, we of the West became cognizant of these happenings or alleged happenings only very recently. Also, it now transpires, detailed and more properly critical information on the subject was even being published in eastern Eurasia centuries ago—for instance in Tibet, China, Mongolia, and Manchuria—and some reflections of this had filtered through to Europe as early as Renaissance times, as is exemplified in certain curious statements in the works of Marco Polo. Millions of people were then taking all this as a matter of course but, the whole thing being completely foreign to European conditions or even thought, it made no impression upon what we now call the Western World until our fourth period—namely that of the mountaineers.

Just how foreign it was prior to that period is clearly demonstrated by the reception, or lack of it, given to a report published in a scientific journal (*Proceedings of the Zoological Society of London*) in the year 1915, and the brief comments upon it made at the time. The report was read before the society by a very well-known botanist and scientific explorer named Henry J. Elwes, and consisted of portions of a letter received by that gentleman from a Forestry Officer by the name of J. R. O. Gent who was stationed in Darjeeling. This read as follows:

I have discovered the existence of another animal but cannot make out what it is, a big monkey or ape perhaps—if there were any apes in India. It is a beast of very high elevations and only goes down to Phalut in the cold weather. It is covered with longish hair, face also hairy, the ordinary yellowish-brown colour of the Bengal monkey. Stands about 4 feet high and goes about on the ground chiefly, though I think it can also climb.

The peculiar feature is that its tracks are about 18 inches or 2 feet long and toes point in the opposite direction to that in which the animal is moving. The breadth of the track is about 6 inches. I take it he walks on his knees and shins instead of on the sole of his foot. He is known as the *Jungli Admi* or *Sogpa*.* One was worrying a lot of coolies working in the forest below Phalut in December; they were very frightened and would not go into work. I set off as soon as I could to try and bag the beast, but before I arrived the Forester had been letting off a gun and frightened it away, so I saw nothing. An old choukidar of Phalut told me he had frequently seen them in the snow there, and confirmed the description of the tracks.

It is a thing that practically no Englishman has ever heard of, but all the natives of the higher villages know about it. All I can say is that it is *not the Nepal Langur,* but I've impressed upon people up there that I want information the next time one is about.

This report, which would today probably cause quite a stir in certain circles, though for various and quite opposed reasons,

* This is also the name of a known tribal group of people in a remote valley of the Himalayas. (For fuller details see Chapter 19.)

seems hardly even to have been commented upon. It would probably have been dismissed altogether—and, most likely not published in the *Proceedings*—had it not been read by such a person as Elwes. As it was, the general impression left was that perhaps a new species of monkey had been found and some local folklore embellished. But, unexpectedly, Henry Elwes then saw fit to make a statement of his own to the effect that in 1906 he had himself seen the same or a similar creature in another part of the Himalayas. Most aggravatingly, he either did not give further details or they were not recorded at the time, and after he died his notes were lost while no mention of the incident was to be found in any of his published writings. Zoologists were apparently quite impressed at the time because of the standing of Elwes, but the matter never got further than the closed confines of professional zoology.

It was, moreover, not until 1920 that the English-speaking public, outside of the limited audience earlier served by the writings of travelers in the Orient, was in any way made aware of this whole business, and, as is so often the case, it was even then more by accident than by design. This part of our story is most intriguing as well as being a sort of turning point in Western thinking, and not only upon this but upon many other matters. But before telling you the details of this little comedy, I just want to diverge a moment to impress upon you once again the fact that what then took place, while a revelation, was more particularly so to the Anglo-Saxon world. A decade before (1907), a certain then young zoologist named Vladimir A. Khakhlov started an extended survey of similar matters throughout central Eurasia and submitted a long report on it to the Imperial Academy of Sciences in Russia; Netherlands authorities had been pestered with annoying (to officialdom) reports of a like nature emanating from Sumatra; the French had undergone the same in Indo-China; and the Brazilians in their country; while even in British Columbia both the courts and the Crown itself had long been bothered by citizens seeking to make depositions on closely related matters. Thus, in retrospect, the happenings of 1920 lose a great deal of their import if not of their impact.

[9]

In that year an incident occurred that was impressive enough but which might have been either wholly or temporarily buried had it not been for a concatenation of almost piffling mistakes. In fact, without these mistakes it is almost certain that the whole matter would have remained in obscurity and might even now be considered in an entirely different light or in the status of such other mysteries as that of "seamonsters." This was a telegram sent by Lt. Col. (now Sir) C. K. Howard-Bury, who was on a reconnaissance expedition to the Mt. Everest region.

The expedition was approaching the northern face of Everest, that is to say from the Tibetan side, and when at about 17,000 feet up on the Lhapka-La pass saw, and watched through binoculars, a number of dark forms moving about on a snowfield far above. It took them some time and considerable effort to reach the snowfield where these creatures had been but when they did so they found large numbers of huge footprints which Colonel Howard-Bury later stated were about "three times those of normal humans" but which he nonetheless also said he thought had been made by "a very large, stray, grey wolf." (The extraordinarily illogical phrasing of this statement will be discussed later on, but it should be noted here that a large party of people had seen several creatures moving about, not just "*a* wolf," and that it is hard to see how the Colonel could determine its color from its tracks.) However, despite these expressions, the Sherpa porters with the expedition disagreed with them most firmly and stated that the tracks were made by a creature of human form to which they gave the name *Metoh-Kangmi*.

Colonel Howard-Bury appears to have been intrigued by this scrap of what he seems to have regarded as local folklore, but, like all who have had contact with them, he had such respect for the Sherpas, that he included the incident in a report that he sent to Katmandu, capital of Nepal, to be telegraphed on to his representatives in India. And this is where the strange mistakes began. It appears that Colonel Howard-Bury in noting the name given by the Sherpas either mistransliterated it or miswrote it: he also failed to realize that he was dealing with one of several kinds of creatures known to the Sherpas and that they, on this occasion, apparently both in an endeavor to emphasize this and for the sake of clarity used as a generic term for all of them, the name *kang-mi,* which was a word foreign to their language. This is a Tibetan colloquialism in some areas, and is itself partly of foreign origin even there, in that *kang,* is apparently of Chinese origin while *mi* is a form of Nepalese *meh.* The combination thus meant "snow creature." His *metoh* would better have been written *meh-teh,* a name of which we shall hear much, and which turns out to mean the *meh* or man-sized *teh* or wild creature. However, the Indian telegraphist then got in the act and either he dispatched this word as, or it was transcribed in India, as *metch.*

The recipients in India were unfamiliar with any of the languages or dialects of the area but they were impressed by the fact that Howard-Bury had thought whatever it might be, important enough to cable a report, so they appealed to a sort of fount of universal wisdom for help. This was a remarkable gentleman named Mr. Henry Newman who has for years written a most fascinating column in the *Calcutta Statesman* on almost every conceivable subject and who has the most incredible fund of information at his finger tips. This gentleman, however, did not really know the local languages or dialects of eastern Tibet and Nepal either, but this did not deter him from giving an immediate translation of this *metch kangmi* which, he stated categorically, was *Tibetan* for an "abominable snowman." The result was like the explosion of an atom bomb.

Nobody, and notably the press, could possibly pass up any such delicious term. They seized upon it with the utmost avidity, and bestowed upon it enormous mileage but almost without anything concrete to report. The British press gulped this up and the public was delighted. Then there came a lull in the storm. During this time, it now transpires, a number of eager persons started a fairly systematic search for previous reports on these abominable creatures, and they came up with sufficient to convince their editors that the story was not just a flash in a pan, but a full-fledged mystery that had actually been going on for years.

Thus, the "birth" of the Abominable Snowman per se may be precisely dated as of 1920. And once it was launched it gathered momentum. As we shall see later when we come to examine the actual reports from the eastern Himalayan region, almost everybody who went there, and notably the mountaineers, reported either seeing "snowmen," their tracks, or hearing them; finding cairns and other objects moved by them; or relating information secondhand that they had gleaned from the native population. The business reached a crescendo in 1939 with the publication of several quite long accounts in books by well-known and much respected explorers such as Ronald Kaulbach. Then came World War II and the matter faded into limbo. But it did not by any means stop.

No sooner was the war over than the onslaught on Mt. Everest was resumed and along with this came a new approach to the ABSM affair. Everybody appears to have felt it incumbent to at least mention the matter even if he could not contribute anything new or material to the story. Yet, there were very few who did not have something concrete to offer and indeed, I am unable to name one who didn't. What is more, prior to World War II, this was an almost exclusively British affair, though there was a book on the first American Karakoram Expedition, entitled *Five Miles High,* that was most pertinent. It has now become international as a result not only of expeditions going to the area from many nations and of multinational composition, but also because of reports that came to light but which were originally made during the war. Also, for the first time, reports by what may be called native foreigners began to appear.

The whole subject of "natives" is a sorry one and it is rather muddling to Americans because, to them, it has several meanings, none of which is exactly synonymous with the term as developed and understood among the British. It was the declaration of independence by a number of Asiatic nations that brought confusion, in that, while these peoples were manifestly native to their own countries, they suddenly be-

[12]

came no longer "natives" in the precise British sense, so that what they said had to be accepted and assessed in an entirely new light. Whereas, while anything stated by such people prior to the war could be passed off as a mere "native tale" or a story "by some benighted native," it had now to be treated with respect as a statement by a responsible citizen. What is more, an Indian traveling through Nepal to Tibet also became just as much a "foreigner" as any Britisher—and, in some cases, actually more so, because there were places where more Britishers had been living longer than any Indians. This proved extremely awkward to the British at first and it took about a decade even for their phlegmatic genius for compromise along with a fairly genuine common decency and belief in good manners, to gain the upper hand.

Despite the international scramble, it was again the British who attracted world attention to the matter of ABSMs and it was still their mountaineers who did this. The most notable was Mr. Eric Shipton who on still another reconnaissance of the Everest Bloc came upon a long set of tracks—not by any means for the first time in his life—and, after following them for some distance, noting they were definitely bipedal but negotiated almost impossible obstacles that would be hard for even an experienced mountaineer to do, took a series of clear photographs of them. These were published in the much respected *Illustrated London News,* not a publication given to elaboration, irresponsible reportage, or the mounting of international jokes. This time everybody had to take the matter seriously; and they did, but in a variety of ways. The public, as is its pragmatic wont, took it at its face value. The press literally howled. The explorers cheered a bit. But the scientists flew into a positive tantrum; an altogether undignified performance, the effects of which have not yet worn off and will not do so for many years. This was in 1951 and it marked the next turning point in the history of ABSMery.

Up till then the matter had been primarily a "Western" and notably a British perquisite; it had also been a child of the popular press with a sort of minor cold war going on between the mountaineers and the zoologists. Now, however, a new agency entered the picture, a polyglot assortment of people of various bents that can only be termed "The Searchers."

Since the turn of the century there had continued to be out-right explorers as well as putterers and sportsmen in the field and not a few of these continued to stumble upon ABSMs, or tracks and other evidence of their passing. None of these, how-ever, had any prior interest in the matter and, like the moun-taineers, had been in the Himalayas primarily for other pur-poses. On the other hand, the whole affair was, until Eric Shipton published his photographs, really nothing more than a news-gimmick though the press had had to tread warily with the reports made by prominent persons and especially the mountaineers engaged in the attack on Everest, which had official backing. The scientific world had not been quite so circumspect. At the outset, it denounced the whole thing as, first, a fraud, and then a case of mistaken identity, and it stuck to this story: and it still in large part sticks to it today, even to the extent of deliberately ridiculing such men as Shipton and Kaulbach. But after their completely unsuccessful attempt to set Shipton's 1951 findings at nought, which backfired with considerable public impact, a sort of revolution began within the ranks of science.

Some topnotch scientists—not just technicians and self-ap-pointed experts who happened to be employed by scientific organizations—started to investigate the whole matter upon truly scientific principles. What is more, these scientists were primarily anthropologists [as opposed to zoologists] and this was of the utmost significance, for the latter had permanently closed the door on the whole question when they could not prove that it was a hoax, stating flatly that all ABSM tracks were made either by bears or monkeys. Also, there were an-thropological expeditions actually going into the field and these too began to report discoveries similar to those of the mountaineers. Notable among the fieldworkers were Dr. Wyss-Dunant of a Swiss expedition, Professor von Fürer-Haimendorff of the School of Oriental and African Studies, and in particular Prof. René von Nebesky-Wojkowitz. Among those not engaged in fieldwork were Dr. W. C. Osman Hill of the Zoological Society of London in England, Dr. Bernard Heuvelmans, Belgian zoologist, in Paris, and latterly a whole group of Russian scientists led by Prof. B. F. Porshneyev.

It was the press, however, that was in the end first in the field with an expedition aimed primarily at the ABSMs. This was organized by the *Daily Mail* of London and went to the Himalayas in 1954. It was a curious outfit and it was not very successful but it initiated a new—and, to date, the last—phase in the history of this mystery. It was led by a reporter, Ralph Izzard and had among its members a professional zoologist, Dr. Biswas of Calcutta and also a man named W. M. (Gerald) Russell, whose experience was of great significance though nobody seems to have realized it at that time. However, it was once again directed by mountaineers. The significance of this escaped everybody then and to a very great extent still does. The universal impression had been gained over the years that the Abominable (as then supposed) Snowman, whatever it might be, was a denizen of the snowfields and therefore inhabited the uppermost slopes of the Himalayas. As a result, its pursuit was looked upon primarily as a mountaineering job and was therefore given to the professionals and the experts in that field of sport. The idea of including a scientist and especially a zoologist, had never occurred to anybody previously. The idea of including a man with the particular skills and experience, as well as training, of Gerald Russell has not even yet, it seems, dawned upon anybody.

Russell alone among the whole army of investigators is really the only man qualified to tackle the problem, for he is a professional collector, which is something absolutely different from either hunters or sportsmen on the one hand, or research scientists on the other. Then again, no ABSM is a denizen of any snowfield—naturally; and as should be obvious to any sane person on a moment's consideration, for in such places there is nothing to eat. All turn out to inhabit dense mountain forests. Thus, just about the last persons suited to search for them are mountaineers (who have a positive passion for climbing mountains above all else, it should be pointed out), while sportsmen and hunters are little better for other and even more obvious reasons.

This is a somewhat sensitive question but one of first importance. The techniques developed over the ages for hunting are basically aggressive, be they noisy as in "beating," or silent as in "stalking." Further, the dog—which is not only a domestic but actually an artificial animal—has been extensively used in hunting. These methods obtain the quickest results, in the largest amounts, of what is specifically desired. Collecting, on the other hand, should best be almost entirely passive. Silence is one of its features in certain of its aspects but almost as much noise is permissible as in hunting in certain circumstances. To obtain animals not normally hunted, the less ground covered the better but the longer the collector must sit and wait for the animals to become used to his presence, the noises he makes, and the effluvia he gives off in the normal course of "living." As many artificial things as possible must be eliminated; and most notably dogs, metal (especially metal cleaned with mineral oils), and suchlike that are not indigenous to the wild. Given time, any wild creature, however timid, will come to investigate the collector, whereas it will fly before the hunter long before it is detected.

Even zoologists, unless they have had extensive *collecting* experience in the field, are little better, for they, poor souls, are hustled about by everybody else into and out of the least likely areas for proper investigation, and are in any case supplied in advance with a sort of "book of rules" that goes far to negating the search for anything that is not already known.

The *Daily Mail* expedition did, nonetheless, include among its ranks, and deliberately, a very experienced zoologist with field experience in the form of Dr. Biswas, and, quite fortuitously in the person of Gerald Russell, the first and only man on any ABSM expedition trained to tackle such a collecting problem. It also accomplished something else, in that it publicized the whole matter and served notice on everybody that the press was no longer overawed by what they had termed "scientific opinion," but from then on took the affair for granted as having graduated from the category of the "silly season filler." In fact, it pointed the way to some serious endeavor designed to try to solve the mystery. This challenge was taken up by quite a new type of operator.

The *Daily Mail* expedition returned in 1955, and in that same year an Argentine mountaineering expedition and another British party (of Royal Air Force alpinists) reported having encountered tracks and other evidence of ABSMs. The following year the young man, John Keel, already mentioned, made his trip through the country and, as stated in his book published in 1958, tracked and sighted an ABSM. At the same time, the Russians were conducting investigations and getting ready to make a concerted attack upon the problem. There were also quite a number of others in the field, while the few serious students at home began to bring to light all manner of related items from the past.

The busiest of these scientific sleuths and the most openminded and best-informed was the zoologist, Dr. Bernard Heuvelmans, who had for long specialized in the collection and examination of evidence for the existence of any creatures as yet unknown to and unidentified by zoologists. It was he, moreover, who first brought the findings of the Hollanders in the East Indies, the French in Indo-China, and to a very considerable extent that of the South American explorers to light. The American edition of these findings by Heuvelmans, *On the Track of Unknown Animals*, was published by Hill and Wang of New York, in 1958. However, the most significant personality to enter the field was the prominent Texan, Mr. Thomas B. Slick.

Tom Slick, as he is known to everybody and all over the world, is a most remarkable man. To Americans he is probably best known because of the airline that carries his name, which is itself a natural advertisement with amusing connotations in the English language. Then, in the world of commerce he is widely known for his position in the mysterious world of oil and the very down-to-earth world of beef; but, his international reputation is based on his extraordinary efforts in the cause of world peace. Tom Slick has done many other things and is not only a patron of but a driving force in many purely scientific endeavors. He established the second largest privately endowed research unit in the world, in the form of the Southwest Research Institute near his home town of San Antonio, and adjacent to this another large organization for educational promotion. I am often asked to describe this man, and my response is invariably the same; namely, to say simply that, for all his activities and the vastness of his outlook and effort, he is less like the popular conception of a Texan than anybody I have ever met. Tom Slick *does* things and very fortunately he became intrigued with the business of ABSMs. Despite ridicule, especially among many of those closest to him, he set to work upon it with the determination that he, almost alone in the Western World it seems, was capable of and willing to apply. And, being a bulldog, he has kept quietly at it ever since.

I speak of Tom Slick at length because it is he, and he almost alone, who has by his quiet persuasion heaved this whole irksome business out of a sort of ten-ring, international circus, into the realm of serious scientific endeavor; while he has also stimulated others in England, France, Italy, India, and elsewhere who are working on the problem, by means of personal contacts and by the exercise of sympathetic encouragement. Finally, he did one more thing. This was to break out of the confined limits of the Himalayan area of the Oriental Region and direct attention and proper effort to other parts of the world, such as California, which are proving to be every bit as important in regards to ABSMs, if not much more significant than even the uplands of Eurasia. He began his own personal investigations by a trip to the Himalayan region in 1957.

In 1957, Tom Slick, together with A. C. Johnson, mounted the first full-fledged expedition to the Himalayas for the specific and sole purpose of investigating ABSMs. This saw the extremely fortuitous bringing together of Gerald Russell and the brothers Peter and Bryan Byrne, and was the happiest event that had until then—and still has been until the time of writing—happened to ABSMery. For the first time in history the leadership was not given to mountaineers or hunters, but to persons with collecting experience who believed that the quarry was real, was multiple in form, and that, in all its forms, it lived in the forests as opposed to on the upper snowfields. As a result, this expedition came closer to obtaining concrete results than any other before or since, and produced more straight evidence of the existence of such creatures than all other expeditions put together (for details see Chapter 12).

In the same year, however, the Soviet Academy of Sciences had established a special commission to co-ordinate the findings of several groups who had been working on the problem in countries within the Soviet sphere. These workers had brought to light the astonishing reports of Khakhlov made to the Academy in 1914, but which had been shelved; they had before them the current report of a Dr. Pronin, a hydrologist of Leningrad University who alleged he had seen an ABSM in the Pamirs, they had a wealth of material from the Mongolian Peoples' Republic and a lot from China; and they had decided to mount proper scientific expeditions to investigate. These were four in number and were put into the field in 1958 —one to the Caucasus where a creature named the "Wind Man" had been rumored for centuries; one to the north face of the Everest Bloc; one to the Mongolian region; and one to the Pamirs, which, for certain odd reasons they considered to be the breeding ground of the ABSMs. Meantime, they started the publication of their over-all findings in the form of booklets (see Chapters 13 and 14) and concurrently with this, a series of studies on fossil men, and particularly the Neanderthalers. Also, a wealth of previously unpublished material, some historical and some current, appeared in certain Russian magazines—notably, *Tekhnika Molodyozhi.*

These Soviet activities shed an entirely new light on the whole business, and also put it on such an altogether higher plane that Western scientific circles were obliged to change their attitude toward the matter quite drastically. No longer could they simply avoid the issue by saying that it had been explained or that its protagonists were merely a bunch of amateur enthusiasts pursuing a fantasy. At the same time, a certain nervous irritation was to be detected in their pronouncements, because the press just then began harping on the case of the Coelacanth fish discovered off the southeast coast of South Africa. This had at first been called a hoax but had finally had to be accepted as living proof of the fact that not everything about the life of this planet is known. Obviously, creatures confidently thought to have been decently extinct for tens of millions of years can still be around.

Further, it was the Russians who first stressed, though perhaps more by inference, something that those scientists in the West who *had* been taking the matter seriously had been harping on for some time. This was that the whole problem is an anthropological rather than a zoological matter. In other words, all the Sino-Soviet evidence pointed to ABSMs being primitive *Hominids* (i.e. Men) rather than *Pongids* (i.e. Apes) or other nonhuman creatures, thus linking them with known fossil forms such as *Gigantopithecus*, the Pithecanthropines, and especially the Neanderthalers. And, in doing this, they also emphasized another point.

That was the now very obvious but totally ignored fact that there is not just one creature called *The* Abominable Snowman, but a whole raft of creatures distributed almost all over the world, of very considerable variety, and of as many as three distinct types in the Tibetan-Himalayan area alone. This suggestion was of course not merely obnoxious but positively horrific to the orthodox scientists who were still vehemently denying even the possibility of the existence of even *one* such entity. Then, the final bombshell landed. At this point in my narrative I must confess to a considerable embarrassment since I must speak in the first person and I do this with much diffidence.

In 1958, I received a number of reports of an ABSM in California. At first, this sounded quite balmy even to us—and we are used to the most outrageous things—and got itself filed among what we call *Forteana*, which is to say those damnable and unacceptable items of the categories collected by the late Charles Fort. However, it so happened that I was privileged to spend the year 1959 touring the North American continent gathering material for a book on its geology, structure, vegetational cover, and wildlife. Before leaving, I had a research specialist—Stanley I. Rowe, with whom I had long been associated—prepare for me from his files, from ours, and from other sources, the details of any and all oddities and enigmas reported from this continent, by states and provinces. These I investigated as a news-reporter as I went along; and when I came to northern California I fell into the most extraordinary state of affairs that I have ever encountered in my life. This was no idle rumor but a full-fledged mystery and a straight-down-the-line, hard-boiled news-story.

This I tell in detail in Chapter 6, so suffice it to say here that I found there clear and most convincing evidence of the existence of a form of ABSM of most outstanding qualities. But worse was to follow for, prompted by this astonishing discovery, I went aside in British Columbia to investigate their long-renowned *Sasquatch,* only to find that it was just as definite, and apparently identical to these *Oh-Mahs* (or "Bigfeet") of California. Subsequent research has, what is more, brought to light a mass of other reports of similar things from Quebec, the Canadian Northwest Territories, the Yukon, the Idaho Rockies, Washington, and Oregon.*

This brings us up to the date of writing, except to note that a large Japanese expedition went in 1959–60 to the Himalayas specifically to search for ABSMs; while there were other expeditions in that area, in Sumatra, and in California, fitted out for the purpose. Finally, later this year (1960), Sir Edmund Hillary, backed by American sponsors and with Marlin Perkins, Director of the Lincoln Park Zoo of Chicago accompanying him as zoological expert, conducted an expedition to the eastern Himalayas with this pursuit as his second major objective.†

* These affairs in our Northwest were summarized in two articles in *True Magazine* for October, 1959, and January, 1960, and set a whole new phase of ABSMery in motion.
† The results of this effort are described in Appendix E.

2. Ubiquitous Woodsmen

May we suggest that laughing at "Indians" is rather old-fashioned while calling a Paleface a liar can be a very dangerous procedure.

In my opening remarks in the previous chapter I said that I was going to tell this story according to the chronology of discoveries made by the Western World, starting about the year 1860, rather than according to straight historical chronology. Having briefly outlined these discoveries from that date up to this year, I landed up in the northwestern corner of North America. I now find that this is just the place where I have to commence my detailed reporting and for several reasons. By way of explanation I resort to a map (Map I); a procedure that, I am afraid, you will discover I nearly always do.

ABSMs have now been reported from several dozen areas scattered all over five of the continents.* At first sight this distribution does not appear to make any sense at all. This is a misconception but to go into the whys and wherefors thereof at this juncture would not only be exhausting but more or less incomprehensible. Nonetheless, one cannot just go barging off all over the world reporting on this and that, both in time and space, without some ordered plan. Skipping around and back and forth over oceans just to point out similarities would be altogether aggravating. Some orderly procedure is therefore called for; and very fortunately there is a ready-made one that will serve many purposes. This is to adopt the travelogue ap-

* For the definition of the continents and their delimitation in accordance with the distribution of land-masses, as well as an explanation of the misconceptions about their identity, see Chapter 18.

proach, starting out from some specific point, visiting all the other necessary points, and ending up where we began. Doing this in the pursuit of ABSMs just happens to be most convenient, and for a number of reasons. If we take northwestern North America as our starting point, we will be able to dispense with a great deal of verbal garbage and duplication.

I therefore propose to take you on a journey starting from western Canada, south through the Americas to Patagonia, then back up to the southern edge of the Amazon Basin; then hop over the Atlantic to West Africa, proceed through or rather around the Congo Basin and over the eastern uplands to the forested coastal land of East Africa. From there, we will jump over the Indian Ocean to the island of Sumatra, proceed from there up the Malay Peninsula to the main body of the great Indo-Chinese peninsula, then turn sharp left in Assam and travel along the Himalayas to the vast Pamirs, and on southwest through Persia to the Caucasus. This will be a turnabout point from which we will return east to the Pamirs, on to the Kunluns, then to the Tien-Shans, Ala-Tau, Altais, and Sayans. From there we will go south through the Khangais and over the Ala-Shan Desert to the Nan-Shans and on to the mountains of Szechwan. Here will be another turnabout point from where we will go north again through the Tsin-lings and the Ordos to the Khingans. In this last lap on our way home we will be following a lot more than ABSMs, and in following these we will cross over the Bering Straits to and through Alaska and the Yukon back to our starting point in British Columbia and specifically to a small place named Yale, on the middle Fraser River.

It was near this place that something frightfully important happened in the year 1884; on the morning of July 3, as a matter of fact. The gorge of the Fraser narrows along this stretch so that rock walls tower on either side. Today, two railroads and the main west-to-east Canadian highway squeeze through this point and the little township of Yale clings to the bank of the river on one side, and is dotted about a narrow meadow on the other. Since I beg to be regarded exclusively as a reporter for the duration of the forthcoming journey, the

MAP I. CENTRAL WESTERN NORTH AMERICA

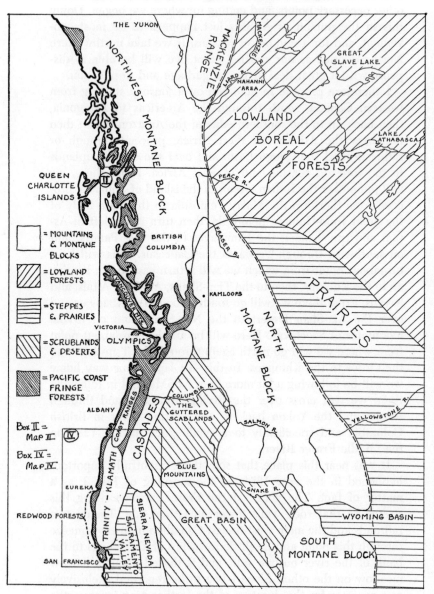

See facing page for legend

best thing for me to do is to quote the original report on what happened there on that day. This goes as follows, as taken from the Victoria newspaper, *The Daily British Colonist:*

Yale, B.C., July 3, 1884—In the immediate vicinity of No. 4 tunnel, situated some 20 miles above this village, are bluffs of rock which have hitherto been unsurmountable, but on Monday morning last were successfully scaled by Mr. Onderdonk's employees on the regular train from Lytton. Assisted by Mr. Costerton, the British Columbia Express Company's messenger, a number of gentlemen from Lytton and points east of that place, after considerable trouble and perilous climbing captured a creature who may truly be called half man and half beast. "Jacko," as the creature has been called by his capturers, is something of the gorilla type standing about 4 feet 7 inches in height and weighing 127 pounds. He has long, black, strong hair and resembles a human being with one exception, his entire body, excepting his hands (or paws) and feet are covered with glossy hair about one inch long. His fore arm is much longer than a man's fore arm, and he possesses extraordinary strength, as he will take hold of a stick and break it by wrenching or twisting it, which no man living could break in the same way. Since his capture he is very reticent, only occasionally uttering a noise which is half bark and half growl. He is, however, becoming daily more attached to his keeper, Mr. George Telbury, of this place, who proposes shortly starting for London, England, to exhibit him. His favorite food so far is berries, and he drinks fresh milk with evident relish. By advice of Dr. Hannington, raw meats have

---

MAP I. CENTRAL WESTERN NORTH AMERICA

This is an arbitrarily chosen area, designed to bring out a number of different physical features. It represents an area of some 1,900,000 square miles, of which some 1,650,000 are land. This is cut diagonally by the Great Barrier—here represented by the Rockies—that extends from the Arctic coast to Vera Cruz on the Gulf coast. To the east of this are lowlands covered, in the north, by the great boreal coniferous forests and, to the south, by the prairies. In the south lies the Great Basin, actually an upland, desert plateau covered with parallel ranges of modest mountains. Between the Sierra Nevada and the Southern Coastal Ranges there is the flat gutter known as the Sacramento Valley. The rest is subdivided into a series of mountain blocks as shown. Each is quite distinct in form, composition, flora, and fauna. It is around the peripheries of these that ABSMs have been reported. The coast, from the Olympics north, is mostly precipitous and without any coastal plain at all.

[25]

been withheld from Jacko, as the doctor thinks it would have a tendency to make him savage. The mode of capture was as follows: Ned Austin, the engineer, on coming in sight of the bluff at the eastern end of the No. 4 tunnel saw what he supposed to be a man lying asleep at close proximity to the track, and as quick as thought blew the signal to apply the brakes. The brakes were instantly applied, and in a few seconds the train was brought to a standstill. At this moment the supposed man sprang up, and uttering a sharp quick bark began to climb the steep bluff. Conductor R. J. Craig and Express Messenger Costerton, followed by the baggage man and brakesmen, jumped from the train and knowing they were some 20 minutes ahead of time, immediately gave chase. After 5 minutes of perilous climbing the then supposed demented Indian was corralled on a projecting shelf of rock where he could neither ascend nor descend. The query now was how to capture him alive, which was quickly decided by Mr. Craig, who crawled on his hands and knees until he was about 40 feet above the creature. Taking a small piece of loose rock he let it fall and it had the desired effect of rendering poor Jacko incapable of resistance for a time at least. The bell rope was then brought up and Jacko was now lowered to terra firma. After firmly binding him and placing him in the baggage car, "off brakes" was sounded and the train started for Yale. At the station a large crowd who had heard of the capture by telephone from Spuzzum Flat were assembled, and each one anxious to have the first look at the monstrosity, but they were disappointed, as Jacko had been taken off at the machine shops and placed in charge of his present keeper.

The question naturally arises, how came the creature where it was first seen by Mr. Austin? From bruises about its head and body, and apparent soreness since its capture, it is supposed that Jacko ventured too near the edge of the bluff, slipped, fell and lay where found until the sound of the rushing train aroused him. Mr. Thomas White, and Mr. Gouin, C. B. E., as well as Mr. Major, who kept a small store about half a mile west of the tunnel during the past 2 years, have mentioned having seen a curious creature at different points between Camps 13 and 17, but no attention was paid to their remarks as people came to the conclusion that they had either seen a bear or stray Indian dog. Who can unravel the mystery that now surrounds Jacko? Does he belong to a species hitherto unknown in this part of the continent or is he really what the train men first thought he was, a crazy Indian?

Now, whatever you may think of the press, you cannot just write off anything and everything reported by it that you don't like, don't believe in, and don't want. Further, to a newspaperman, this report is excellent, being factual, giving names that were obviously carefully checked even to titles such as the C. B. E. of Mr. Gouin, and hardly being at all speculative. In fact, it is really a model report and one that some present-day newsmen might well emulate. Then, the persons concerned were not a bunch of citizens with names only to identify them; they were mostly people with responsible positions who must have been widely known at that time throughout the area, for the railroad played a very important part in the opening up and development of lower British Columbia. The reporter, moreover, himself took a very common-sense view of the business when he inquired what manner of creature this might be and stated flatly that it was completely human but for being covered with silky black hair and having exceptional strength in its arms. The asinine opinions of others—such as, that the similar if not identical creature seen before might have been a bear or a "stray Indian dog"—are recorded "straight" and without facetious comment. The whole thing cannot, in fact, be lightly dismissed. It therefore has to be most seriously considered.

The story has been publicized for some 50 years now, so that afficionados of ABSMery can often almost quote it verbatim but, although I must here class myself among these reportorial limpets, I wish to put on record one thought about it that has always stayed with me. This stemmed from a comment made in another paper shortly after the original story was published, and which asked quite without facetiousness also but with a slight air of mystification, how anybody could suggest that this "Jacko" could have been a chimpanzee that had escaped from a circus. This little aside puts the whole affair in a remarkably vivid light, for we tend to forget that it was penned 75 years ago in a country that was then only recently connected with the rest of the world. Also, it was written before palaeontologists had demonstrated that true monkeys and, more so, the apes (i.e. Pongids), never have existed in the Western Hemisphere.

This creature was captured, and it is absolutely sure that it existed in "captivity" for some time (a reporter in 1946 interviewed an old gentleman in Lytton, B.C. who remembered having seen it): it was not human, yet it was more so than it was anything else; it had definitely been captured on the Fraser River; therefore, there had to be some explanation of how it got there and what it was. The standard answers to these questions today would undoubtedly be that it was (1) a hoax, or (2) a "cross"—though between what and what would doubtless not be suggested, (3) a throwback—and probably an "Indian" one, (4) a little boy who had been lost years before on a hunting trip and either managed to survive all on his own or been fed by wolves, (5) a mentally defective glandular case from an "institution," or (6), and most likely of all, an ape escaped from a bankrupt circus. Surprisingly, the locals and even hard-boiled newspapermen of the time did not indulge in any of these latterly foibles: rather, they asked a straight question and poo-poohed any outlander's suggestion that it was a chimp escaped from a circus. They even inquired as to whether it might be a very primitive form of human or an as yet unidentified species of great ape, and in either case indigenous to the area.

I may be properly accused of harping on this case, but I think that of almost all ABSM reports it is perhaps the most cogent. It took place just within the "age of reason" (today, perhaps, rather a misnomer) in a country then inhabited and being opened up by the most extremely pragmatic Westerners of predominantly hard-headed Anglo-Saxon stock, at a time when there was little call for phoney sensationalism. It was not just a report of tracks or other secondary items, nor even of an alleged sighting; it was a clear and definite account of a capture by known people with all the witnesses needed for confirmation. Quite apart from anything else, it alone sets at nought the constant refrain "Well, why haven't we ever caught one?"

This is by no means the only ABSM that has been caught, but it is the only one that I know of that was caught by what we must call for lack of a better phrase "Westerners," and it is this culture that is the most skeptical, the most stubborn, and at the same time the most interested. Of course, the more aggravating part of the business is that there is no proper end to the Jacko story, and no physical evidence of his existence has come down to us—at least as far as anybody so far knows. What actually happened is not recorded; the only inkling that I have traced being a remark by Mr. Stephen Franklin, staff writer of *Weekend* magazine, in his excellent article dated April 4, 1959, in which it is stated (and I quote) that "The editor of the *Inland Sentinel* inopportunely chose this month (the one in which Jacko was captured) to hump his newspaper and his presses up the canyon from Yale to Kamloops, and didn't publish an edition for several weeks."

This statement is itself a kind of *non sequitur* since the original reports come from *The Daily British Colonist*, of Victoria. I made somewhat extensive search for any series on the forlorn Jacko in a Yale paper of old, but was unable to unearth even the morgue of the *Inland Sentinel* which moved to Kamloops. Jacko, sad to tell, just "dropped out of the news" without apparently further comment; perhaps the most enigmatic figure ever to appear on the pages of history and potentially one of the most important.

Would that we could unearth the end of this story and learn what did happen to him, for he must have either (1) escaped, (2) died, or (3) been killed, and in the two last events it is possible that some part of him may have been preserved and be lying either in somebody's attic trunk, or even in a museum. And do not for a moment get the idea that the latter is impossible. (See Chapter 20.)

Jacko, however, is not just an isolated imp that suddenly appeared upon the scene and then disappeared. Before his capture either he or one of his species had been reported from the same area by Mr. Alexander Caulfield Anderson, a well-known explorer and an executive of the Hudson's Bay Company, who was doing a "survey" of the newly opened territory and seeking a feasible trade route through it for his company. He reported just such hairy humanoids as having hurled rocks down upon him and his surveying party from more than one slope. That was in 1864. Many years later, Mr. J. W. Burns (now retired and living in San Francisco) who had devoted a lifetime to the study of this business, unearthed an old Amerindian woman from Port Douglas at the head of Harrison Lake (see Map II) who alleged, and brought some seconders to confirm, that she had been kidnapped by one of these creatures in the year 1871, kept by it for a year, but finally returned by it to her tribal homestead because she "aggravated it so much" (though, she said, it had treated her with every consideration). This old lady died in 1940 at the age of 86. When abducted she was 17 years old and was, she stated, forced to swim the Harrison River by the ABSM and then carried by him to a rock shelter where its aged parents dwelt. This account comes from Mr. Burns who had for years enjoyed the confidence of this retiring Amerind. It has been embellished in various ways by others to the effects that the girl had rosin plastered over her eyes by the creature; that she became pregnant by it; and that she subsequently gave birth to a half-breed that either was stillborn, died shortly after birth, or is still hidden by her people from the eyes of the white man. She never said any of these things to Mr. Burns but adhered to her straightforward story till her death.

Nor is this woman's story unique. All the Amerinds of southern British Columbia, Washington State, Oregon, parts of Idaho, and the Yuroks and the Hüppas of northern California not only have similar tales to tell but a history of these creatures so complete and extensive that it would take a volume to tell in itself. The poor Amerinds have always been and still are regarded by Americans and Canadians as "natives," which indeed they are, but in the same light as the British used to regard the inhabitants of all countries other than their own or at least beyond the confines of western Europe. The stories told by, and the traditions of, Amerinds are not, therefore, regarded as of much worth or reliability. Nonetheless and despite the fact that these peoples did not previously write and have had even today little if any contact among themselves over any distance, their reports upon these local ABSMs are absolutely the same all the way from the Mackenzie Range of Alaska through the Yukon and British Columbia, down through Washington and Oregon to California, and back to the western flank of the Rockies in Idaho. There are traditions and folk-tales spread over an even wider area among these people, but this is another matter. I am here speaking of perfectly straightforward, up-to-date accounts of encounters with such creatures that have been made by them ever since the white man first got to speak with them and which have come in from one source or another annually every year since the capture of Jacko. I will interject some of these as I go along.

Before doing so, however, I must put on record that I do not share the old British or what seems to be the current American opinion of "natives" and never have. Further, as a working reporter, having now been privileged to travel extensively throughout just the five continents with which we are concerned in this story, I would state that I find the so-called "native" in some respects on the whole more reliable than the foreigner, and the white foreigner in particular. First, they seem to me to know their country better; secondly, those of them that are country folk are almost invariably consummate naturalists and know their local fauna inside out (and much better than we do); third, if they like you and feel that you are not going to laugh at everything they say, they are very pragmatic and are willing to tell you, straight, what is what in their opinion; fourth, provided one appreciates the very basic fact that to many non-Europeans there is a nonmaterial world that is just as real as the material one, one can readily distinguish between stories of one and the other, and may even without giving offense ask the teller to which category any story belongs. When my job was collecting animals for scientific institutions in out of the way parts of the world—a profession I pursued for two decades—I always asked the natives for information on their local fauna. While all people may display, and often do so, lapses or gaps in their knowledge, and so just do not know an animal that has always been right under their noses, what they *do* tell has, I have found, invariably turned out to be the truth. More than this, some peoples, such as the Mayas of Yucatan, are absolutely incredible "taxonomists" in that they differentiate, and have names for every type of animal, so that in one case I found out after long and patient

recording phonetically that they even had the spiders of their
country classified, all in just the same way as does our modern
zoology. Then finally, I would also put on record that I have
a particular respect for the nonprofessional American "Indian"
as he is so incorrectly and lugubriously called.

My wife and I have lived with various of these peoples—and
they are as varied a lot as "Europeans" if not more so—off and
on for many years; we did so in rather exceptional circumstances in that we were neither their employers nor employees,
were not interested specifically in their "culture," art, or anything else, but had several mutual interests with them in their
crops, stock, local wildlife, and plants. My wife has an exceptional knack of learning languages by ear and under appropriate circumstances and in local costume she can look like
almost any race on earth while I, as a "doctor" or "medicine
person" was on the one hand unobtrusive and inoffensive to
them while, on the other, having my wife with me I could
browse around in the obscurer corners of life without giving
concern to the elders or alarm to my male contemporaries.
Thus, by simply living alongside these people—and going to
their dances only for the fun of it, instead of to study their
alleged implications, and so forth—we came to chat around
the evening fire of many things. While I have found the African the most enjoyable company at such times of genuine
relaxation, and the Malayan peoples the most informed (sometimes terrifyingly so to a European), it has been the Amerinds
that I have found to be the most down to earth and pragmatic.
Many of these peoples—and they are the first to admit it; roar
with laughter at the fact; and will not be offended by a sincere friend saying so—love to drink alcohol and sometimes indulge in stimulants that we class as narcotics, and when they
do so they can very readily become uproarious in all manner
of ways. At these times they will concoct the most delicious
imagery compounded of mysticism, ancient tradition, and personal whim, and, while there may be all manner of historical
gems to be gleaned from such outpourings, none of it should
be taken as "exact science." When, however, they are stone-
cold "sober," in the strictest sense of that loose term, they can

give out information of a caliber that would do justice to a Yale professor. Don't ever underestimate the Amerind or his knowledge! I shall not forget a remark made to a partner of mine, who has also lived with these people and likes them very much, so that they seem to like him. He was making exhaustive inquiries into this very matter of ABSMs, when an old gentleman—a doyen of his tribal unit and a pillar of the local church —suddenly burst out with "Oh! Don't tell me the white men have finally gotten around to *that?*"

Let us, nonetheless, ignore the Amerinds for the moment and concentrate on the unfolding of ABSMery in and about British Columbia as reported by "white men" or allegedly witnessed by them. This history is now just about 100 years old, starting with Mr. Anderson of the Hudson's Bay Company. During this period some paleface appears to have reported an ABSM incident almost every year and they are now doing so in droves, to such an exaggerated extent that even Chambers of Commerce (*vide* that of Harrison Lake, the leading resort area for the vast city of Vancouver) have gotten into the act, and one sees large cutouts of the creatures along highways advertising everything from motels and garages to bakeries, cleaning services, and speedboats. Most notable contributions to this tradition have been made in the years 1901, 1904, 1907, 1909, 1910, 1912, 1915, 1924, 1936, 1939, 1941, 1948, 1954, 1955, 1956, and 1959. And all but two of these were "sightings" or rather personal encounters, but usually confirmed by more than one witness—not just dreary footprints found in snow or mud, hanks of hair, overturned barrels, or piles of excrement. This is really a pretty astonishing picture and makes affairs even in Nepal look somewhat picayune.

All of this centers around the lower Fraser River and notably around Lake Harrison. Therefore, I resort, as usual, to a map in order to cut down verbiage. All of these reports have been published before, and often so many times that there are those who feel that the process has been protracted *ad nauseam.* Nevertheless, I am, as I have said, myself reporting and I do not know of any one place where all of them have been brought together in chronological order. That anything like this could have been going on for a century right in our front yard—it being politically in Canada—is amazing enough but we are to get an even more profound jolt when we come to see that the very same thing has been going on in our own *back yard*—to wit, in Washington, Oregon, California, and, according to none other than Theodore Roosevelt, at one time at least, in Idaho.

The opening gambit was a sworn statement made by a highly respected lumberman who had also been most successful as a timber-cruiser and prospector, named Mike King. This gentleman had had to penetrate an isolated area in the north of Vancouver Island in 1901 alone, because his Amerindian employees refused even to enter it on any account but mostly because they said that it was a territory of the "Wildmen of the Woods." From other accounts of Mr. King it seems that he was not a man to be diverted from essential business routine by such stories, but that he had a profound respect for the local "natives" because they had guided him to a reasonable fortune on more than one occasion simply by their real knowledge of the country and the timber that grew in it. Some days after penetrating this wild area, Mr. King topped a ridge and spotted below a creature squatting by a creek washing some kind of roots and arranging them in two neat piles beside him, or her, on the bank. This should be compared with the specific remarks made by Mr. Ostman (Chap. 3) on the same subject. In my interview with Mr. Ostman, he stressed the collection of roots by the creatures and even named the plant most chosen, also the careful washing and stacking of these. Perhaps he got the notion from reading this account, but personally I doubt it. King's natural instinct was to raise his rifle and sight, for the creature was large, covered in reddish brown fur, and thus potentially dangerous. By the time the fact that brown bears don't wash roots and stack them up had penetrated, he realized that he had some kind of humanoid in his sights and he lowered the rifle. The creature took off, running like a man and, as Mr. King later reported: "His arms were peculiarly long and used freely in climbing and bush-running [i.e. scrambling on all fours through scrub]." King descended the slope and inspected the spoor left by the departed one, and noted that it was a distinctly "human foot but with phenomenally long and spreading toes."*

On reading the original account from an old clipping to a company of easterners some years ago, I heard somebody murmur, "And so endeth the first lesson." And so indeed! For, although that statement has been repeatedly recounted and Mike King has been repeatedly said to have elaborated, no further direct quotes appear to be extant. This is the way that unexpected things happen. I know from the few that I have experienced. You are not prepared for them; by the time you have managed to bring your senses to bear upon them, they are up and away; and you are left gaping, with a blurred impression all around a single vivid centerpiece. What more can you add unless you want to be a tattler? Mike King apparently had both the decency and the common sense to say what he had to say and then shut up.

The next lot to have a similar encounter (in 1904) were out hunting near Great Central Lake on Vancouver Island. Their names were J. Kincaid, T. Hutchins, A. Crump, and W. Buss, four citizens of Qualicum. They were apparently beating the bush, and put up what they afterward described as a boy ABSM that was covered with brown hair but had long *head-hair and a beard*. This is a very odd report in that it otherwise crops up only once or twice in all the accounts of ABSMs, and is, categorically, contrary to all the other reports by everybody who has alleged that he or she has seen these creatures at close range.

The third classic report is dated 1907 and was made by the Captain and crew of the coastal steamer *Capilano* on their return from a routine cruise during which they had called at a small landing named Bishop's Cove. There, they said, the entire Amerindian population had come charging aboard begging for asylum or outright emigration due to a huge monkey-like, human-shaped creature that had been clam-digging

---

* This remark, and particularly the word *"long"* used to describe the toes, rather than the whole foot, is most pertinent as we shall see when we come to examine the tracks of the *Oh-Mahs*.

along their beach for a number of nights in succession, and which gave vent to most disturbing high-pitched howls. These people readily identified the creature but insisted that it had moved into their territory with its family, if not its whole clan, and that it would not brook any interference by a few poorly armed humans. The comments on this report are rather illuminating as they display a curious acknowledgement of the presence of such "Wildmen" and the fact that, while they are accepted as being basically peaceable and known to mind their own business, and while they avoid organized men in masses, they tend to adopt a nasty tone when it comes to hunting and collecting rights, and appear then to regard the Amerinds as interlopers and a nuisance. In 1907, however, the attitude of even the British toward real primitives was going through a peculiar phase; halfway between the concept of the "worthless native" and that of the "noble savage." The Amerinds had proved an unreliable labor force, while certain other non-Europeans had turned out to be far too civilized for rank exploitation. The idea of really primitive creatures had not yet been abandoned and everybody was still undecided just how to behave toward them. The thought that we might be dealing with sub-hominids did not, of course, occur to anybody professing any education (after all, Darwin was hardly cold as of then) but it remained in no way illogical to the uneducated, and it was played on by the press.

This may in some measure account for the solemnity with which a discovery made in 1912 was greeted. I got this report from Mr. Burns, mentioned above. It came to him from the principal, a Mr. Ernest A. Edwards, who states that he was residing at Shushwap, B.C., at that date, and that he and his wife had unearthed on the small island of Neskain a little way off the coast, a human skeleton that they found protruding from the bank of a river. The location was noted for its abundance of "arrowheads" of Amerindian origin. This skeleton is stated to have measured "from skull to ankel-joints—7 feet 6 inches, so with feet and scalp, the person must have been 8 feet tall." Mr. Burns received this information in a letter from Mr. Edwards in 1941, and this included the further comments that "I, together with my wife, examined the jaw. The teeth were of huge size, but in perfect condition—no cavities noticeable. The jawbone was so large it would span my face easily at the cheek bones. Together with the help of Indians, I crated it and shipped it to Rexham Museum, North Wales, England, where I believe it still is. In his acknowledgment, the Curator of the museum was greatly astonished, remarking among other observations, that it was hard to believe such jaws and teeth 'existed' in human beings."

The receipt of such intelligence as this naturally prompts an almost fiendish "Ho-ho! what is this?" on the part of any reporter, so I wrote to the Curator of the museum specified and got the following reply from the Librarian of the town of Wrexham (not Rexham, there being no such town in Wales or anywhere else in Great Britain): "With regard to your query, I have checked the Minutes of this establishment [i.e. museum and public library] for the years 1912, 1913, and 1914, and there is no mention of the receipt of a skeleton. Yours sincerely, Clifford Harris, F.L.A."

Reports of the discovery of the skeletons of giant humans or humanoids are extremely numerous, and have been coming in from all over this continent for many years. They constitute a subject of their own which I have endeavored to pursue for a long time now but, I regret to have to say, without any success. One and all have just "evaporated" like this, but, I must admit, very often within the portals of some museum which had acknowledged receipt of the relic. There is the famous story of the forty mummified giants in Mammoth Cave, Kentucky; of the giants in giant coffins in some unnamed cave in Utah; of others dug up in a peat bog in West Virginia and allegedly shipped to the Smithsonian; and of others "preserved" in sundry small county museums in Nevada. I have voluminous correspondence on file on these items but I have never yet managed to obtain sight of any single bone. This is odd because human giants are not really terribly rare [I have seen it stated that there are several thousand men over 7 feet tall living today in the United States] whereas such persons in the past would probably have been regarded with some awe and might be expected to have been accorded rather special burial, so augmenting our chances of unearthing them.

The matter of skeletal remains of ABSMs is, of course, of first importance and second only to the procurement of a whole living specimen. The chance of unearthing a skeleton of one is not quite so unlikely as one might suppose, for it now transpires that very primitive peoples indeed seem to have performed deliberate interments, if only to clear away refuse from a cannibalistic meal in a cave. Some ABSMs might well be or have once been at such a level of "cultural" development and it is constantly reported by the Amerinds in this area that their particular local variety indulge something akin to hibernation, or at least winter inactivity equivalent to that of the local bears, and that they do this in caves. This presents a dubious aspect of these traditions however, because, in the absence of limestone strata in the area, caves are rarities. Nonetheless, there are caves in volcanic rocks of certain kinds and some have been alleged to have been found in the mountains around Harrison Lake. There is one story of such that pertains to ABSMs. This again I got from Mr. J. W. Burns. It goes as follows and comes from an Amerind named Charley Victor, a resident of Chilliwack on the lower Fraser:

The first time I came to know about these people [the local ABSMs, now named *Sasquatches*], I did not see anybody. Three young men and myself were picking salmonberries on a rocky mountain slope 5 or 6 miles from the old town of Yale. In our search for berries we suddenly stumbled upon a large opening in the side of the mountain. This discovery greatly surprised all of us, for we knew every foot of the mountain, and never knew nor heard there was a cave in the vicinity. Outside the mouth of the cave there was an enormous boulder. We peered into the cavity but couldn't see anything.

We gathered some pitchwood, lighted it and began to explore. But before we got very far from the entrance of the cave, we came upon a sort of stone house or enclosure. It was a crude affair. We couldn't make a thorough examination, for our pitchwood kept going out. We left, intending to return in a couple of days and go on exploring. Old Indians, to whom we told the story of our discovery, warned us not to venture near the cave again, as it was surely occupied by a *Sasquatch*. That was the first time I heard about the hairy men that inhabit the mountains. We, however, disregarded the advice of the old men and sneaked off to explore the cave, but to our great disappointment found the boulder rolled back into the mouth and fitting it so nicely that you might suppose it had been made for that purpose.

This story seems to me to have a certain ring of truth about it, and the idea of using a boulder as a door, either for protective purposes or for concealment of a breeding-chamber, is not in any way illogical or impossible, There is, however, it should be pointed out, a modern tendency to, as it were, chase anything elusive back into caves, and especially wild men; probably because of all that has been written, from archaeological texts to comic books, about "Cave Men." The majority of primitive hominids did not live in caves; simply, because the number of caves available was, except in a few special areas, very limited. [Further, they may have first entered them to get away from either heat or rain as much as from cold.] Yet, the remains of early men and animals are better and more readily preserved in cave floors than out in the open, while locating open-air camping sights is very chancy. The idea that men went through a cave-living phase, all over the world, has therefore gained wide credence. *Sasquatches* could just as well hole up in ice-caves made by themselves in deep snow, as some bears do. But caves should be searched most diligently for remains or other evidence of their occupation.

It was not too far away from this alleged cave site that the next encounter of which we have record and that is documented, sworn to, and witnessed by more than one person, took place in 1915. A Statutory Declaration of this was sworn to in September of 1957 by one of the participants, Mr. Charles Flood of Westminster, B.C. This goes as follows:

I, Charles Flood of New Westminster (formerly of Hope) declare the following story to be true:

I am 75 years of age and spent most of my life prospecting in the local mountains to the south of Hope, toward the American boundary and in the Chilliwack Lake area.

In 1915, Donald McRae and Green Hicks of Agassiz, B.C. and my-self, explored an area over an unknown divide, on the way back to Hope, near the Holy Cross Mountains.

Green Hicks, a half-breed Indian, told McRae and me a story, he claimed he had seen alligators at what he called Alligator lake, and wild humans at what he called Cougar Lake. Out of curiosity we went with him; he had been there a week previous looking for a fur trap line. Sure enough, we saw his alligators, but they were black, twice the size of lizards in a small mud lake.

Awhile further up was Cougar Lake. Several years before a fire swept over many square miles of mountains which resulted in large areas of mountain huckle-berry growth. Green Hicks suddenly stopped us and drew our attention to a large, light brown creature about 8 feet high, standing on its hind legs (standing upright) pulling the berry bushes with one hand or paw toward him and putting berries in his mouth with the other hand, or paw.

I stood still wondering, and McRae and Green Hicks were arguing. Hicks said "it is a wild man" and McRae said "it is a bear." As far as I am concerned the strange creature looked more like a human being. We seen several black and brown bear on the trip, but that thing looked altogether different. Huge brown bear are known to be in Alaska, but have never been seen in southern British Columbia.

This document brings up two questions that I should dis-cuss briefly forthwith. The first is the matter of the Law. As I have already said, we in this country do not have much re-spect for this aspect of human organization and often tend to the observation that "laws are only made to be broken." This is not so in some other countries however, and the Cana-dians have an intense respect for their laws and for authority in general. Canadians will scoff at the suggestion that one of their countrymen is more likely *not* to lie before a justice of the peace than an American, but it is nonetheless a fact that a Canadian is more likely to make such a deposition if his veracity has been called in question and/or he wants to as-sert his sincerity. Also, he will think longer and more care-fully about his statement if made before established authority because, should anything he say therein be mendacious and thereby cause any distress or harm to others, he will be held fully accountable. Thus, these sworn statements and others that follow have a rather strong implication. The other matter is the introduction of an almost classic red herring.

As I explain at greater length in Chapter 19, an inexplicably high percentage of all esoteric investigations turn up other unexpected and apparently unrelated matters that are often just as weird, if not more so, than the original object of pursuit. In this case, the matter of "alligators" is quite extraordinary and quite beyond my comprehension. Alligators, per se, are only two in number, one species being indigenous to the Mississippi Valley and around the Gulf coast to Florida; the other to the Yangtse-Kiang Valley of China. The term "alligator" has, however, become a colloquialism for all the crocodilians, and it is also applied in some countries to various lizards that spend most of their time in fresh water. Popular names are also very dangerous in that they become displaced in the most outrageous manner, such as the designation of a species of tortoise in Florida as a "gopher," when that is the name for a group of small mammals otherwise called groundsquirrels. Reptiles are, however, cold-blooded, and the existence of an aquatic one in even southern British Columbia would be unlikely, to say the least. Yet, there is a species of salamander [an amphibian named *Batracochoseps*] found in Alaska, and the giant salamander of the mountain streams of Japan is customarily iced in every winter. The mere mention of such a creature as an alligator in this story tends to cast doubt upon its other features, but then who is to say what can or cannot be. There is volcanicity in the area, and there might thus be hot or warm springs and lakes there. Also, at some time, one or other of the present-day species of alligator *must* have gotten either from China to the Mississippi, or vice versa. The only route for such an emigration is over the Bering Straits; thus passing through what is now British Columbia along the way.

This matter of volcanicity and hot springs brings us to another really quite fabulous item of Canadian ABSMery. This is the matter of the lower Nahanni area of the Northwest Territories. If you go to the western part of the Northwest Territories you will sooner or later be told about the place where banana trees have been grown. This sounds quite wacky but, if you pursue the matter diligently, you will learn that in the area of the junction of the Liard and South Nahanni Rivers (see Map I), lying against the vast mountain barrier which cuts our entire continent from the mouth of the Mackenzie River on the Arctic Sea to Vera Cruz on the Gulf of Mexico, abutting on to the central plains like a monstrous wall, there is a volcanic area where hot springs are found. There have been mission stations along the Liard for over a century and it is quite true that at these, magnificent vegetables are grown out in the open in the brief but intense summer. Also, they have been raised indoors, and among these vegetables have been a number of banana trees. However, this area, which lies at the south end of the vast Mackenzie Range, has long been one of myth and fantasy. The reports emanating from there cannot better be summed up than by quoting a column from a publication named *Doubt*, the periodical of the Fortean Society of New York. It was founded by the late author, Tiffany Thayer, in conjunction with several other notable persons such as Ben Hecht, in memory of, and to carry on the work of Charles Fort, that assiduous collector of borderline reports for so many years. This reads in part, when speaking of an expedition said to have been organized to visit the area:

This Valley, number one legend of the Northlands, has as its background, stories of tropical growth, hot springs, head-hunting mountain-men, caves, pre-historic monsters, wailing winds, and lost gold mines. Actual fact certifies the hot springs, the wailing winds, and some person or persons who delight in lopping off prospectors' heads. As for the pre-historic monsters, Indians have returned from the Nahanni country with fairly accurate drawings of mastodons burned on raw hide. The more recent history began some 40 years ago (circa 1910) when the two MacLeod brothers of Fort Simpson were found dead in the valley, and reportedly decapitated. Already the Indians shunned the place because of its "mammoth grizzlies" and "evil spirits wailing in the canyons."

Canadian police records show that Joe Mulholland of Minnesota, Bill Espler of Winnipeg, Phil Powers and the MacLeod brothers of Ft. Simpson, Martin Jorgenson, Yukon Fischer, Annie La Ferte, one O'Brien, Edwin Hall, Andy Hays, an unidentified prospector and Ernest Savard have perished in the strange valley since 1910. In 1945 the body of Savard was found in his sleeping bag, head nearly severed from his shoulders. Savard had previously brought rich ore samples out of the Nahanni. In 1946 Prospector John Patterson disappeared in the valley. His partner, Frank Henderson, was to have met him there, but never found him.

The "head-hunting mountain-men" are alleged locally [and for a great distance around, stretching to the limits of the mountain forest toward Alaska,* east to northern Manitoba, and south all the way to the lower Fraser and beyond], to be ABSMs of the *Sasquatch* type and with all its characteristics, such as winter-withdrawal, occasional bursts of carnivorousness, and so forth. I also have reports in the form of private letters of similar creatures from all across the Northwest Territories just south of the tree-line, and again in northern Quebec Province.

This is a somewhat irksome matter as I have been unable to obtain any casts of footprints or other physical evidence from these regions nor even sworn statements as yet. The reports are categoric and specific. Those from northern Manitoba are second hand only, and from Amerindian informants via white men who have hunted there for many years in succession. Those from Quebec have puzzled me for years. I have constantly heard about them but have only three pieces of paper to show for my exhaustive and prolonged inquiries and appeals. These are all letters from American summer visitors on serious hunting and camping trips by canoe, guided by professional Amerindian trappers and hunters. All three are substantially identical and all give somewhat similar accounts of events in widely separated places. One is from a lone man, a business executive from Chicago; one is from a party of four men of assorted professions who have hunted for years on their annual vacations together; the third is from the father of

* The dividing line between two major types of vegetation forms a great curve to the north close to this area, and then bends down to the south, and even southeast for a stretch, along the Pacific coast. The southernmost of these is a type of forest that grows far up mountainsides; the northern type grows only in valleys, leaving the upper slopes bare. ABSMs are reported from all over the former in the mountains but not from the latter. (See Chapter 18.)

a family of four—three grown sons and a (then) teenage daughter.

In each case, a tall, very heavily built, man-shaped creature with bullet-head and bull-neck, and clothed all over in long shiny black hair, with very long arms, short legs and big hands, is said suddenly to have appeared on the bank of a river in which the party was quietly fishing. On one occasion, the creature is said to have carried off some fish left on a rock on the bank; on another it chased the Amerindian guide out of the woods and into his canoe and then waded some distance out into the water after him. The family party seem to have become fairly familiar with two of the creatures over a period of several days. They say they constantly prowled around their camp, and showed themselves among the trees whenever they went out in the canoes. One seems to have shown signs of chasing the girl on one occasion but, the father told me, they gained the impression that this seemed to be more through curiosity than menace. Two of the Amerinds are said to have asserted that they and their people knew the creatures quite well and that there were quite a lot of them in those forests. The other guide, who was chased, appeared to be scared almost witless and swore that the thing was some form of spirit or devil. However, it smashed branches and hurled stones, it is reported.

I am frankly stymied over these reports. Two of the writers asked that I withhold their names *in perpetuo* as they did not want the reports to become known to their business associates. The third man I never traced. It was many months before I could get to the places from where these people wrote and although I traced two of them, they all stopped answering my letters and I am left with nothing to follow up. This is an almost chronic condition of laborers in the vineyards of ABSMery. People almost all just dry up in time. Of course, many probably write in the first place by way of a joke or just to see how gullible the inquirer is; but not all are of this ilk. Many people also, I believe, take fright at the possibility of ridicule, or even become alarmed about their own sanity, after they have once gotten something so unusual off their chests. Others again, either consider the matter explained or just don't want it explained. It takes years of work to get at the facts and this is rendered almost futile when one is dealing with a new locale that is only just being penetrated by civilized people.

The ABSM tradition extends all across Canada but is concentrated in southern British Columbia; probably because that was the first area opened up and is still being probed from all around.

3. Further Sasquatchery

What are you going to do with a new story
when you've got one? How do you know it is not
an old one plastered over with new facts?

Just because I have skipped over some 60 years by the recounting of only 8 stories, is not to be taken to mean that these were the only reports current during that period. Quite to the contrary, almost every year somebody or some group of people in southern British Columbia stated that they had either run into a *Sasquatch*, been chased by one, shot at one, or seen its foot-tracks. Many of these accounts are from our friends, the Amerinds, and many of them are not specifically dated. They begin "Some years ago . . ." or "Early last year . . ." but fail to state which year, or how many years ago. A lot of these have become garbled because of loose reporting or because they were made to specialists in local languages, each of which has a different name for its local ABSMs. The very name, *Sasquatch*, now so widely disseminated and known in Canada, is actually of partially artificial construction and was first, I understand, coined by Mr. Burns in an effort to obviate some of this muddle and to draw attention to the fact that throughout a very wide area—from the Yukon to California—all the names refer to the same creature. This name is derived from the Salish Amerindian word for "wildmen of the woods" which may be transliterated as *Te Smai'Etl Soqwaia'm*, also written as *Sami 'Soq' wia'm*, the form used by the Chehalis tribal group. Farther south among the Pugets, the name was *Hoquiam*, now the name of a flourishing small town on the Chehalis River south of the Olympic Mountains in Washington State. However, many of the locals had a habit of prefixing almost everything with a sibilant so that this name also came as *S'oq'wiam*. In the Cascades the name was *See-ah-tik* but down around Mt. Shasta it was *See-oh-mah*. In the Klamaths we note that it is still *Oh-Mah* among the Hŭppa, while the Yuroks call them *Toki-mussi*. On Vancouver Island, and north up the inlets of the mainland, the sound changes to something more like "*Sokqueatl*" or "*Soss-q'atl*" and it was from this that Mr. Burns derived the anglicized "*Sasquatch*," or "*Susquotch*" as Americans have usually written it.

I mentioned above that all these names refer to a single kind of creature. This is so, as far as the Amerinds are concerned; but, you may well ask as you read on, how come these creatures are stated to vary so much in appearance. On analysis, it will be noted that this variation is almost exclusively in two features—length and quality of hair and its disposal about the body, and color of skin and fur. Further analysis will also show that these differences seem to be due to age and sex. The young ones, like Jacko and the one shot by a local hunter and to be described in a moment, are said to have had light faces and yet black, shiny, straight, and apparently orderly hair all over (one imagines like that of a chimpanzee), but the adults are invariably said to have black faces and skin, and reddish-brown fur, often shaggy, and sometimes washed with white or silver-tipped. The matter of long head-hair is variable but most of the close-up sightings speak of very short head-hair, no beard, but a curiously forward, upward, and finally backward curl of longer hair all across the brow like that seen on certain Spider Monkeys (genus *Ateles*). I reproduce a photograph of a sketch that I made under the direction of Mr. Ostman during our interview, that emphasizes this strange feature. (See Fig. 41.)

The growth and rearrangement of body hair with age is absolutely consistent with what is known among other mammals and notably primates and particularly apes. Further, the changes in color are exactly what we would expect and are very similar to those to be noted among gorillas and some gibbons. Baby chimpanzees often start off with faces and

MAP II. BRITISH COLUMBIA

See facing page for legend

hands the color of those of white men but end up with complexions as dark as Dravidians or Wolofs. Some gorillas develop a distinct gingery tinge—the "black" of mammalian hair being only melanin, and really a very dark red—and almost all of them go silvery gray with age. Some gorilla families have bright red topknots just like some human beings. Some gibbons vary in a most bewildering way in coat colors. They may be black, gray, chocolate, white, or beige to start with and throughout life, or they may change from one color to another with age. Different races of the different species do all manner of different things in this respect. It is therefore quite consistent that these large ABSMs should start off with jet black hair and light skins, and end up hoary old black-faced creatures with silver-tipped reddish fur. The females might lack the gray and might be less shaggy. There may also be family likenesses to start with.

Let us assume that we are now chronologically at the turn of the year 1920 to 1921 but still in British Columbia. As I said in the brief historical review of world-wide ABSMery, this was a most important date in that it saw the birth of the term "Abominable Snowman" and really kicked off the whole

---

MAP II. BRITISH COLUMBIA

This represents an area of some 270,000 square miles. Ninety per cent of this is uninhabited, despite the enormous conglomeration of the City of Vancouver, the old capital of Victoria on Vancouver Island, and the somewhat extensive cultivated areas on that island and about the lower reaches of the Frazer River from Agassiz west. The coastal plains of Puget Sound add only 2 per cent. The whole of it, apart from Vancouver Island, the Frazer delta, and the Puget Sound area, is mightily mountainous and great parts are not truly explored, though there are now excellent large-scale maps resultant from aerial surveys. The Olympic Mountains and the coastal fringe northward around Vancouver Island and north of the lower Frazer River are clothed in an immensely tall, several-layered "Rain Forest" with conifers predominating (the largest trees in the world are found here) and choked with mosses, ferns, and a broadleafed undergrowth. The other areas are heavily forested but for their peaks.

thing. I have often wondered what would have happened if the Squamish word for these creatures in their country, instead, had happened to have been mistranslated as something equally fetching. I suppose we would then, in time, have witnessed a *New York Journal American* Expedition to Harrison Lakes, and Admiral Byrd flying skin-trophies to Chicago from the hamlets of the Alaskan panhandle. It is nothing more than a quirk of history and a series of harmless mistakes that has put Nepal instead of Vancouver Island on the map in this respect; though it has to be admitted that Mt. Everest has played its part.

It was about this time, moreover, that an incident is alleged to have occurred in this area that is in many ways perhaps one of the most fantastic ABSM stories ever told. It only came to light in 1957 but concerns happenings alleged to have taken place in 1924 in the mountains behind Toba Inlet, which is on the coast of British Columbia (see Map II). It came to light through a letter (written to John Green, owner of *The Advance*, published in Agassiz, a small town near Harrison, some 70 miles from Vancouver) by a retired prospector and lumberman of Swedish origin named Mr. Albert Ostman. This letter was a result of the publication by Mr. Green of an affidavit sworn to by a Mr. William Roe (now of Edmonton, Alberta) concerning certain experiences he had in the year 1955 on Mica Mountain on the Alberta border. (This latter statement is reproduced in full in the next chapter and concerns Mr. Roe's meeting with a female *Sasquatch.*) Reading this, Mr. Ostman apparently decided to break more than a quarter century of silence and relate what had happened to him. Mr. Ostman now lives at Fort Langley outside Chilliwack, and John Green, who for years has gathered information on the *Sasquatches,* sought him out and persuaded him to write his full story. This Mr. Ostman did—painstakingly, and in two large notebooks. John Green published this in his newspaper along with a photostat of a sworn affidavit testifying to its truth by Mr. Ostman.

I had the pleasure of meeting both gentlemen in company

with a partner of mine, Robert Christie, who was traveling with me at the time, and a Mr. and Mrs. René Dahinden. He is Swiss; his wife, Swedish, as is Mr. Ostman. As I already had Mr. Ostman's story both on paper and on tape from an interview between him and a reporter from a local radio station, I confined my questions to trying to recall his memory about certain zoological or anthropological details. I fully admit to having loaded these questions with snares and abstruse technical catches, and to having been rather rough in my approach. I know that I thereby incensed John Green and the Dahindens, who not only have a very great affection and respect for Mr. Ostman but feel that, with his still slight language difficulty, outsiders such as I tend to rattle him. I do not agree, in that Mr. Ostman has the wisdom of age as well as long experience, and a sense of humor that cannot be downed; and I don't think that he was annoyed with me then, or will be hurt if he reads this. In fact, I felt that he was twinkling at me all the time; and I fancy that, if he ever thought of me after I left, it was simply as a "very funny fellow," as he might say. This is more the case since I went away a very puzzled reporter.

This story, when read cold, sounds utterly preposterous. If one has read a great deal on ABSMs in general and on the *Sasquatch* in particular it also, at first, appears highly suspicious because it seems to knit together just about everything else that has ever been published on the matter. In fact, given some firsthand experience of the country, I could have written just that story myself. The world is full of good weavers of yarns and some of them, who are not professional writers of fiction, can be so damnably convincing that they have fooled not only the press but governments and even peoples, if not the whole world. Fabrications, if well enough done, consistently adhered to, and big enough lies, can, as has so often been pointed out (e.g. the case of Hitler) be utterly convincing. However, in technical matters, and most notably in the biological sciences, there are subjects that just cannot be imagined or thought up by anybody, unless they have learned of them

[51]

specifically in advance and, what is much more important, their exact significance relative to a whole host of other technicalities is appreciated. Anybody can read everything that has been published on *Sasquatches* and yet still attribute to them some trivial biological character that really is *impossible*. In the case of ABSMs there are a large number of very abstruse matters of this nature that may be slipped in casually. Only one answer to these can be right, while an endless string of other answers will be wrong, and conclusively so. I put about two dozen of such, directly and unexpectedly, to Mr. Ostman and, of all those for which he had a reply, he did not miss once—not one impossible answer; not a single uncalled-for elaboration; and not one unrequested fact that did not have a possible and quite logical place in the general picture. What is more, when we got off on the sketching of the creature's head, there emerged several points that were *not* then in published *Sasquatch* literature, nor in that on any ABSM, nor even in textbooks of physical anthropology. Yet, subsequent to that interview, some of these points (such as the odd head-shape) *have* appeared in the last type of publication.

This is really rather alarming and has given me many sleepless nights. Some things I just cannot bring myself to take at their face evaluation; and, frankly, Mr. Ostman's story was at first one of these. Besides, he even included some gross fallacies such as that he became poisoned through eating a broody grouse—an old wives' tale, if ever there was one. But then, I have to admit to myself now, that this fact is still believed in parts of his home country—namely, that one does get poisoned by eating birds taken sitting on eggs—and that he probably believed this; while he was in poor enough condition at the time of his adventure to be made sick by almost anything. Also, I ask myself, why tell this story? Mr. Ostman is not an uneducated country bumpkin. He is well read, speaks three languages, has traveled quite a lot, lives very much in the world, and knows quite well what ridicule is, and all about its deadly efficacy. He is retired, owns his own property, has many friends, and does not need publicity; nor does he welcome it, though he is extremely long-suffering and most gracious in discussing his experiences with newsmen and others who call upon him. He never told his story in his youth for fear of ridicule, knowing what effects it might have. He doesn't care now: he is still sincerely puzzled; and he is eager to do anything he can to help clear up the mystery. Mr. Ostman is, in fact, sick and tired of skeptics.

After a strenuous year on a job, he decided to take a part vacation with some prospecting on the side. He chose a wild area at the head of this Toba Inlet which is the first substantial fjord north of Powell River. This is on the mainland opposite the middle of Vancouver Island. There was allegedly a lost gold mine thereabouts and he decided to take a crack at finding it. He hired an old Amerind to take him up the fjord and he says that he first heard from him on that journey of the existence of the giant hairy "Wild Men of the Woods." He had supplies for three weeks, plus rifle, sleeping bag, and other basic equipment. The local man left him alone on shore and he proceeded inland and found a good campsite.

This he fixed up very comfortably, making a thick bed of small branches on which to place his sleeping bag, and hung his supply bags well off the ground on a pole. The next morning he, nonetheless, found his things disturbed, though nothing was missing. Being a knowledgeable woodsman, he assumed that a porcupine was responsible, so the following night he loaded his rifle and placed it under his bed flap. The next morning he found, to his dismay, that his packsack still hung from the pole well off the ground but that its contents had been emptied out and some items of food taken. Strangely, his salt had not been touched. This surprised him not a little, because porcupines have an insatiable craving for salt and always go for it first. At the same time, he did not think that it was a bear because, although he admits to having been a very heavy sleeper, bears usually make a great rumpus and smash up everything. Albert Ostman did not like these events one bit, so he stayed rather closely around camp in the hopes of catching the marauder in the act. On the third night he took special precautions; intending to stay awake all night, he did not undress but merely removed his boots and left them at the bottom of his sleeping bag, put his geological pick to hand, and took his loaded rifle into the bag with him. But he did fall asleep.

The next thing he knew he was being picked up like a puppy in a paper bag, and felt himself heaved, as he at first thought, on to a horse's back. Bemused and half awake, he tried to get at his knife to cut his way out of his sleeping bag, but he was wedged down into the bottom in a sitting position and could not reach it. Then he felt his packsack bumping against him with the hard cans within clearly discernible by their sharp impact. As far as I was able to ascertain in my interview with him, he was completely in the bag, as one might say, and its opening was being held shut above his head. How he managed to breathe in such circumstances, and for over an hour, puzzled me until he explained that he was slung over the back of something walking on two legs and that its hand was not big enough to go all around the bag-opening. I never heard Mr. Ostman say that he was scared, but he admits that he was terribly hot in there and that his cramped legs were extremely painful. Don't forget, moreover, that he hadn't a clue at that time as to what was going on or what had got him.

He says that he was carried up hill and down dale, when he was dragged along the ground, and that his carrier even jog-trotted over level places. This is some going for anything carrying a man of Ostman's size plus a knapsack full of supplies and other equipment. But, this is by no means the strangest part of the proceedings; yet it is still at least possible. Another aspect seems quite impossible; namely, that Ostman estimates—and sticks to it—that this trip in the bag took three hours. In an interview with a commentator from a radio station (a tape of which I have), but made, of course, a quarter of a century later, he says thirty miles. Personally, I fail to see how he survived such an ordeal, stuffed up in a bag, but that is not so much the point: what is are the time, the distances, and the speed of travel implied. These are not easy things to estimate at the best of times, and they are among the first to become exaggerated in the mind with the passage of time. I wish that Mr. Ostman had not tried to give any estimates at such a late date since it causes the eyebrows of all who read or hear his story to go up sharply.

Anyhow, at the end of what must have been an ordeal, however brief it really was, he was dumped unceremoniously on the ground. He heard some voices gibbering but not using true speech as far as he could ascertain. He apparently got his head out of the bag for air and then tried to crawl out, but his legs had rather naturally gone numb and it was some time before he could emerge and rescue his boots. It was still dark and starting to rain. He then tells, in various characteristic ways, what happened when it began to dawn and he could see the outlines of four large creatures on two legs around him. I don't know if his native Swedish wit got the better of him, but he says that when he could stand up he asked the somewhat banal question: "What do you chaps want with me?" I find this most refreshing.

He found that his captors consisted of two big ones (a pair), and two youngsters, also a male and a female. He stresses that the two latter seemed thoroughly scared of him, and that the "Old Woman," as he rather delightfully called the elder female, seemed very peeved with her mate for dragging such an object home; but, he then goes on to say—and this I find very interesting, if odd—that the "Old Man" kept gesticulating, and *telling* the others all about it. In other words, their gibbering *was* speech. All of them were hairy and without clothing; and Ostman estimates the "Old Man" to have been between seven and eight feet tall. When the sun was fully up, they all left him.

He says that he found himself in a ten-acre bowl high in the mountains, its edges so steep as to be unscalable, and with only one outlet—a V-shaped cut with walls about twenty feet high and about eight feet wide at the bottom. It is not quite clear why, at this point, he did not try to make a break for this gap, but this was possibly because of his still wobbly legs. Later on, he made several attempts, both frontally at speed and by subtlety, but the "Old Man" kept a weather eye on him and invariably cut off his approach, making *"pushing"* motions with his hands, and a sound that Ostman invariably describes as something like "sooka-sooka." However, when he first arrived, he moved over to the opposite side of the bowl and set up camp under two small trees. I find the inventory that he says he took of his possessions most interesting. Prunes, macaroni, his full box of rifle cartridges, and his matches were missing; so was his pick. Otherwise, all was intact. He had an emergency waterproof box of matches in his pocket but says that there was no dry wood in the valley, which seems to have been open and grassy with a few scattered junipers. All his cooking utensils had also been left, but he opened a coffee-can and went to look for water.

I will now complete the story as best I can from the various versions that I have heard, though I would stress that Mr. Ostman is remarkably consistent however many times he tells his story. Each interviewer, however, manages to ask a new question and elicit from him some scraps of information that the teller had not thought of or mentioned before. As I don't know the sequence in which the various versions were recorded, I have no way of differentiating between inconsistencies and mere additions. It would seem that Ostman made his first attempt to get out on the second day but was driven back by the "Old Man." The young male kept coming closer to him and he finally rolled his empty snuff box to him. The *Sasquatch* grabbed it, showed it to his sister, and then took it to his father. Somehow, Ostman got it back, because he used it later. During the next five days nothing much seems to have occurred except that the young male gave Ostman some grass with sweet roots to eat and got some snuff in return, which he chewed. The "Old Man" then also developed a liking for snuff; and this finally did the trick.

On the seventh day, as far as I can make out, the boy and the "Old Man" came right up to Ostman and squatted down watching him take a pinch of snuff. Ostman held out the box to him (the "Old Man") who, instead of taking a pinch in imitation, grabbed the box and emptied its whole contents into his mouth and swallowed it. In a few minutes his eyes began to roll, he let out a screech, and grabbed a can half full of cold coffee and coffee-grounds, which he drank. This made him worse; and, after rolling about some more, he charged off to the spring. Ostman gathered up his possessions and made a dash for the opening in the cliff. The "Old Lady" tried to intercept him and was very close on his heels, but he fired a shot at the rock above her head and she fled back again. Ostman found himself in a canyon running south, down which he made record time, as he put it. Then, he climbed a ridge and saw Mount Baker way off to the side, so that he knew which way to go to hit the coast. He was not followed.

He rested for two hours on the ridge, then started down again. That night he camped near heavy timber and shot a grouse sitting on eggs; he roasted and ate the bird. The next morning he was very groggy and stomachically upset, which he attributed to eating the grouse, since it was his belief that a broody bird was poisonous. Finally, he heard a motor running and made for it, coming out at an advance logging operation. The foreman, seeing that he was just about at the end of his tether, took him in and fed him and let him rest up for a couple of days. Ostman then made his way down to a camp on the Salmon Arm Branch of the Sechelt Inlet, where he got a boat back to Vancouver.

This is Mr. Ostman's story and you may make what you will of it. As I have said, there are some curious discrepancies in it but not even these are impossibilities, with the exception of the times and the distances as mentioned above. The grouse, broody or not, could quite well have upset his stomach. Mr. Ostman seems to be a straightforward and honest man. But, it is the facts that he gave me about the ABSMs themselves that go farther than anything else to convince me of the validity of the whole thing—unless, of course, as I have also said above, he read all of these elsewhere.

His descriptions of the creatures are considerably detailed. What is more, the sexual and age differences he describes are very reasonable, and do not in any way insult such variations as found among men or other primates. Of the adult male, he says that he was about eight feet tall, barrel-chested, with powerful shoulders and a very pronounced and large "hump" on his back, causing his head to be carried somewhat forward. This is exactly in accord with the posture of some sub-hominids as deducted from the angle at which the condyles are set to the back of the skull. The biceps were said to be enormous but to taper to the inside of the elbow; the forearm to be disproportionately (to a human) long but well proportioned. The hands were wide but the palm long and curved permanently into "a kind of a scoop"; the fingers short, and the nails flat, broad, and "shaped like chisels." Mr. Ostman mentioned to me quite casually that they were copper-colored. This is most significant, as we shall see later (Chapter 14). He estimated the neck to be about thirty inches around. The whole body was covered in hair, somewhat longer on the head; shorter but thicker in other parts. It covered his ears. Only the palms of the hands and the soles of the feet, which had pronounced pads, were naked and a dirty dark gray in color. The "top" (i.e. bridge) of the nose and eyelids alone were naked. The big male's canine teeth were longer than the others but not sufficiently so to be called tusks.

The adult female he described as being over seven feet tall and weighing between 500 and 600 pounds. He said that she could have been anywhere between forty and seventy years old, using humans as a criterion; but, she was apparently very ugly, with an enormously wide pelvis that caused her to walk like a goose. She had long, large, and pendent breasts.

The young male spent the most time near Ostman and was thus most closely observed. Ostman says that he could have been anywhere from eleven to eighteen years of age, but was already seven feet tall and weighed about 300 pounds. His chest would have measured between fifty and fifty-five inches around and his waist some twenty-six to thirty-eight inches:

and don't forget that Mr. Ostman was a lumberman and better at estimating the girth of things (like trees) than the average person. He had wide jaws and a narrow, sloping forehead. The back of his head, as in all of them, apparently rose some four or five inches above the brow-line, and was pointed. Mr. Ostman went to great pains to explain this, and to get the shape just right, as shown in the sketch that I made under his direction (see Fig. 41). The head-hair was about six inches long; that on the body shorter but much thicker in some areas.

The young female was very shy; she did not approach Ostman closely but kept peeking at him from behind the bushes. He could not estimate her age, but remarks that she was without any visible breast development and was, in fact, quite flat-chested. Like her mother, she had a very pronounced up-curled bang across her brow-ridges. This was continuous from temple to temple. Curiously, no amount of questioning would prompt Mr. Ostman to elaborate any further on this individual, which may in part be psychological since it seems to be his conviction that he had been kidnaped as a potential suitor for her, and I think he has a sort of subconscious and rather touching modesty about her shyness. Mr. Ostman maintains a delightful old-world delicacy about the proprieties and neatly turned aside some purely biological questions with such noncommittal phrases as "I wouldn't know about that." But he did tell us of a few most interesting observations on the behavior of the group.

First and foremost was this gibbering in which they indulged. As his story progresses, it becomes quite clear that he assumed in the end that they were actually communicating intelligently, since they made a variety of noises befitting special situations and seemed to discuss the objects they carried one to the other. There was also the delightful expression "ook" that the young male made on one occasion. Then, almost equally significant, was the fact that the old female and the young male went regularly to gather vegetable foods; the former going out of the gap and returning with armsful of branches, including fresh spruce and hemlock tips, grasses, and ferns. These, he told me, she washed and stacked up. She also brought quantities of a certain kind of "ground nut" of a kind that Mr. Ostman had often seen in abundance on Vancouver Island. (Shades of Mike King!) Inquiry elicited the fact that this is a root-nodule of a herbaceous plant related to the Hemlock of Europe (not the tree called by that name in this continent), one form of which grows such nutlike growths that are edible and, in fact, delicious. The young male used also to go every day and return with bundles of a kind of grass with a "sweet root."

Mr. Ostman stressed the incredible climbing ability of the male youngster and remarked on the form of his and his father's feet as having an enormous big toe. At one point he states that, in order to get a purchase in climbing, all he would need would be to find a resting place for this toe alone.

One of Mr. Ostman's observations is very peculiar, and is one which can be taken either as evidence that the whole thing is a wild fabrication or as glowing testimony to the recorder's veracity and powers of observation. It brings up some very fundamental matters with regard to the history of culture among early hominids—if it proves to be true, that is. This was that, according to Mr. Ostman, the four creatures slept and lived for the most part under a rock-ledge like the rock-shelters known to have been favored by many Stone Age men. In this, which was some ten feet deep and thirty feet wide, he says that they had regular beds of branches, moss, and dry grass, and that they had coverlets of woven strips of bark, forming great flattened bags, *and stuffed with dry grasses and moss.* However, I could not elicit from Mr. Ostman any facts as to whether he visited the shelter and examined these objects or, if not, how he knew so much about their construction and composition. This worried, and still worries, me.

Should such items have existed, combined with the primitive speech, the collection of food and its washing, we are faced with a pretty problem. Are we to suppose that, prior to the use of bone and horn tools (such as the little very primitive Australopithecines of South Africa are now thought to have used) and the discovery and control of fire, hominids (man or otherwise) went through a prior period of food-gathering but still knew weaving? This would seem not to be unreasonable or illogical, though even crude weaving calls for considerable dexterity. Be it noted at the same time that Orangs, Chimps, and Gorillas tie true knots when making their sleeping platforms on occasion, while some Gorillas do so regularly. Weaving in its most primitive form, moreover, is little further advanced than excessive knot-tying; besides, some birds do the most incredibly accurate jobs of weaving, even with different colored wools, on a piece of small-mesh wire. Also, animals, and particularly the primates, definitely do communicate. (I may say that even I can speak fairly good Rhesus!)

Thus, there is nothing really outrageous about Mr. Ostman's statements about these creatures nor about the whole concept of some of the *Sasquatches* (Neo-Giants, as we shall eventually come to call them) being food-collectors, with a primitive speech but lacking fire, clothes, and tools. And, it is even more interesting to note that Mr. Ostman states clearly that he never saw them bring to their camp, or eat, any animal food. The most primitive sub-sub-hominids were probably, like their close congeners (the apes), fruit- and leaf-eaters. Only when some of them were forced out onto the savannahs, scrublands, and deserts did they have to take up animal-hunting and become partially or wholly carnivorous, as, apparently, did the Australopithecines of South Africa. If the Great Apes, still living today, have continued to be pure vegetarians, there is no reason why some of the most primitive Hominids could not also have so continued to be. This gives us a somewhat new concept of our own background and of the possibilities for ABSMs.

This brings up several questions that, if it were possible, ought to be discussed concurrently with any straight reportage on the ABSMs themselves. The details of a report on any such alleged creature cannot be evaluated properly without prior knowledge or exposition of certain aspects, on the one hand, of vegatology and, on the other, of palaeanthropology, both physical and cultural. Our whole outlook on the last of these fields has undergone a complete revolution in the past two decades. The old idea was that sub-Hominids had bent knees, a stooped gait, ape-like faces and teeth and tiny brains, and no "culture" at all, in that they had no speech, no fire, no tools. Then, it was also previously believed, sub-Men came along that stood more upright and were bigger-brained and less ape-like about the muzzle. These creatures were assumed to have invented tools by bashing at things with stones, which often cracked, giving them cutting edges. The usual idea was that they were hunters and lived in caves, and progressed steadily toward Man, though taking an inordinately long time about it. Finally, some of them developed such big brains and pushed-in faces that they became true Men.

Meantime, their tools got better and better, finer made, and more diversified. Also, the great growth in certain parts of their brains made cogent speech possible. Then, the theory went, they somehow got on to fire and its uses as opposed to its dangers, developed "society," developed the art of pottery, and finally realized that from tiny seeds tall grasses grow, so that they gave up hunting and settled down to agriculture. And, in time, came the wheel, writing, money, and all the other improvements that inevitably contributed to their downfall. Be that as it may, the development of Hominid *mentality*, as opposed to mere brain capacity and structure, was not much considered, being assumed simply to have advanced along with his gray matter, since, it was then believed, you could not be expected to assess the psychology of any extinct creature and especially one with a brain no bigger than an ape's.

The first real break through this massive theoretical structure was really made by a rather dubious antiquarian named Mr. Dawson, who foisted upon science not only the now infamous Piltdown cranium, teeth, and mandible, but also a

[62]

fraudulent tool that he himself appears to have made from a semi-fossilized bone of some elephantine. Piltdown Man never did look quite right but was fully accepted by physical anthropologists as a very early and primitive man-thing but with a very large brain. Thus, his grotesque tool was also accepted. Then, there was also some suspicion that tools had been found in the same strata in which Dr. DuBois found his genuine "Apeman" in Java, but the matter was rather hurriedly suppressed. Acceptance of tools along with sub-Humans finally came with the diggings in north China that produced Pekin Man. This was rather a rude shock, but did not grossly disturb the neat historical sequence then believed in. It simply meant moving tool-making back some way. The real shocker came with Dr. Raymond Dart's discovery of enormous quantities of bone and tooth tools most obviously and carefully worked, which had to have been made by none other than the little Australopithecines that were at first classed as Apes, and only grudgingly accepted as most primitive sub-Men after the discovery that they walked erect. Worse still, there was a strong plea made for acceptance of the fact that they used fire as well. It then was decided (by most, but not all, anthropologists) that the Hominids went through what is called an odontokeratic tool-making phase before they came to use stones.

This picture has now been considerably muddied by Dr. Leakey's discoveries of early Chellean Man in East Africa, an appalling-looking chap with positively immense brow ridges, but who made splendid hand-axes of stone. Nevertheless, it is only now slowly dawning on anthropologists that the first tools were more probably sticks, and otherwise wooden; for the earliest Hominids were definitely vegetarians and forest dwellers. The bone-tooth toolmakers were carnivorous. The use of wood implies pulling twigs and branches from trees and the discovery of the many uses of strips of bark. From this to primitive weaving is but a step. Thus, it is quite probable that the earliest Hominids were vegetable gatherers, using sticks and possibly the crudest weaving, and that they so equipped themselves long before they got around to breaking stones, using fire, or even developing a true language. It is therefore most interesting to note, as our story continues, that the only tools ever reported in use by ABSMs have been sticks.

4. The Appearance of Bigfeet

If you want to find out how crimes are really
solved, ride around with a police patrol for a few
nights. The same little things, happening time
and time again, always bring the culprits to book.

Mr. Ostman's story was related to Queen Elizabeth II when
she visited British Columbia in 1959. The story is said to have
been submitted to Her Majesty by an official, along with other
Sasquatchery, in a remote vacation cabin at a lake near Kamloops on August 28. By coincidence, I was on that same day
closeted in a small railroad shack with a charming Amerindian
couple named Mr. and Mrs. George Chapman, at Jacko's old
retreat of Yale, some miles lower down the Fraser River. I
also was hearing a story, but firsthand, and in what turned out
later to have been rather extraordinary circumstances.

We had crossed the log-filled Fraser in a small boat, rowing
first away upstream, then very rapidly a long way downstream
broadside, and then finally a long way back upstream again
on the other side in the lee of a tall bank. Scrambling to the
top of this we struck a railroad along which an Amerindian
family were straggling in from the hills. By some strange quirk
of fate, this turned out to be the Chapman family for whom
we were looking. They hospitably invited us in to the freight
office, behind which they had a small house.

That could have been a very tense or even profitless interview for several reasons. Here we were, two palefaces with
locally odd accents—Robbie Christie, though born in New
Jersey, has ranched in Colorado, wears a Texan-type hat, and
has a vaguely British accent; while I talk a sort of bastardized
Anglo-Saxon with an American intonation and a British ac-

cent, neither of which are popular in Canada—who had met up with a reticent Amerindian couple, apparently quite by chance on a railroad track, and who now had suddenly demanded to hear the facts of a series of incidents that had happened to these good people 18 years before. Somehow, however, and perhaps due mostly to a kind of mild shock, we all got off on the right foot and within a surprisingly short space of time Mrs. Chapman was recounting those terrible hours with complete clarity, only every now and then being mildly corrected by her husband, or having her account augmented by details which she had not witnessed.

We had heard their story from several sources and had read it in several printed versions, but I wanted to get it firsthand and I wanted to be able to shoot my particular glossary of awkward biological questions at principals, who were alleged eyewitnesses of a living *Sasquatch* in daylight. It is just as well that we crossed the Fraser River just when we did, and so met the Chapmans, because about a month afterward they were drowned crossing at the same spot late one night. The irony and tragedy of this event upset me greatly for, as I have said, I have a great liking and respect for the Amerindian peoples and I not only found this couple graciously natural and friendly but they also impressed me, as very few other people have ever done, with their sincerity and honesty. The Chapman family at the time of the incident consisted of George and Jeannie Chapman and three children. Mr. Chapman worked on the railroad. They lived near a small place called Ruby Creek, 30 miles up the Fraser River from Agassiz. It was about 3 in the afternoon of a cloudless summer day when Jeannie Chapman's eldest son, then aged 9, came running to the house saying that there was a cow coming down out of the woods at the foot of the nearby mountain. The other kids, a boy aged 7 and a little girl of 5, were still playing in a field behind the house bordering on the rail track.

Mrs. Chapman went out to look, since the boy seemed oddly disturbed, and then saw what she at first thought was a very big bear moving about among the bushes bordering the field beyond the railroad tracks. She called the two smaller children who came running immediately. Then the creature moved out onto the tracks and she saw to her horror that it was a gigantic man covered with *hair*, not fur. The hair seemed to be about 4 inches long all over, and of a pale yellow-brown color. To pin down this color Mrs. Chapman pointed out to me a sheet of lightly varnished plywood in the room where we were sitting. This was of a brownish-ocher color.

This creature advanced directly toward the house and Mrs. Chapman had, as she put it, "much too much time to look at it" because she stood her ground outside while the eldest boy—on her instructions—got a blanket from the house and rounded up the other children. The kids were in a near panic, she told us, and it took 2 or 3 minutes to get the blanket, during which time the creature had reached the near corner of the field only about 100 feet away from her. Mrs. Chapman then spread the blanket and, holding it aloft so that the children could not see the creature or it them, she backed off at the double to the old field and down on to the river beach, out of sight, and ran with the kids downstream to the village.

I asked her a leading question about the blanket. Had her purpose in using it been to prevent the children seeing the creature, in accord with an alleged Amerindian belief that to do so brings bad luck and often death? Her reply was both prompt and surprising. She said that, although she had heard *white men* tell of that belief, she had not heard it from her parents or any other of her people, whose advice regarding the so-called *Sasquatch* had been simply not to go farther than certain points up certain valleys, to run if she saw one, but not to struggle if one caught her, as it might squeeze her to death by mistake.

"No," she said, "I used the blanket because I thought it was after one of the kids and so might go into the house to look for them instead of following me." This seems to have been sound logic as the creature *did* go into the house and also rummaged through an outhouse pretty thoroughly, hauling from it a 55-gallon barrel of salt fish, breaking this open, and scattering its contents about outside. (The tragic irony of it is that all those original three children *did* die within 3 years, while, as I have said, a month after we interviewed them, the Chapmans and their *new* children drowned as well.)

Mrs. Chapman told me that the creature was about 7½ feet tall. She could easily estimate the height by the various fence and line posts standing about the field. It had a rather small head and a very short, thick neck; in fact really no neck at all, a point emphasized by William Roe and by almost all others who claim to have seen one of these creatures. Its body was entirely human in shape except that it was immensely thick through its chest and its arms were exceptionally long. She did not see the feet which were in the grass. Its shoulders were very wide and it had no breasts, from which Mrs. Chapman assumed it was a male, though she also did not see any male genitalia due to the long hair covering its groin. She was most definite on one point: the naked parts of its face and its hands were much darker than its hair, and appeared to be almost black.

George Chapman returned home from his work on the railroad that day shortly before 6 in the evening and by a route that bypassed the village, so that he saw no one to tell him what had happened. When he reached his house he immediately saw the woodshed door battered in, and spotted enormous humanoid footprints all over the place. Greatly alarmed —for, like all of his people, he had heard since childhood about the "big wild men of the mountains," though he did not hear the word *Sasquatch* till after this incident—he called for his family and then dashed through the house. Then he spotted the foot-tracks of his wife and kids going off toward the river. He followed these until he picked them up on the sand beside the river and saw them going off downstream *without any giant ones following.*

Somewhat relieved, he was retracing his steps when he stumbled across the giant's foot-tracks on the river bank farther upstream. These came down out of the potato patch, which lay between the house and the river, milled about by the river, and then went back through the old field toward the foot of the mountains where they disappeared in the heavy growth.

Returning to the house, relieved to know that the tracks

of all four of his family had gone off downstream to the village, George Chapman went to examine the woodshed. In our interview, after 18 years, he still expressed voluble astonishment that any living thing, even a 7-foot-6-inch man with a barrel-chest could lift a 55-gallon tub of fish out of the narrow door of the shack and break it open without using a tool. He confirmed the creature's height after finding a number of long brown hairs stuck in the slabwood lintel of the doorway, above the level of his head. George Chapman then went off to the village to look for his family, and found them in a state of calm collapse. He gathered them up and invited his father-in-law and two others to return with him, for protection of his family when he was away at work. The foot-tracks returned every night for a week and on two occasions the dogs that the Chapmans had taken with them set up the most awful racket at exactly 2 o'clock in the morning. The Sasquatch did not, however, molest them or, apparently, touch either the house or the woodshed. But the whole business was too unnerving and the family finally moved out. They never went back.

After a long chat about this and other matters, Mrs. Chapman suddenly told us something very significant just as we were leaving. She said: "It made an awful funny noise." I asked her if she could imitate this noise for me but it was her husband who did so, saying that he had heard it at night twice during the week after the first incident. He then proceeded to utter exactly the same strange, gurgling whistle that the men in California, who had told us they had heard an *Oh-Mah* (or "Bigfoot") call, had given. This is a sound I cannot reproduce in print, but I can assure you that it is unlike anything I have ever heard given by man or beast anywhere in the world. To me, this information is of the greatest significance. That an Amerindian couple in British Columbia should give out with exactly the same strange sound in connection with a *Sasquatch* that two highly educated white men did, over 600 miles south in connection with California's Bigfoot, is incredible. If this is all a hoax or a publicity stunt, or mass hallucination, as some people have claimed, how does it happen that this noise—which defies description—always sounds the same no matter who has tried to reproduce it for me?

A somewhat more colorful story was told by a well-known old Amerindian "medicine man" named Frank Dan. (This I reproduce by the kind permission of Mr. J. W. Burns.) This, he says, occurred in July, 1936 along Morris Creek, a small tributary of the Harrison River (see Map II). J. W. Burns writes of Frank's story:

It was a lovely day, the clear waters of the creek shimmered in the bright sunshine and reflected the wild surroundings of cliff, trees, and vagrant cloud. A languid breeze wafted across the rocky gullies. Frank's canoe was gliding like a happy vision along the mountain stream. The Indian was busy hooking one fish after another; hungry fish that had been liberated only a few days before from some hatchery. But the Indian was happy as he pulled them in and sang his medicine song. Then, without warning, a rock was hurled from the shelving slope above, falling with a fearful splash within a few feet of his canoe, almost swamping the frail craft. Startled out of his skin, Frank glanced upward, and to his amazement beheld a weird looking creature, covered with hair, leaping from rock to rock down the wild declivity with the agility of a mountain goat. Frank recognized the hairy creature instantly. It was a Sasquatch. He knew it was one of the giants—he had met them on several occasions in past years, once on his own doorstep. But those were a timid sort and not unruly like the gent he was now facing.

Frank called upon his medicine powers, sula, and similar spirits to protect him. There was an immediate response to his appeal. The air throbbed and some huge boulders slid down the rocky mountain side, making a noise like the crack of doom. This was to frighten away the Sasquatch. But the giant was not to be frightened by falling rocks. Instead he hurried down the declivity carrying a great stone, probably weighing a ton or more [sic], under his great hairy arm, which Frank guessed—just a rough guess—was at least 2 yards in length. Reaching a point of vantage—a jutting ledge that hung far out over the water—he hurled it with all his might, this time missing the canoe by a narrow margin, filling it with water and drenching the poor frightened occupant with a cloud of spray.

Some idea of the size of the boulder may be gained from the fact that its huge bulk blocked the channel. Later it was dredged out by Jack Penny on the authority of the department of hinterland navigation. It may now be seen on the 10th floor of the Vancouver Public Museum in the department of "Curious Rocks." When you're in Vancouver drop in to the museum and T. P. O. Menzies, curator, will gladly show it to you. The giant now posed upon the other ledge in an attitude of wild majesty as if he were monarch of these forboding haunts, shaking a colossal fist at the "great medicine man" who sat awe-struck and shuddering in the canoe, which he was trying to bail out with his shoe. The Indian saw the Sasquatch was in a towering rage, a passion that caused the great man to exude a repugnant odor, which was carried down to the canoe by a wisp of wind. The smell made Frank dizzy and his eyes began to smart and pop. Frank never smelt anything in his whole medicine career like it. It was more repelling than the stench of moccasin oil gone rotten. Indeed, it was so nasty that the fish quitted the pools and nooks and headed in schools for the Harrison River. The Indian, believing the giant was about to dive into the water and attack him, cast off his fishing lines and paddled away as fast as he was able.

I include this story not so much for anything it might add to the general picture of ABSMs in the area—there is ample evidence of that in any case—but to exemplify the type of tale told by the Amerind that cause the white man to doubt his veracity. Frank Dan was an old and respected medicine man living by the precepts and beliefs of his ancestors. Thus, his interpretation of events had to be in accord with his position in the community. I believe that facts colored by these precepts may be readily spotted in his account and just as readily eliminated. If this is done, we are left with a pretty straightforward account; namely, that while fishing, a *Sasquatch* appeared, hurled some rocks at the old gentleman, and stank like hell. The induced landslide and the weight of the second rock hurled, or perhaps merely dislodged into the river, as well as the giant's implied curse, are pure embellishments. Even the mass exodus of the trout might well be perfectly true and due to a cascade of boulders rather than to a stink in the air that they could of course not smell in the water. Besides, Frank Dan's "medicine" came off second best and he had manifestly fled. He couldn't explain this fact away, so he just did the best he could so not to show up in too poor a light. As a matter of fact, Mr. Burns records that he gave up being a medicine man from then on, saying that his powers had been finally defeated. That would seem to be the act of an honest man.

During this decade the Amerinds of this area appear, by all accounts, to have suffered quite a spell with their *Sasquatches*. One by the name of Paull, in company with others returning from a lacrosse game, met one on the main road near Agassiz; another party only a few miles away ran into one on a mountain, and one of the men fired at it in pure fright, whereupon it pursued them to their canoe, in which they just managed to escape. Another local man, when dressing after a swim in a river on a hot summer day was confronted by one near a rock, and was just about to address it in his language when it rose to its full height and nearly scared him out of his wits. Still another group told Mr. Burns that they had watched one fighting a large bear for a long time and finally killing it by strangulation. In another place, an old man said that a party of *Sasquatches* used to watch loggers at work and then, after they had gone home for the evening, come out and imitate their activities as if playing a game. But, perhaps the most curious is an incident told to the same indefatigable investigator, Mr. Burns, by the same Charley Victor of Chilliwack already mentioned, and which I herewith reproduce with the former's permission. Charley speaks, and says:

I was hunting in the mountains near Hatzic. I had my dog with me. I came out on a plateau where there were several big cedar trees. The dog stood before one of the trees and began to growl and bark at it. On looking up to see what excited him, I noticed a large hole in the tree 7 feet from the ground. The dog pawed and leaped upon the trunk, and looked at me to raise him up, which I did, and he went into the hole. The next moment a muffled cry came from the hole. I said to myself: "The dog is tearing into a bear," and with my rifle ready I urged the dog to drive him out, and out came something I took for a bear. I shot and it fell with a thud to the ground. "Murder! Oh my!" I spoke to myself in surprise and alarm, for the thing I had shot looked to me like a white boy. He was nude. He was about 12 or 14 years of age.

[In his description of the boy, the Indian said his hair was black and woolly.]

Wounded and bleeding, the poor fellow sprawled upon the ground, but when I drew close to examine the extent of his injury, he let out a wild yell, or, rather a call as if he were appealing for help. From across the mountain a long way off rolled a booming voice. Less than half an hour, out from the depths of the forest came the strangest and wildest creature one could possibly see. I raised my rifle, but not to shoot, but in case I would have to defend myself. The strange creature walked toward me without the slightest fear. The wild person was a woman. Her face was almost negro black and her long straight hair fell to her waist. In height she would be about 6 foot but her chest and shoulders were well above the average breadth.

The old man remarked that he had met several "Wild Persons" in his time but had never seen anyone half so savage in appearance as this woman. The old brave confessed that he was really afraid of her and that he had fled.

This story does add some significant facts to the over-all picture because of the details given of the youngster's fur color compared to that of the female, and the curious statement about the length of her head-hair. The former agrees with the accounts of Jacko and some other reputed ABSM youngsters: the latter is, as far as I know, a completely unique item. I wonder about this latter because I have noted a distinct tendency, perhaps psychological, for people to assume that the head-hair of wild people would be of the Lady Godiva type. A good friend of mine, a well-known artist who has illustrated many scientific works and natural history books, once sent me his "impression" of a Californian *Oh-Mah* which greatly surprised me. Despite this man's extensive knowledge of mammalian anatomy and long experience in drawing animals to the specifications and approval of zoologists, he had depicted just a great big white-type man with long flowing hair and an immense beard. This seems, indeed, to be the popular conception of an ABSM; yet, everybody who claims to have seen one makes special mention of the small pointed heads, small round eyes close together and directed straightforward, extra long arms, and *short* head-hair, a naked face *without* beard and prognathous jaws but no lips (i.e. no eversion of the lips). The picture given of all of them

[73]

by those who claim to have seen them, is of creatures with several distinctly nonhuman characters, especially about the head and face. However, the same witnesses everywhere, and all natives who say they know of the existence of ABSMs— and this goes for the Central Asiatics, as well as Malays, African, and North and South Americans—insist just as vehemently that the creatures are human rather than animal. Quite where various people draw the dividing line between the two presents other puzzles, but the Kazakhs of the U.S.S.R. who caught one of their *Ksy-Giiks,* thought it was a man wearing a disguise, while the Soviet Army medical officer who examined a *Kaptar,* pronounced it so human that it should be released. Even the Hill Batuks of Sumatra, who are themselves just about at the bottom rung of the cultural ladder, call their local *Orang Pendeks* and *Orang Gadangs* by a name that denotes "wild *men.*" The Malays of the same country, however, call even the *Mias* (their great ape), the Orang-utan (i.e. "hutan"*), which simply mean wild (*utan*) man (*orang*). The Amerinds of our Northwest insist that the *Sasquatches* are very lowly forms of men, so lowly that they, Amerinds, do not want to associate with them in any way; preferring not to talk about them and especially about the possibility of mating with them. That would lead to contamination of their race, and, if the very idea got into the white man's head, it would lead to a further degradation of their status by the implication that they might be partly wild themselves.

The basic "humanity" of ABSMs is perhaps understandable as regards the pigmy and the giant types, for both leave what at first sight look exactly like either very small or very large human footprints, as most certainly do the Eurasian *Almas.* The man-sized *Meh-Teh* type, on the other hand, leave a most unhuman type of footprint (see Appendix B). Encounters with *Sasquatches* are really so common that they become boring in the telling. I could give dozens more, all

---

* This is the correct spelling in Malay, and "orang" really means "person," not "man."—Author.

[74]

of which were allegedly witnessed by more than two people and occurred between 1930 and 1960, but I shall refrain and confine my concluding remarks to three cases that for some reason created great stirs and which appear to have finally convinced the general public that something was going on.

The first would not appear to have been any more outstanding than dozens of others, but the personalities of the couple concerned played a considerable part in the formulation of public opinion. These were two young people named Adeline August and her boy friend William Point. They happened to be particularly popular and attractive, and were then attending the local high school. They had been on a picnic and were walking home along the Canadian Pacific Railroad track right by Agassiz when a large *Sasquatch* stepped out of the woods ahead of them. Adeline sensibly bolted, but young William stood his ground to cover her flight and grabbed up two rocks with which to defend himself. However the ABSM kept steadily advancing and when it was only 50 feet away William Point decided to retreat. He said that it was about twice the size of an ordinary, large, well-built man, covered with hair, and had arms so long that they almost reached the ground. William Point also said, "It seemed to me that his eyes were very large, and the lower part of his nose was wide, and spread over the greater part of his face." Locally, the account of this young couple was fully believed, and *despite* the fact that they were Amerinds.

This was in 1954. The following year the most outstanding of all Canadian cases occurred. This was related by one William Roe, mentioned above, and is succinctly and amply covered in the following affidavit:

Deposition by Mr. William Roe

From the City of Edmonton, Alberta. An affidavit by William Roe. To the Agassiz, Harrison Advance, Printers & Publishers, Drawer O, Agassiz, B.C.; Attention Mr. John W. Green. From the legal Department of Allen F. MacDonald, B.A., L.L.B., City Solicitor., H. F. Wilson, B.A., Asst. City Solicitor and R. N. Saunders, Claims Agent.

Dear Sir:

Re Affidavit of Mr. William Roe, on August 26th, 1957. Mr. Wm. Roe approached the writer requesting the swearing out of An Affidavit in regard to a strange animal he had seen in British Columbia.

The affidavit was drawn up by a member of our legal department and sworn to in the usual manner by the writer.

I cannot state as to the creditability of the story.

We trust the foregoing information will be of assistance.

Yours truly,

(signed)   W. H. Clark

Asst. Claims Agent

WHC:ek.

Affidavit.

I, W. Roe, of the City of Edmonton, in the province of Alberta make oath and say,

(1) That the exhibit A attached to this, my affidavit, is absolutely true and correct in all details.

Sworn before me in the City of Edmonton, Province of Alberta, this 26th day of August, A.D. 1957,

(signed) Wm. Roe and then
signed by Clark under a
numbering D.D. 2822

EXHIBIT A

Ever since I was a small boy back in the forests of Michigan, I have studied the lives and habits of wild animals. Later when I supported my family in northern Alberta by hunting and trapping, I spent many hours just observing the wild things. They fascinated me. The most incredible experience I ever had with a wild creature occurred near a little place called Tete Jaune Cache, B.C., about 80 miles west of Jasper, Alberta.

I had been working on the highway near this place, Tete Jaune Cache, for about 2 years. In October 1955, I decided to climb five miles up Mica Mountain to an old deserted mine, just for something to do. I came in sight of the mine about 3 o'clock in the afternoon after an easy climb. I had just come out of a patch of low brush into a clearing, when I saw what I thought was a grizzly bear in the brush on the other side. I had shot a grizzly near that spot the year before. This one was only about 75 yards away, but I didn't want to shoot it, for I had no way of getting it out. So I sat down on a small rock and watched, with my rifle in my hand.

I could just see part of the animal's head and the top of one shoulder. A moment later it raised up and stepped out into the opening. Then I saw it wasn't a bear.

This to the best of my recollection is what the creature looked like and how it acted as it came across the clearing directly towards me. My first impression was of a huge man about 6 feet tall, almost 3 feet wide, and probably weighing near 300 pounds. It was covered from head to foot with dark brown, silver-tipped hair. But as it came closer I saw by its breasts that it was female.

And yet, its torso was not curved like a female's. Its broad frame was straight from shoulder to hip. Its arms were much thicker than a man's arms and longer, reaching almost to its knees. Its feet were broader proportionately than a man's, about 5 inches wide in the front and tapering to much thinner heels. When it walked it placed the heel of its foot down first, and I could see the grey-brown skin or hide on the soles of its feet.

It came to the edge of the bush I was hiding in, within 20 feet of me, and squatted down on its haunches. Reaching out its hands it pulled the branches of bushes towards it and stripped the leaves with its teeth. Its lips curled flexibly around the leaves as it ate. I was close enough to see that its teeth were white and even. The head was higher at the back than at the front. The nose was broad and flat. The lips and chin protruded farther than its nose. But the hair that covered it, leaving bare only the parts of its face around the mouth, nose and ears, made it resemble an animal as much as a human. None of this hair, even on the back of its head, was longer than an inch, and that on its face much shorter. Its ears were shaped like a human's ears. But its eyes were small and black like a bear's. And its neck also was unhuman, thicker and shorter than any man's I have ever seen.

As I watched this creature I wondered if some movie company was making a film in this place and that what I saw was an actor made up to look partly human, partly animal. But as I observed it more I decided it would be impossible to fake such a specimen. Anyway, I learned later there was no such company near that area. Nor, in fact, did anyone live up Mica Mountain, according to the people who lived in Tete Jaune Cache.

Finally, the wild thing must have got my scent, for it looked directly at me through an opening in the brush. A look of amazement crossed its face. It looked so comical at that moment I had to grin. Still in a crouched position, it backed up three or four short steps, then straightened up to its

[77]

full height and started to walk rapidly back the way it had come. For a moment it watched me over its shoulder as it went, not exactly afraid, but as though it wanted no contact with anything strange.

The thought came to me that if I shot it I would possibly have a specimen of great interest to scientists the world over. I had heard stories about the Sasquatch, the giant hairy "Indians" that live in the legend of the Indians of British Columbia and also, many claim are still, in fact, alive today. Maybe this was a Sasquatch, I told myself.

I levelled my rifle. The creature was still walking rapidly away, again turning its head to look in my direction. I lowered the rifle. Although I have called the creature "it," I felt now that it was a human being, and I knew I would never forgive myself if I killed it.

Just as it came to the other patch of brush it threw its head back and made a peculiar noise that seemed to be half laugh and half language, and which I could only describe as a kind of a whinny. Then it walked from the small brush into a stand of lodge-pole pines.

I stepped out into the opening and looked across a small ridge just beyond the pine to see if I could see it again. It came out on the ridge a couple of hundred yards away from me, tipped its head back again, and again emitted the only sound I had heard it make, but what this half laugh, half language was meant to convey I do not know. It disappeared then, and I never saw it again.

I wanted to find out if it lived on vegetation entirely or ate meat as well, so I went down and looked for signs. I found it* in five different places, and although I examined it thoroughly, could find no hair or shells or bugs or insects. So I believe it was strictly a vegetarian.

I found one place where it had slept for a couple of nights under a tree. Now, the nights were cool up the mountain, at this time of year especially, and yet it had not used a fire. I found no signs that it possessed even the simplest of tools. Nor did I find any signs that it had a single companion while in this place.

Whether this creature was a Sasquatch I do not know. It will always remain a mystery to me unless another one is found.

I hearby declare the above statement to be in every part true, to the best of my powers of observation and recollection.

Signed William Roe

Witnessed

* [I add here the following note that I presume he is referring there to droppings or faeces of this animal of which he says he found evidence in five different places.]

1: Track of *Meh-Teh* on upper snowfield of Southern Tibetan Rim. (Eric Shipton & the Mt. Everest Foundation)

2 & 3 (*above*). Desiccated hand of alleged ABSM from Pangboche, Nepal; Fig. 3 as seen from below. (Slick-Johnson Exp.)

4 (*left, below*): Another desiccated hand from Pangboche. (Prof. Teizo Ogawa)
5 (*right, below*): Desiccated forearm of Snow Leopard from Makalu village, Nepal. (Slick-Johnson Exp.)

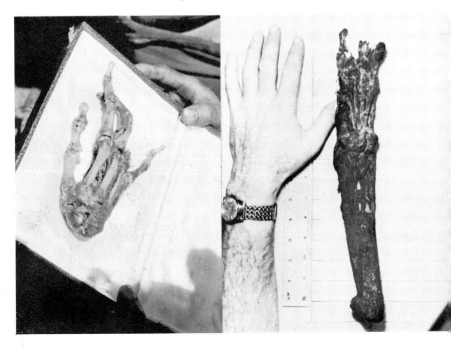

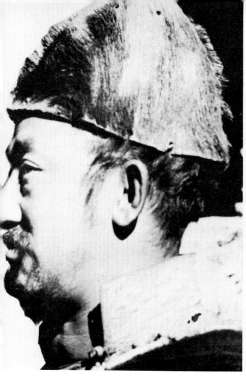
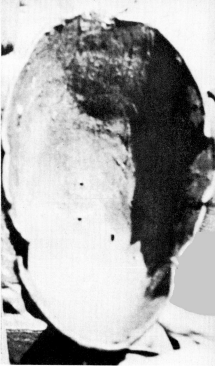

6 (left, above): A Sherpa Headman wearing a cap made in imitation of a Meh-Teh scalp. (Slick-Johnson Exp.)

7 (right, above): Same scalp, seen from inside. Preserved at Pangboche. (Slick-Johnson Exp.)

8 (left, below): Same scalp, showing holes for insertion of tassels. (Navnit Parekh, Bombay)

9 (right, below): Another fur cap. These are used for traditional pantomime. (Slick-Johnson Exp.)

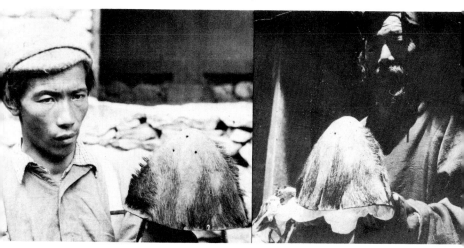

 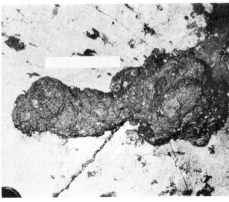

10: Himalayan Black Bear.   11: American (Kodiak) Brown Bear.

FAECAL MASSES

12: Giant Panda.   13: Californian ABSM (Oh-Mah).

(All photos by Prof. W. C. Osman Hill)

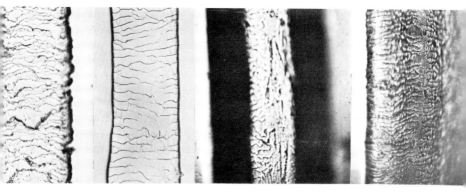

14: Himalayan Black Bear (near tip) ($\times$400).   15. Same Himalayan Black Bear (near root) ($\times$400).   16: Lowland Gorilla ($\times$250).   17: Orang-Utan ($\times$250).

HIGHLY MAGNIFIED HAIRS

18: Caucasoid Human head-hair ($\times$550).   19: Tibetan Langur Monkey ($\times$470). 20: Tibetan Blue Bear (*Ursus arctos pruinosus*)—fine hair ($\times$400).   21:Tibetan Blue Bear—coarse hair ($\times$400).

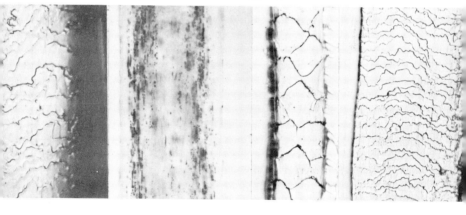

(All photomicrographs by Prof. W. C. Osman Hill)

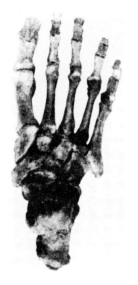

22: A Neanderthaler-type Hominid from the Crimea (from above). (Dr. W. Tschernezky)

23 (left, below): A Human. (American Museum of Natural History)
24 (right, below): A Lowland Gorilla. (American Museum of Natural History)

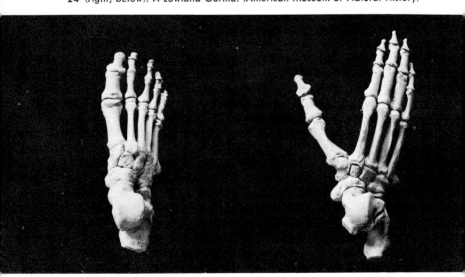

← 25: Feet of Lowland Gorilla in quadrupedal stance. (University Museum, University of Pennsylvania)

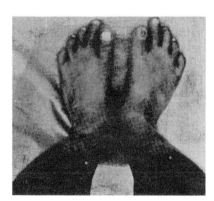

26: Abnormal (Human) feet of an Australoid. (Dr. W. Tschernezky)   →

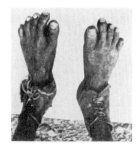

← 27: Abnormal feet of a Caucasoid. (Freiherr E. von Eickstedt)

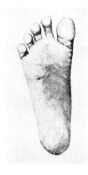

28: Sole of foot of an African Negrillo (Pigmy). (Freiherr E. von Eickstedt)   →

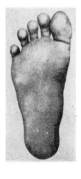

← 29: Sole of foot of adult Negroid man, used to going barefoot. (Dr. W. Tschernezky)

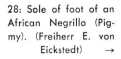

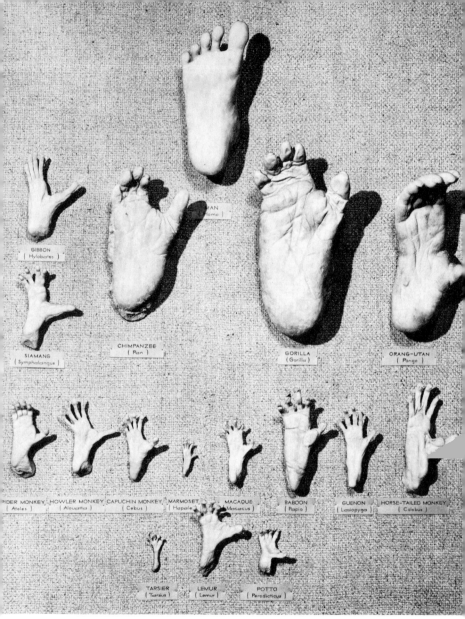

GIBBON
(Hylobates)

SIAMANG
(Symphalangus)

CHIMPANZEE
(Pan)

MAN
(Homo)

GORILLA
(Gorilla)

ORANG-UTAN
(Pongo)

PIDER MONKEY
(Ateles)

HOWLER MONKEY
(Alouatta)

CAPUCHIN MONKEY
(Cebus)

MARMOSET
(Hapale)

MACAQUE
(Macacus)

BABOON
(Papio)

GUENON
(Lasiopyga)

HORSE-TAILED MONKEY
(Colobus)

TARSIER
(Tarsius)

LEMUR
(Lemur)

POTTO
(Perodicticus)

30: Casts of soles of hind feet of Man and various other Primates. (American
Museum of Natural History)

IMPRINTS OF: 31 (*left*) Fore and hind right feet of Eurasian Brown Bear, in snow; 32 (*center*) Hind right foot of Himalayan Langur, in snow; 33 (*right*) Right foot of Gorilla, in snow (made from a cast). (All photos by Dr. W. Tschernezky)

IMPRINTS OF RIGHT FOOT OF: 34 (*above*) *Meh-Teh*-type ABSM from Nepal, in snow. (Eric Shipton & the Mt. Everest Foundation); 35 (*below*) Californian *Oh-Mah,* in soft clay. (Author)

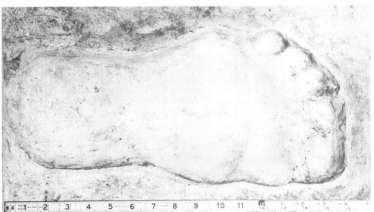

← 36: Adult male and female Lowland Gorillas. (Philadelphia Academy of Natural Sciences)

37 (*below*): Wow-Wow Gibbon walking. (Roy Pinney)

38 (*above*): Corpse of Sloth Bear killed in Nepal and at first alleged to be that of an ABSM. (Slick-Johnson Exp.)

← 39: Reconstruction of *Meh-Teh* (and photo) by Dr. W. Tschernezky.

40: Artist's conception of a female Sasquatch. (Morton Kunstler) →

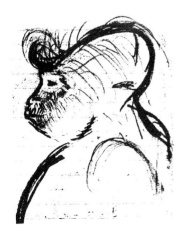

← 41: Field-sketch of head of male Sasquatch (and photo) by the author, under the direction of Mr. Albert Ostman.

← 42: Drawing of a *Gin-Sung* (giant ABSM type of Szechwan) from an 18th-century Mongolian manuscript. (Prof. Emmanuel Vlec)

43: Reproduction of the above in a later Chinese manuscript. (Prof. Emmanuel Vlec) →

44 (left, above): Reconstruction of an Australopithecine. (M. Wilson, 1950)
45 (right, above): Reconstruction of head of *Zinjanthropus*. (World Wide Photos)

46 (left, below): Reconstruction of head of *Pithecanthropus*. (University Museum, University of Pennsylvania)
47 (right, below): Reconstruction of head of a Neanderthaler. (University Museum, University of Pennsylvania)

48: Head of an Australoid.
(Author)

49: African Negrillos (Pigmies). (University Museum, University of Pennsylvania)　　　　→

← 50: Negrito girls, Philippine Islands. (University Museum, University of Pennsylvania)

51: Head of girl, Negroid type.
(Quentin Keynes)

52: Head of girl, Caucasoid type.
(Photo Library, Inc.)

53: Head of girl, Mongoloid type.
(Philip E. Pegler, Inc.)

54: Man with (abnormal) tail (the Philippines). (Author)

55 (below): The famous Tensing Norgay, Conqueror of Everest, and his family at home. It is his people, the Sherpas of Nepal, who first led the world to the ABSMs. (Information Bureau, Government of India)

56 (above): The author with a family of Mayan friends—the Het Zooz-Mukuls of Tekom, Yucatan. The mother is holding one of the author's god-children, Manuelita. Note: all are standing on the same level. The author is 6 feet tall. (Author)

This priceless document was also unearthed by the indefatigable John Green of the *Agassiz-Harrison Advance,* upon
whom the mantle of *Sasquatch* research, nobly worn by Mr.
J. W. Burns for so many years, seems to have fallen. He published it in his paper and the results were electric. Not only
did it bring Mr. Ostman's story to light; it got the whole
neighborhood on its toes, including even the Chamber of
Commerce of the resort town of Harrison which made moves
to advertise a *Sasquatch* hunt as a come-on for its centenary
celebrations! Fortunately, and decently, this idea was dropped
but $5000 is said to have been offered for the capture of a
Sasquatch. This was not, of course, collected but it brought
forth another rash of encounter stories. Notable among these—
and most noted in the world press—was a story reported by a
Mr. Stanley Hunt of Vernon, B.C., a respected and widely
known auctioneer, who, when driving at night along the
Trans-Canada Highway near a place called Flood on the lower
Fraser River south of Yale, on May 17, 1956, had to slow
down to permit one of them to cross the road. It was immense
and covered with "gray hair," and, waiting for it on the other
side of the road, there was, Mr. Hunt relates, another one
"gangly, not stocky like a bear."

According to C. S. Lambert, writing in 1954, the situation
changed considerably in 1935 when:

After a series of alarming reports that these giants were prowling
around Harrison Mills, 50 miles East of Vancouver, disturbing the residents
by their weird wolf-like howls at night, and destroying property, a band
of vigilantes was organized to track the marauders down. However, no
specimen of the primitive tribe was captured, and many white people became openly sceptical of the existence of the giants.

According to Allen Roy Evans, in the *Montreal Standard* ("B.C.'s Hairy
Giants"), the Indians are now very sensitive to any imputations cast upon
their veracity in this matter. During the 19th century they were ready to
tell enquirers all they knew about the Susquatch men; but today they
have become more reserved, and talk only to Government agents about
the matter. They maintain that the "Wild Indians" are divided into two
tribes, whose rivalry with each other keeps their number down and so
prevents them becoming a serious menace to others.

[79]

Expeditions have been organized to track down the Susquatch men to their lair in the mountains; but the Indians employed to guide these expeditions invariably desert before they reach the danger zone. However, certain large caves have been discovered, with man-made walls of stone inside them, and specially-shaped stones fitted to their mouths, like doors. The difficulty in the way of penetrating to the heart of the Morris Mountains district is very great. The terrain is cut up by deep gorges and almost impassable ravines; it is easy to get lost, and hard to make substantial progress in any one direction for long.

In the fall of the following year large human-like footprints turned up overnight all over the place in this area. Throughout a hundred years of *Sasquatchery*, footprints are often mentioned casually, but nobody seems to have been particularly impressed by them or to have done anything about them. Suddenly they took over the front pages.

5. Footprints on the Sands of . . .

Some things people accept; some they reject.
Others, they will accept as long as they have a
ready-made answer, but—certain things they sim-
ply don't know what to do about.

If you look out of your window one morning to find that it
has snowed during the night, you may be happy or you may
be sad. If then, while contemplating this quite natural phenomenon, you perceive upon its pristine surface a number of
marks of regular shape, forming a set of tracks, the sundry
relays, feedbacks, and synapses in your brain may snap open
or shut in ordered patterns, causing you to register almost sub-
consciously such concrete items as man, dog, car, snowplow,
or suchlike. You may even go so far as actually to think, saying to yourself "That's funny, Mary went out already." Foot-
tracks are commonplace, and quite logical, and we consider
them as objects. Yet they are not even quasi-objects; they are
entirely negative physically; are purely subjective concepts;
and in almost all cases are ephemeral things. Nevertheless,
they are quite acceptable, provided we have a ready-made
answer for them, ranging from vague terms such as "dog,"
all the way to "Mary wearing a particular pair of shoes."
When, however, a set of foot-tracks turns up on snow, or any
other surface for that matter, to which people cannot immediately put a label, they become quite hysterical, and in
their frantic efforts to explain this appalling thing, they will indulge in the most terrifyingly illogical actions. They also say
the silliest things.

Simple logic demands that a foot or any other print must
have been made by something, and something which must

MAP III. NORTH AMERICA

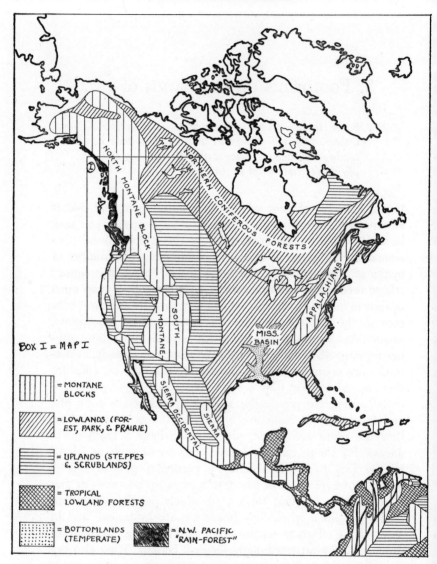

See facing page for legend

have been at the point where the imprint was made. But sometimes, unfortunately for humanity, matters don't always work out that way, in either one or both of these respects. The second class of problems is the less awful. For instance, "How on earth did Mary get up on the barn roof?" may jolt you but can have all sorts of logical explanations. If one is sufficiently concerned about Mary's welfare, it is the common practice to investigate these in order of likelihood, starting by asking Mary, if she is around; and ending by calling in the long-suffering police if she has disappeared. Even in this class, however, there can be nasty ones. We once found a set of what looked like our tame porcupine's tracks, inside an empty cage, which was constructed of heavy wire in the form of a cube on all six sides, and had a firmly locked door. That took some investigation and it reduced a number of normally sane citizens to gibbering idiots in the meantime.

(Said porcupine had once been housed in that cage for an hour or so, while its own cage was cleaned and repaired, by an assistant who was not present when the bizarre discovery was made. The earth floor inside the cage had been wet at the time and the animal had left deep tracks in the claylike mud. This dried solid. The assistant had then, in accord with his routine duties, put a 2-inch covering of fresh earth over this. The night before the uproar there had been about 15

MAP III. NORTH AMERICA

This continent should be regarded as reaching from the Arctic Ice-Raft to the Isthmus of Tehuantepec. It is divided into three parts: first, into a western and an eastern, by the Great Barrier, the dividing line running roughly down the 110th Meridian. Secondly, the eastern half is sub-divided latitudinally about the 45th parallel; to the north being closed forest and tundra; to the south, open forest (parklands) and prairies. The midwest, southwest, and Mexico are arid and covered with scrub and desert. The rest is mountainous, and forested almost exclusively with conifers. In the Mexican Sierras there are some tropical forests. Along the eastern fringe of the continent lie the Appalachians, and there is another upland area in Labrador. The valley of the Mississippi and its tributaries form extensive, swampy bottomlands.

minutes of torrential rain, which had washed all this top layer exactly off the old hardened one and the tracks had appeared looking just as if they were fresh and, of course, once again in damp earth.)

The more abominable class is that of individual prints or sets of tracks—and the two items are quite different and should be at all times most carefully defined by the use of the appropriate term—for which there is not a ready-made explanation. A *print* (or imprint) is an individual item such as that of one foot. A set of *tracks* (or a track) is, on the other hand, a series of prints, either interrupted as in animals, or continuous as made by wheeled machines, left by some moving object. There are quite a lot of reports of single prints being found both in such positions as may be explained—as in a small patch of mud on a rocky path—but on occasion in places that cannot be explained. These last are, of course, very unnerving.

Sasquatch imprints and tracks, along with those of their relatives or congeners, by whatever name they were known, were perfectly all right by the Amerinds because they had just such a ready-made answer, all of them, as they readily tell one, knowing perfectly well that they were made by the big, wild, hairy men of the woods: or by their wives and children. As the Amerinds gave up being Americans and started to become, or were forced to become sort of bogus Europeans, they forgot to tell their own children about these personages. The result was that in time we even have Amerinds becoming for a time slightly disturbed. [Amerinds *never* under any conditions become "hysterical."] When, however, white men first saw these large ABSM tracks they invariably went into a fairly advanced trauma. This habit was apparently universal among Europeans and people of European origin, right up until the time when a ready-made answer became disseminated—namely, *Sasquatches, Oh-Mahs,* etc.—whereupon a happy reaction set in. This was simply to say: "Oh, those! Don't worry, they're made by runaway Indians; they have huge feet, you know, and sometimes grow hair to keep out the cold." (Amerinds, I should point out here, are either wholly or substantially of Mongoloid ancestry, the group of the human race that is defined as being the most glabrous [almost without body hair], and having particularly small, neat feet.

It is rather interesting to note in passing that persons of African ancestry have behaved quite otherwise throughout. They possess ancestors who have always recognized a non-material world just as widespread and as real as the material one. This is probably why they are such great pragmatists. What is more, according to them, entities in both worlds customarily muck about in the other, so that men's souls can range around "elsewhere" and *chumbis*—or what we in our innocence call ghosts, poltergeists, and spirits—can, in their estimation, quite well leave imprints and foot-tracks. Africans of the Negroid branch of humanity and their descendants are, therefore, the greatest skeptics throughout our story, they have never really been interested in or even much surprised about the matter, for they have a sort of built-in answer; and while they have always thought Europeans to be stupid for not carrying on with disembodied entities, they usually think the Amerinds quite batty for needing an embodied entity to explain these tracks. The few people of African origin whom I have met in the course of this business in North America, as well as in Africa appear, furthermore, to have accepted the physical appearance of ABSMs that they themselves have witnessed, with the utmost equanimity and simply as lucky or dangerous happenstances.

I bring all this up now because it has to be aired in any case sooner or later, and because from now on we are going to have all three major branches of the human race involved in the matter. Their reactions are indeed different, whatever anybody may say about generalizations. All three "races" are present in the United States, where our story now takes us, and since we are going to follow the foot-tracks of the ABSMs, clear through this country to tropical America, we are going to have to be prepared for some real surprises—both ways. You will see what I mean by this in a minute.

At this point I would ask you to glance at Maps III and

XVI, before proceeding, because, without some idea of the facts of vegetational distribution, very little of what I have to say in this and the next chapter will make much sense. I know by experience that it is quite all right for me or anybody else to say almost anything about foreign lands, and the farther away and thus foreign they are, the more outrageous the claims may be. This is the reason why such a high percentage of "explorers" are found, on proper investigation (if that is possible, which it seldom is), to be phonies, even if only mildly and innocuously so. When, on the other hand, anybody makes even slightly unusual remarks about the country *in* which he is speaking and to citizens *of* that country, he is almost certain to be disbelieved, probably ridiculed, and oftimes harassed for his pains. This applies to statements as innocent as "You know, the hillbillies down there don't wear shoes." Try it sometime, *down there,* but don't wait to see what happens, for you'll have the local State Department on your back if you have published your statement, and you'll find yourself excluded from private swimming pools if you have merely said it in family circles.

Since I have a private swimming (duck) pond of my own, and seldom wear shoes indoors in winter or either in- or out-of-doors throughout the whole summer and early fall as well as, for other reasons that I will not go into, I have made a profession of saying things about the country I am in. I am, in fact and as I said at the outset, a reporter and as I don't give a damn whether anybody wears shoes or not, nor what their opinions are on that or any other subject, and am interested only in facts, I am constantly saying things that annoy people. What I have to say now is going to annoy some types very much. Moreover, if you haven't as yet glanced at these maps, you may be so annoyed that you will just stop reading. I don't want you to do this, but for purely altruistic reasons— namely that these facts are such fun. To keep you reading, therefore, let me just tell you that, if you do so, you are going to get a really good laugh, specifically at the expense of just those people whom you have always thought were idiots in any case. [Admittedly, this includes almost everybody other than yourself, which makes it all the more pleasant.]

Animals (and ABSMs) take no account of political boundaries even when they are physically erected by people in the form of barbed-wire fences or iron curtains. They do, on the other hand, not only take into account but conform absolutely to certain boundaries and dividing lines set up by Nature. No animal ever, it seems, transgresses such a boundary and these boundaries may often be so precise that you can stand with one foot in one great natural province and the other foot in another. There are animals that range over more than one and sometimes over half a dozen provinces. These are called *catholic* species; but most animals stay within the confines of just one province. Within the provinces, moreover, there are a number of natural niches or environments. Nature abhors a vacuum (as we have been repeatedly told) and she fills all her niches with an appropriate animal species. If any one dies out or is exterminated, some other animal will come in to inhabit its niche. As an example, the South American aquatic porcupine called the Coypu (*Myopotamus coypu*) the fur of which is called *nutria*, was introduced into North America 50 years ago and immediately started to fill up the niche previously occupied by the Beaver which had, at that time, been largely exterminated in this country by fur trappers.

Sometimes a species of animal will introduce itself into an area and do battle with the established occupants of the particular niche that it likes. Then again, men have introduced animals from one country to another and started virtual animal wars, usually with fatal consequences to one or the other party. In Australia introduced European animals, like the dog, cat, fox, and rabbit, have committed mass mayhem on the indigenous fauna: on the other hand, attempts to introduce the pheasant in certain parts of North America have repeatedly failed. The whys and the wherefores of these results have proved very puzzling in that there seemed to be no rhyme or reason for them. There is, nonetheless, a law governing the matter, and a very precise one. This is a botanical matter.

The whole earth is portioned out into different types of plant growth—different in the way the vegetation grows (in height, density, and so forth) rather than in what particular types of plants it contains—and these form great belts around the earth regardless of oceans, seas, and mountains. These belts, which meander about and broaden out or wither down sometimes almost to nothing, are also subdivided into blocks or provinces going from east to west, like the cross-stripes on a banded snake. Each one of these provinces has its own history, climate, weather, soils, flora, and fauna. What is more, it has now been discovered that all faunas are wholly dependent upon vegetation but not so much upon the constitution of that vegetation as upon the way in which it grows. Human beings are animals and they conform to these general principles too, even down to national types. So it seems, do ABSMs. (For fuller details of all this, see Chapter 18.)

Man, however, is what is called an adaptable animal. He is also incredibly tough, and can survive in more types of vegetation and in a wider variety of environments than most animals, being surpassed in this ability by only a few other animals, such as the spiders and their allies, which live in water and in air, and range from icecaps to still hot lava flows, and to the tops of mountains where even plants give up. Nevertheless, when man comes to settle down and try to earn a living and breed, even he conforms to the old pattern. Hollanders gyrate to coastal flats, and Norwegians to warm, wet fiords. However, man *can* survive an ousting from his natural environment and he has often done so. The Neanderthalers appear to have been driven back into the hills by the folk of Cromagnon-culture; and the Jews were blasted all over the lot, and have survived.

ABSMs, it seems, have also been driven back into certain environments. By the time my story is told, you will see why I say this and why it happened. There is nothing mysterious about it. It is simply that ABSMs are Hominids or, just as every benighted native has always asserted, *human* rather than animal, and thus are endowed in one degree or another

[88]

with human attributes, and most notably their powers of survival, their adaptability, their toughness, and their acuteness. The Pongids, or apes, on the other hand, though looking so like humans, are the lousiest adapters, are completely stuck with their special environments and in their particular provinces. They can hardly breed outside them, even with the very best and most modern human medical assistance—as witness the tiny number of gorillas born in captivity. In other words, about 50 million years ago, Nature started an experiment with a couple of Primate types now called the Hominid and the Pongid. The first made the grade, and mostly through the efforts and discoveries of ABSMs; the latter failed, and are doomed.

If there are ABSMs in North America, as well as Central and South America (as would appear from what follows), and they are Hominids, they must have come here from somewhere else, for we can say with almost absolute certainty that neither Man nor the Hominids was evolved in the New World. What is more, not so much as a single bone or other indication has ever been discovered suggesting that either the Pongids or any of the true Monkeys ever even got here. On the other hand, men got here, and at a rather early date. Bones of the animals he brought back from hunting forays have been dated certainly back to before the last ice-advance; some are claimed to be more than 40,000 years old. We have not yet obtained the bones of the earliest of these men themselves, but, if some anthropologists are right, there are some extremely old and quite primitive stone implements at the lowest levels, and we now know that a creature (such as East Africa's *Zinjanthropus*) was a toolmaker but most certainly would be called an ABSM if he were found running around today. Failure to find the bones of ABSMs is no cause for stating that they never existed. Tools of the types known as Chellean and Acheulian have been known from all over southern Europe and Africa since men started collecting such items, but it was not until the last decade that we found a single bone of the men who made them—if we have yet done so, as a matter of fact.

However, ABSMs seem once to have roamed much of North America. Why, then, should those alleged still to do so, although really very hominoid in form, appear to be without tools, fire, or speech? We have to look at it this way. They were probably here in the purely "animal" stage of their development, and they kept coming in waves [over the Bering Straits, if you like] at ever increasingly efficient levels of tool-making and development, until they were replaced by their cousins who were so *"something-or-other"* that we, upon digging up their remains, call them Men. [Lots of these came too, making ever better tools, until the misguided Amerinds made the mistake of tagging along. At this point we enter history and the domain of other specialities.] As brighter and better ABSMs turned up, however, the previous occupants had to move out into less desirable environments—nasty places like deserts and mountains—and by the time proper Men arrived, these places were getting quite crowded. At that point another factor became operative.

ABSMs, both here and all over the world, had been getting "better"—which is another way of saying more complicated or mixed up—and, thus, in certain ways less efficient again. The more complex their culture became—and don't think that they didn't have a culture for Nutcracker-Man (*Zinjanthropus*) of 600,000 years ago in East Africa made splendid tools but had a brain somewhat more paltry than the average chimp—the more dependent they were upon an easy environment, which means one where it was easy to obtain a living. Chased out into a rough one by still more cultured chaps, they began to find the going very hard. In fact, the more "cultured" they were, the worse they fared when pushed up into the mountains; and the more advanced they were, the more easily and rapidly they gave up and became extinct. Thus, we have the extraordinary spectacle of the more primitive surviving and the more advanced wilting away. Today, only the most primitive have apparently survived, and in the remotest and ruggedest places where any other ABSM less rugged could not get along; where Man, however tough,

[90]

failed; and where even Modern Man, who has really got some-where with his culture, finds it hard going. And just where is this?

The answer is very simple and absolutely definitive. It is what is called by botanists *The Montane Forests*. This is why I suggested that you take a look at the maps and see where such forests are, especially today, on our continent. From these you will note that their distribution coincides exactly with that of the reports of our ABSMs; as it does on all the other continents with *their* ABSMs. There is only one excep-tion, from the botanical point of view, and this I would like to dispose of forthwith.

The last retreat on land of anything is a forest. In North America between those latitudes occupied by the United States, most lowland forests are woodlands, and anything un-wanted in them has long ago been eliminated. [One can't speak of feral dogs because we introduced them.] In Canada, of course, such forests are still virtually impenetrable. There remain then the montane forests [which are not quite the same thing as mere forests on mountains] and one other type of vegetational growth. This is what are called technically the Bottomlands. By this is meant swamps at low level but mostly in river valleys and deltas, that are covered with a closed-canopy forest of some kind however short in stature, and which are either flooded all the time, seasonally, or from time to time, so that they are unpleasant for man to live in and a lost cause to try and clear, drain, and farm. It so hap-pens that we have a very great acreage of just such country in the United States that is tacitly ignored by everybody and frankly unknown to most. This is concentrated along the Mississippi Valley and up the valleys of the tributaries of that great river.

The best road maps of the states that straddle these Bottom-lands look perfectly OK at first sight, being covered with roads of various grades, having names of counties, townships, and so forth scattered all over them and seeming, when viewed individually, to be quite consistent with all other road maps of our country. If, however, you look more carefully at them, take a pair of dividers, consult the scale at the foot of the map, and then select your areas carefully you can isolate almost endless parts of the map that look like this:

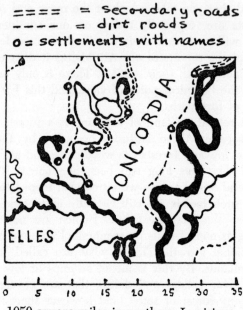

1050 square miles in northern Louisiana.

This you will not, of course, believe. It will also probably make you very annoyed. You might therefore assuage your fury by going out and buying or writing to one of the oil companies to obtain maps of such states as Missouri, Illinois, Kentucky, Tennessee, Arkansas, Alabama, Mississippi, and Louisiana, and spend a moment or two with a rectangle of the dimensions and the scale of the above. It will probably make you even more angry, but I said that I would name names, even if I am "down there."

The reason I bring this obnoxious subject up at this time is that, before we can get back to the main road of our travelogue, there is something that is really unpleasant that has to be taken care of. This is the "Little Red Men of the Trees." How aggravating can I get; and how far out on what limb can I wriggle? You would be surprised indeed; but I warn you in the most friendly fashion, please don't forget that I am a reporter and, as of now, nothing else. It is therefore my duty to report to you; so here goes:

Dear Sir,

My name is James Meacham, I read the article that you wrote for *True* Magazine.* I have been planning on going to California in the same area that your article was about. I was a little surprised to read about such a creature as an abominable Snowman living so close to where I intended to visit. I have always liked to explore places that other people care little about. I would like to know all you can tell me about this creature if you can tell anything more than you did in the article. I am sure a man of your standing must have more information about this subject than was in those few pages. I will gladly pay the postage on the information you can send. I cannot offer more because I am not working at the present.

I have met a few strange things in my life; as I am still young, there are many more I will probably see. I would like to know if you can tell me anything about a creature that looks like a small ape or a large monkey that has hair the color of fur a reddish orange color. I saw such a creature when I was 15. A friend was with me but did not see it. Whatever it was did not have a tail like a monkey but it did swing like one by its arms. This may sound like something that I thought I saw but really didn't which I would believe except for a few details.

I had a .22 calibre semi-automatic with me. I watched this thing for about 5 minutes so I have to believe it. I put fourteen .22 long-rifle shells into whatever it was. From where I was standing I couldn't have missed. We found 1 bullet in the tree trunk so 13 of them hit it. The part that sounds more impossible is that whatever it was, did not even move while 13 bullets went into it. If I had missed all 14 bullets would have gone into the tree trunk.

I have told many people about this but nobody believes it. We found a few hairs where I had shot, but nothing else except the bullet. There was not a trace of blood. My partner thinks it was a squirrel but no squirrel grows that big. If it had been one, 2 of those bullets would have stopped it dead. Whatever it was did not even move till I headed for the

* *"The Strange Story of America's Abominable Snowman," True, The Man's Magazine,* Vol. 40, December, 1959.

tree. It traveled through those trees like an express train. I could hear the leaves rattle but could not see it.

I searched for it for a long time after that but never saw it again. No one in that area knows anything about it or has ever seen it. It had a cry that was enough to drive a person crazy. That was almost 3 years ago [1957] and I still wake up in my sleep sometimes when that sound comes back to me. If you can give me any advice as to what it could have been I will greatly appreciate it. If I had not shot it myself I would not believe it, not being able to find any blood. I know you must receive a lot of letters about this sort of thing, but all I want to know is what animal in a marsh near Jackson, Tenn. could hold 13 long-rifle shells without even moving till you start to come after it? That is what started me looking for things most people think cannot possibly exist.

Yours truly,
James M. Meacham

In 1954 a young Orang-utan escaped from a shipment of apes to a well-known Florida organization, took off into the woods, and has never been seen again. I refrain from giving further details because the valuable ape was paid for, but reported as DOA, a trade term for "dead on arrival," and someone still might get in trouble. The incident is fairly widely known in certain circles, and has been a perfect nuisance because when anything like the above is reported, even as far away as Tennessee, it is immediately dredged up by way of explanation. I suppose it is just possible that a healthy young *Mia* [a better name for what we call the Orang-utan] could survive a succession of mild Southern winters and it could travel an enormous distance by trees alone, but what it would eat during most of the year I don't know. Much more important is that a lost ape that has once been in captivity for even a short period would be almost certain to head for the nearest human habitation the moment it got hungry or saw anything novel that frightened it. In all the years that I had a zoo, I never knew an escaped animal [apart from local fauna, and even many of those] *not* to return voluntarily to its own cage during the night. Of course this "ape" might have escaped from some zoo much nearer the place where this correspondent said he saw it, but the loss of a $5000 specimen from a zoo would not go unnoticed; though, it must be admitted, it might well go unreported—to the Directors, that is. There is as much hanky-panky in the animal business as in any other. An escaped Mia is, however, I rather think, itself merely an escape mechanism as it is called, especially when we come to contemplate the following.

From *Hoosier Folklore*, Vol. 5, p. 19, March, 1946:

Another type of story that is of much more concern to us here in Southern Illinois nowadays is the "strange beast" legend. . . . Every few years some community reports the presence of a mysterious beast over in the local creek bottom.

Although it is difficult to determine just where a story of this sort has its beginning, this one seems to have originated in the Gum Creek bottom near Mt. Vernon. During the summer of 1941, a preacher was hunting squirrels in the woods along the creek when a large animal that looked something like a baboon jumped out of a tree near him. The preacher struck at the beast with his gun barrel when it walked toward him in an upright position. He finally frightened it away by firing a couple of shots into the air.

Later the beast began to alarm rural people by uttering terrorizing screams mostly at night in the wooded bottom lands along the creeks. School children in the rural districts sometimes heard it, too, and hunters saw its tracks. . . . By early spring of 1942, the animal had local people aroused to a fighting pitch. About that time, a farmer near Bonnie reported that the beast had killed his dog. A call went out for volunteers to join a mass hunt to round up the animal.

The beast must have got news of the big hunt, for reports started coming in of its appearance in other creek bottoms, some as much as 40 or 50 miles from the original site. A man driving near the Big Muddy River, in Jackson County, one night saw the beast bound across the road. Some hunters saw evidence of its presence away over in Okaw. Its rapid changing from place to place must have been aided considerably by its ability to jump, for, by this time, reports had it jumping along at from 20 to 40 feet per leap.

It is impossible to say how many hunters and parties of hunters, armed with everything from shotguns to ropes and nets, went out to look for the strange beast in the various creek bottoms where it had been seen, or its tracks had been seen, or its piercing screams had been heard. Those taking nets and ropes were intent on bringing the creature back alive.

Usually* this strange beast can't be found, and interest in it dies as mysteriously as it arose in the beginning. . . . About 25 years ago, a 'coon hunter from Hecker one night heard a strange beast screaming up ahead on Prairie du Long Creek. Hunters chased this phantom from time to time all one winter. Their dogs would get the trail, then lose it, and they would hear it screaming down the creek in the opposite direction. It was that kind of creature: you'd hear it up creek, but when you set out in that direction you'd hear it a mile down creek.

And again:

Dear Mr. Sanderson,

I listened to you on Long John Nebel's program last Thursday and was very much surprised that you talked about such things as Abominable Snowmen in America. I am a housewife but I majored in biology, attended our state university and have an M.A. in plain zoology. My husband is an experimental chemist employed by . . . [company name withheld for obvious reasons: *Author*.] and my eldest son is a technician in the Air Force. I come from Mississippi but we have resided here (in Kentucky) for ten years now.

I wonder if you have ever heard of the Little Red Men of the Delta? Nobody thought anything much of them where I was raised except that one had better be careful of shooting one because it might be murder, or so the sheriff might think if anything came of it, but I was surprised to find that the folks hereabout know it too though they took some years to talk about it to me. My husband is a New Englander and these folks don't talk much. They are [the Little Red Men of the Delta] said to be about the size of a ten year old kid and able to climb like monkeys and to live back from the bayous. They talk a lot but keep out of gunshot range and mostly go into the water. They are people and the muskrat trappers say they often wear scraps of discarded lines [linens?] old jeans and such.

If you have heard about them will you talk about them on the air as it puzzles me that nobody has ever talked about them but everybody in some places seems to know about them. There was sure nothing in my biology course about them but there's a lot folks don't know or don't talk about . . .

Yours, etc.,
Mrs. V. K.

And you can say that again! Plain ordinary citizens just don't talk; they are born with too much sense. Ridicule is the

* This is a funny word. Does it imply that sometimes it *can* be found?

[96]

most dastardly thing and can ruin one's whole life in one small jump. It takes real guts to come right out and say you've seen the Loch Ness Monster; and you'd better have private means, if you do. Otherwise, humanity at large will round on one and jump in unison, and they have a collective memory that can last for a century. Don't do it, brethren and sistren! [That's why I always ask specifically whether I may publish a name.]

I could go on quoting tidbits like the above for quite a long time and give transcripts of some tape recordings that I have but what, frankly, is the use? No one will believe either the stories or me. Nonetheless, I would be failing in my duty—which, incidentally, I take very seriously; and please make no mistake about that—if I did not put this outrageous matter before the public. Like many other things "reported" it needs, and can stand, a good airing. I am not saying that there is even so much as a word of truth in any of it but there it is, and it is no good just ignoring it. If people "down there" will persist in penning such tripe, we had better get on with the job of showing it up for what it really is. But just what is it? You tell me: I am merely reporting, and I have not yet had the time, money, nor opportunity to go to those particular places to investigate the matter. Since others apparently have not either, perhaps it would be better that everyone shut up. Meanwhile, however, I refuse to just discount everything anybody from the states listed above says. That would be tantamount to calling them all liars and idiots; and I know for a fact that they are usually neither. What is more, that is their country, and I am prepared to accept the fact that they know more about it than all of us, however whacky what they say may sound. And then there is the matter of the road maps. Just what is anyone prepared to swear under oath he knows about the Bottomlands? I have been a little way into some for brief periods and I must say that I am not prepared to give out much about them at all—they are far too vast, complex, and incomprehensible to any "foreigner." The geodeticists have surveyed them; let them tell us. Their maps are excellent—they are made from points 60 miles apart and from the air. They show everything!

As a sort of parting shot, I quote a newspaper clipping of recent date:

MONSTER AROUND

Reform, Ala.—A mysterious creature is still roaming the woods around nearby Clanton. It eats peaches, makes sounds like an elephant, and leaves footprints like an ape.

This whole bit is really becoming very difficult because little squibs like this should not include so many splendid possibilities. Of course it would eat peaches, who wouldn't? And I must admit that a herd of elephants in a forest can sound exactly like a troop of chimpanzees having a ball. But who in Reform, Alabama [I like that name] is that good on the ichnology of the *Anthropoidea*? There is a sort of chatty approach about this story, giving the impression that among the citizens of Reform and Clanton there is a considerable understanding about this beast, and there is definite indication that its presence is not a new event. In fact then, are the Bottomlands full of runaway apes or do we have an indigenous and most particular abomination thereabouts? I could give an opinion but I shall refrain, for it would be even more loathsome.

Now, and with a certain sense of relief I may say, we can get back on the straight and narrow path, and pick up our foot-tracks again. These we first stumbled upon in southern British Columbia at the end of the *Sasquatch* trail. Thence, they went south over the border and, willy-nilly we have had to follow. This is going to get us into a most unpleasant labyrinth. It is, actually, a maze with several alternate correct routes, all of which cross each other and land us up in seemingly impossible predicaments. I follow the foot-tracks first.

In progressing in space we have first to retrace our tracks in time to even earlier than before—to the "49ers," in fact. It was about that date (1849) that Anglo-Saxon type Americans first descended upon the West in any substantial masses. It was, of course, the gold that did it. Actually, this area was the first to be penetrated and colonized by Europeans on this continent; the Spaniards having made some really astonishing advances north through it from Mexico. Few people realize that these intrepid savages in their clanking armor carrying little more than their lovely and holy crosses, actually got into what is now Canada through the mountainous third of our country. This area is still giving our bulldozer operators trouble in crossing from east to west. But, here again, is another story. The point is that the Spaniards later, and very sensibly, contracted into the more fertile and pleasant areas and just left the rest to the benighted Amerinds.

During this long period of some 300 years no less, things went along much as they had done since the last ice-advance in this area—outside the Spanish Missions. There were, however, some most agile-minded priests who interested themselves enormously in the land and took the Amerinds quite seriously. They left records of some of the legends of their flocks that make most interesting reading. I have to mention the fact of the existence of these now because they constitute the earliest sight of our trail, leading, as always, from the Northlands on toward the salubrious climes of tropical America. They [the records] speak of great wild men of the dry upland arroyos and massed piñon forests, that tramped lugubriously about at night scaring adolescent Amerinds and leaving monstrous footprints on the sands of that time all over the region. But, after these ecclesiastical indiscretions, there is a complete blank as far as I know until the 1849 Gold Rush. Then things began to happen in typically Yankee fashion.

This particular facet or phase of ABSMery has, like the overall picture, to be tackled in retrospect and in the order of its rediscovery. The alleged incidents in some cases occurred over a century ago but the records came to light only in the last few years. They had been lying buried in newspaper morgues. What actually happened—and this is quite apart from the reports on individual incidents—is that a whole mass of Easterners, unacquainted with the Far West, suddenly appeared on the scene and went barging off into the outlands looking for gold. Prior to their arrival there had been plenty of people along the coast and idly dotted about the inner belt of the West, but they had stayed literally around the water holes in the latter, while they had not gone back inland from the coast in the more northern and better watered areas—that is from the north end of the Sacramento Valley to Puget Sound. It was when the Easterners tried to penetrate these lands of mighty forests and seemingly everlasting mountain ranges, one behind the other, that things began to happen. Sometimes, they got a bit out of hand.

We are now back in the montane forests of which we have spoken so firmly, and we are going to stay in them for a very long time. Before we go any farther into them, though, I should state a few basic facts. Such types of forest—and there are actually about a couple of dozen of them between Alaska and Tierra del Fuego—are well-nigh impenetrable. That is why not only just substantial parts, but the greater part of them, even in our own country, are not yet "opened up." This is a loose term; so, to be more precise, let me give one example of the state of current affairs in what is just about the *most* accessible of all of them today. This is the 17,000 square-mile block of territory centered around the Klamath River area in northern California.

The extent, position, and boundaries of this area may be seen on Map IV. You may calculate its dimensions for yourself. This I beg of you to do, rather than writing to me about it.* Please note also that it starts at the bottom about Clear Lake which is just 70 miles north of San Francisco, and it continues on north into Oregon. Actually it is confluent with a much vaster block in the Cascades, and is nowhere completely cut off (by farmland or nonforested land) from other lesser blocks in Oregon and thence on to Washington. I should explain that in delineating these wilderness blocks I do not consider a road, even a main blacktop, to be a boundary for it does not deter any living thing that I know of from passing from one side to the other, provided there is cover on both

* In the article mentioned above in *True* Magazine, an extra zero unfortunately became attached to the area given, a mistake that started with my typing but went clean through to the published story. This resulted in a deluge of several thousand letters. But, when it had been corrected, just as many people wrote scoffing at the true figure. Many of these were Californians; and some even from the counties concerned!

sides right up to its edges. This great area has been surveyed and there are maps of it down to very large scale in conformity with the best series published by the Coast and Geodetic Survey, National Forests, and other official agencies. There are neat county maps covered with names and a grid on a scale of 4/10 of an inch to the mile, that look perfectly splendid at first sight. However, I ran into a Federal agency surveying party when I was deep in the middle of this block in 1959, and spent several evenings with the Chief Surveying Officer who told me things and demonstrated certain facts that, metaphorically speaking, caused myself and my two traveling partners to lose our eyebrows—upward.

It transpired that this area has only once been "surveyed" and that was by unofficial surveyors under contract to the U.S. Government, *in the year 1859!* Further, the survey was ostensibly made on a 1-mile grid; that is to say the surveyor was supposed to walk a mile north, south, east, or west, take a fix and drive a stake, and continue doing this till he reached some previously selected line at the other end that linked up with the next survey. The original notebooks carried by these surveyors of 1859, and in which they recorded the facts and figures of their surveys in the field, a page to a mile, are on file in the Lands Office in San Francisco. They are a revelation. The surveyor whom we met told us that in one notebook he had found no less than 23 pages absolutely blank and without so much as a thumbmark on them, and he told us that all the books covering this area were like that. He stressed that this is no deprecation of the early surveyors as, he said, they actually did a remarkable job on the whole, managing to join up the surveys to the 60-mile triangulation made from mountaintops (and now corrected from aerial photography), but he pointed out that the greater part of the resultant maps are pure conjecture and most of them made by what surveyors call "camp-surveying." What, of course, happened was that the country was so rugged and impassable that the surveyors just went in as far as they could, then came back out, went around to the next possible entrance, and tried again. When they had enough fixes around the edges, they just ruled lines connecting what they had, adjusted a bit for error, and then ruled in the rest of the grid. And this, combined with names given to visible mountains, ranges, Amerindian settlements in accessible valleys, and logging operations, filled the whole thing out nicely, so that on paper it looks almost like the outskirts of San Francisco.

Actually, this great block of territory is quite unknown. Nobody goes into it much except a short way from its edges, and practically nobody has gone through it. I interviewed one experienced locally bred woodsman who took a 3-week summer vacation to attempt this. He did cut across the southwest corner of the square but was a week late getting back to work. A friend of mine working in there at the time of writing did come upon a lone and unknown prospector of the old school some distance in, and he had *a mule* in there. One "scientist" from a "university" in California wrote a furious letter after I had published my report on the ABSMery of this area, stating that he had "collected animals all over every bit of the area during several seasons" and adding gratuitously that "its entire fauna has for decades been well known." This is a point at which I find it very hard to remain civil.

The whole of this country is clothed in a particular kind of montane, closed-canopy, mixed deciduous-coniferous forest, of magnificent proportions and containing some of the finest timber in the world. It grows in three tiers with an undergrowth. The tallest trees such as Sitka Spruce, and Douglas Fir, run up to 150 to 200 feet and stand pretty close together. Under them on the upper reaches there is a closed-canopy of smaller conifers, in the valleys of deciduous trees such as maples, madroñes, etc., and beneath both of these there is usually another closed-canopy of large saplings and smaller trees of mixed constitution. Beneath this again is another layer that is almost impenetrable, being composed of bushes and the dead branches of the spruces and firs which are as strong as spring steel even when leafless, and which persist right down to the ground like a barbed-wire entanglement. It took me half an hour with a sharp machete to get far enough from the one road in the country not to be able to talk to my companions left on that road. I am a fair bushwacker, having been at it all my life, and I am pencil-thin and thus highly suitable for going through and under things.

But this is not by any means all. The whole of this country is constructed like a freshly plowed field on a monstrous scale. While its mountains and peaks are not high by Western standards they are immensely steep, and closely packed so that there is practically no horizontal ground throughout the whole country. The whole thing is a nightmare even to experienced woodsmen, and something much worse to road builders.

This is the real state of affairs throughout a huge block of territory within a hundred miles of one of our greatest cities, although almost everybody in that city would deny it positively, and even the majority of citizens of Eureka, a large and prosperous community right on its edge, have no idea of its true nature. Conditions are even more difficult in other montane areas but from now on I shall simply be saying of them, as we approach them, that they are either better or worse than the Klamath. This is going to relieve me of the necessity for a lot of verbiage. Readers may also find this useful in arguments; while it will give some sort of key to assess other forests in other lands. Actually, though, this Klamath forest is just about as difficult as I have ever run into, and that goes for the tropics too, but it, of course, pales before the British Columbian vegetation on the grounds of topography for, whereas we have here to deal only with little mountains, there we have enormous ones.

It was such topography, moreover, that was tackled by the greenhorns from the East looking for gold. They didn't get very far, but they did, according to the older Amerinds still living, and who got it from their fathers and grandfathers, cause the ABSMs to make a sudden mass withdrawal into the inner recesses of each of the blocks, at that time. This interesting information was first given to me by a Mr. Oscar Mack, doyen of the Yurok clans of the upper Hoopa Valley. The same statement has cropped up again and again during my investigations all over the Puget Sound to California area. If, moreover, you look at Map I you will note an extremely odd fact. This is that early reports (and of various types) came also from what is now Idaho in what is called more technically the North Montane Province. Some very funny things happened there in early days and they seem still to be happening. Most of them center round the real wilderness area about the upper Salmon River which flows into the Snake River as shown on that map. It was in Idaho also that the first foot-track scare took place.

This is an interesting story in several ways, and has naturally been received with whoops of joy by the skeptics. The story is from the *Humboldt Times* of January 3, 1959, and reads:

STORY OF CENTURY OLD BIGFOOT IN IDAHO ADDS COLOR TO LEGEND: by Betty Allen, *Times* Correspondent: Willow Creek—Mrs. Alvin Bortles, Boise, Idaho, discussed an account of a "Big Foot" who lived prior to 1868 in the wilderness of Idaho.

The mother of Kenneth Bortles, vice principal of the Hoopa valley high school, Mrs. Bortles said that mysterious tracks of a tremendous size and human shape stirred the residents of Idaho in the early days. Just as with the "Big Foot" tracks of Northern California's Bluff Creek area, some believed they were genuine, others saw in them a clever hoax.

The "Big Foot" lived in the remote wilderness of Reynold's Canyon now known as Reynold's Creek. A thousand dollars was offered for him, dead or alive. Here the likeness to the local "Big Foot" ended for the "Gigantic Monster," as he was called in Idaho, was a killer. The full extent of the depredations of this Big Foot were never known, for many robberies and murders were attributed to him which he probably did not commit. The sometimes wanton killings that were the work of almost superhuman strength both with stock and humans, brought about his downfall. A thousand dollars was offered for Big Foot dead or alive.

John Wheeler, a former army man, set out to collect the reward. In the year 1868, he came upon Big Foot and shot him 16 times. Both legs and one arm were broken before he fell to the ground. As he lay there he asked for a drink of water and, because of his great fear, Wheeler shot him, breaking his other arm before giving the water to the creature. Before he died, he told Wheeler that his real name was Starr Wilkerson and he had been born in the Cherokee nation of a white father. His mother was part Cherokee and part Negro. Even as a very small boy everyone had called him "Big Foot" and made fun of him. At the age of

19 the white girl he loved jilted him for another. Gathering a small band of men about him he killed then, for the sheer love of killing. Later he killed the girl that he had loved.

The foot length of this great giant of a man was 17½ inches and 18 inches around the ball of the foot. His height was 6 feet, 8 inches, with a chest measurement of 59 inches, and his weight was estimated at 300 pounds. He was all bone and sinew, no surplus flesh. He was known to have traveled as far as 60 or 75 miles in a 24-hour period.

Adelaide Hawes gives an account of Starr Wilkerson or "Big Foot" in her book, *The Valley of the Tall Grass*, written in 1950.

I have other old stories from Idaho, mostly of sheep being torn apart and monstrous human-like footprints by water holes, but nothing ever came of them. There is one story, however, that has always impressed me. This is told by none other than Theodore Roosevelt in a book he published in 1892 entitled *Wilderness Hunter*. Teddy was not a boy to be taken in by anybody much, and he was a great skeptic and debunker, especially in the field of wildlife, being the originator of that most excellent expression of opprobrium, "Nature-Faker." This story seems to have impressed him not a little and mostly because of the still noticeable terror of the teller, half a lifetime later. He was an old man when he talked to Roosevelt and the incident had happened when he was young. His name was Bauman and he was born in the area on the then frontier, and had spent all his life as a hunter and trapper. Roosevelt's account goes as follows:

It was told [to me] by a grizzled, weather-beaten old mountain hunter, named Bauman, who was born and had passed all his life on the frontier. He must have believed what he said, for he could hardly repress a shudder at certain points of the tales.

When the event occurred Bauman was still a young man, and was trapping with a partner among the mountains dividing the forks of the Salmon from the head of Wisdom river. Not having had much luck, he and his partner determined to go up into a particularly wild and lonely pass through which ran a small stream said to contain many beaver. The pass had an evil reputation because the year before a solitary hunter who had wandered into it was there slain, seemingly by a wild beast, the half-eaten remains being afterwards found by some mining prospectors who had passed his camp only the night before.

The memory of this event, however, weighed very lightly with the two trappers, who were as adventurous and hardy as others of their kind. . . . They then struck out on foot through the vast, gloomy forest, and in about 4 hours reached a little open glade where they concluded to camp, as signs of game were plenty.

There was still an hour or two of daylight left, and after building a brush lean-to and throwing down and opening their packs, they started up stream. . . .

At dusk they again reached camp. . . .

They were surprised to find that during their absence something, apparently a bear, had visited camp, and had rummaged about among their things, scattering the contents of their packs, and in sheer wantonness destroying their lean-to. The footprints of the beast were quite plain, but at first they paid no particular heed to them, busying themselves with rebuilding the lean-to, laying out their beds and stores, and lighting the fire.

While Bauman was making ready supper, it being already dark, his companion began to examine the tracks more closely, and soon took a brand from the fire to follow them up, where the intruder had walked along a game trail after leaving the camp. . . . Coming back to the fire, he stood by it a minute or two, peering out into the darkness, and suddenly remarked: "Bauman, that bear has been walking on two legs." Bauman laughed at this, but his partner insisted that he was right, and upon again examining the tracks with a torch, they certainly did seem to be made by but two paws, or feet. However, it was too dark to make sure. After discussing whether the footprints could possibly be those of a human being, and coming to the conclusion that they could not be, the two men rolled up in their blankets, and went to sleep under the lean-to.

At midnight Bauman was awakened by some noise, and sat up in his blankets. As he did so his nostrils were struck by a strong, wild-beast odor, and he caught the loom of a great body in the darkness at the mouth of the lean-to. Grasping his rifle, he fired at the vague, threatening shadow, but must have missed, for immediately afterwards he heard the smashing of the underwood as the thing, whatever it was, rushed off into the impenetrable blackness of the forest and the night.

After this the two men slept but little, sitting up by the rekindled fire, but they heard nothing more. In the morning they started out to look at the few traps they had set the previous evening and put out new ones. By an unspoken agreement they kept together all day, and returned to camp towards evening.

On nearing it they saw, hardly to their astonishment, that the lean-to had been again torn down. The visitor of the preceding day had returned, and in wanton malice had tossed about their camp kit and bedding, and destroyed the shanty. The ground was marked up by its tracks, and on leaving the camp it had gone along the soft earth by the brook, where the footprints were as plain as if on snow, and, after a careful scrutiny of the trail, it certainly did seem as if, whatever the thing was, it had walked off on but two legs.

The men thoroughly uneasy, gathered a great heap of dead logs, and kept up a roaring fire throughout the night, one or the other sitting on guard most of the time. About midnight the thing came down through the forest opposite, across the brook, and stayed there on the hill-side for nearly an hour. They could hear the branches crackle as it moved about, and several times it uttered a harsh, grating, long-drawn moan, a peculiarly sinister sound. Yet it did not venture near the fire.

In the morning the two trappers, after discussing the strange events of the last 36 hours, decided that they would shoulder their packs and leave the valley that afternoon. . . .

All the morning they kept together, picking up trap after trap, each one empty. On first leaving camp they had the disagreeable sensation of being followed. In the dense spruce thickets they occasionally heard a branch snap after they had passed; and now and then there were slight rustling noises among the small pines to one side of them.

At noon they were back within a couple of miles of camp. In the high, bright sunlight their fears seemed absurd to the two armed men, accustomed as they were, through long years of lonely wandering in the wilderness to face every kind of danger from man, brute, or element. There were still three beaver traps to collect from a little pond in a wide ravine near by. Bauman volunteered to gather these and bring them in, while his companion went ahead to camp and made ready the packs.

On reaching the pond Bauman found 3 beavers in the traps, one of which had been pulled loose and carried into a beaver house. He took several hours in securing and preparing the beaver, and when he started homewards he marked, with some uneasiness how low the sun was getting. . . .

At last he came to the edge of the little glade where the camp lay, and shouted as he approached it, but got no answer. The camp fire had gone out, though the thin blue smoke was still curling upwards. Near it lay the packs wrapped and arranged. At first Bauman could see nobody; nor did he receive an answer to his call. Stepping forward he again shouted, and as he did so his eye fell on the body of his friend, stretched beside the trunk of a great fallen spruce. Rushing towards it the horrified trapper found that the body was still warm, but that the neck was broken, while there were four great fang marks in the throat.

The footprints of the unknown beast-creature, printed deep in the soft soil, told the whole story.

The unfortunate man, having finished his packing, had sat down on the spruce log with his face to the fire, and his back to the dense woods, to wait for his companion. . . . It had not eaten the body, but apparently had romped and gambolled round it in uncouth, ferocious glee, occasionally rolling over and over it; and had then fled back into the soundless depths of the woods.

Bauman, utterly unnerved, and believing that the creature with which he had to deal was something either half human or half devil, some great goblin-beast, abandoned everything but his rifle and struck off at speed down the pass, not halting until he reached the beaver meadows where the hobbled ponies were still grazing. Mounting, he rode onwards through the night, until far beyond the reach of pursuit.

Judged by the time of publication of this story and what the old man said, this must have taken place in the early 1800's. Conditions changed radically about those parts in the 1850's, but then, strangely, they lapsed once more into a form of oblivion and, despite the incredible advance of civilization and the complete opening up of the whole West, until it stands today as second to no other area in the Union, parts of it are really less known now than they were a hundred years ago. I have observed this strange progress of progress in action in other lands, notably in the Republic of Haiti. The population of that small Caribbean country is so enormous that the whole of it, and right up its towering mountains, is virtually one great garden-city. You can stand anywhere and spit in three directions and be sure to hit somebody's compound. When the troubles took place in the 1920's and the American Marines took over, they built motorable roads in a network all over the country. Then they left; but at the same time there came the commercial airplane. By 1940 you just couldn't find any of the roads made by the Marines, while a new network was being built that went roaring straight through the country from one important center to another. All in between had gone right back to conditions pertaining before the advent of the Marines, and in some large areas apparently to those pertaining before Columbus. So also with great pieces of our own country.

It was during this initial period of lapse, or collapse, that we once again pick up our tracks. The strangest story is that of Capt. Joseph Walker, an account of which lies in the files of a paper called the *Eureka Daily Leader*, dated February 14, 1879. This recounts that this gentleman, who was then a most renowned mountaineer, trapper, and guide, due to his many exploits in the Rockies, had recently returned from investigating a newly opened territory near the mouth of New York Canyon and had brought to the office of the *Leader* a slab of sandstone about 20 inches long and 14 inches wide and some 3 inches thick. "On the surface of this slab of sandstone was imprinted the clear form of a gigantic footprint [I am quoting here], perfect except for the tip of the great toe. The footprint measured 14½ inches from the end of the heel to the tip of the toe and was 6 inches wide across the ball of the foot. Captain Walker related how he had found the slab of sandstone formation under about 2 feet of sand."

This story has sundry rather odd features. First, a foot 14½ inches by 6 inches across the ball is hardly a gigantic foot compared to what is coming in a moment, but it has a plantar index of 2.42 which is much wider than a human foot and would give an impression of great size. The fact that it was impressed in a slab of sandstone might at first sound more than just suspicious because you can't impress anything into solid rock—you have to chip it out. However, and this should be borne in mind, sandstone can form in a matter of days under certain conditions. A surface of argillaceous sand may dry out under a hot sun and remain baked to the consistency of pottery for months. If then a flood brings down a layer of sand or other material and deposits this on top of it and also immediately dries out, you may get conditions similar to those that pertained in our Porcupine cage at my zoo. More drying, compression and the solution and percolation of, say, lime from the covering layer may then, in only a few years, solidify the bottom layer and turn it to a sandstone. I have seen car tire tracks in sandstone so solid you needed a cold chisel to chip it.

Captain Walker was not a man to be fooled either and he retained a high reputation so that Walker River was named after him by the Federal Government. He was, in fact, solidly on the right track!

6. In Our Own Back Yard

When something happens in the next county, we all get excited, but how many of us go take a look at it? If it is not pleasant or impugns our local community, we usually assert it is a hoax.

Now that we are squarely back on the tracks, we might as well stay on them and skip, for the moment, all chronology. There is a business about giant, humanoid-appearing foot-tracks that has been going on in this country for far too long. It needs examination, and either exposition or debunking. It centers around the Great Basin, which is mostly now the state of Nevada, but it slops over in all directions and, in the form of giants capable of making such tracks, it reaches from Canada to Mexico, from the Pacific coast to Pennsylvania, and right on into the portals of the Smithsonian Institution in Washington, D.C.

As will be seen from the discussion of what I euphemistically call Myths, Legends, and Folklore (Chapter 17) such things are linked up all over a wide area from New Mexico to Puget Sound but center round the Sierra Nevada. They are linked by both traditional, early, and even recent accounts of a giant race of wild people, who inhabited this area in bygone days, and who not only were there before the Amerinds arrived, but persisted for a very long time after they had done so and, it is alleged, still linger on today. In tradition, these personages are not overly exaggerated. They are consistently reported as having been on an average about 7 to 8 feet tall (or its equivalent), but to have included outsized individuals, to boot, that were especially reverenced.

We have seen a record of a skeleton fitting these dimen-

MAP IV. NORTHERN CALIFORNIA

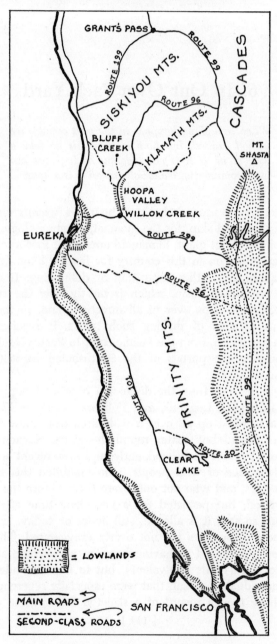

See facing page for legend

sions allegedly found in British Columbia. There are plenty of others but we just don't talk about them. Unfortunately, I have not had the opportunity to follow up the notes given to me some time ago on this subject, regarding Nevada, by a good friend, a man with a genius for bibliographical research, a very wide and real knowledge of American prehistory and folklore, of Amerindian history, and of colonial tradition. It is a voluminous and very startling file, containing what seem to me to be endless references to what are classed as "giant burials" from all over the place. Many of these are said to be housed in small county and city museums dotted about the West, and most seem to have been lodged therein during the 19th century. A few are said to have gone to the Smithsonian, yet all have been totally ignored ever since. The reason for this ig–*nor*–ance, as well as a notable ignorance of the subject, is stated by this indefatigable literary groundhog to be the really abominable story of the foot-tracks of Carson City jail, a most odd affair. It goes as follows:

This business was a *cause célèbre* 80 years ago. It could have passed almost unnoticed but for two facts; first, that said tracks were found in sandstone at a depth of some 25 feet below the surface in the jail compound of Carson City, Nevada. The second thing that stimulated such

---

MAP IV. NORTHERN CALIFORNIA

The land area on this map represents approximately 45,000 square miles. All but a small portion at the extreme south around San Francisco, and a sliver of the upper Sacramento Valley, are mountains. These are not excessively high but are very steep and closely packed, with deep narrow gorges between. However, the various blocks contained within this area are not at all homogeneous. The mighty Cascades are volcanic and much larger than the coastal ranges. The Klamaths are the "oldest" from a faunistic point of view; the Trinities are newer and of somewhat different phyto-geographic constitution. Along the coast, from just south of Cape Blanco, but a little way in from that coast, to a little south of San Francisco Bay is the land of the great Redwoods (*Sequoia sempervirens*). The whole mountainous part is clothed in almost unbroken forest, and ABSMs have been reported from Clear Lake in the south to the northern edge of the Siskiyous and beyond to the northeast.

[113]

wide interest was not that some scientists said that some of these tracks were made by giant men of over a *million* [sic] years ago, but that Mark Twain wrote a positively hilarious story with the discovery as its main theme or *raison d'etre*. These facts have been quoted, misquoted, and just mentioned over and over again. The true story represents one of the finest examples of scientific skullduggery—and vagueness—on record.

First, some of the tracks—there were others of elephantines, deer, cats, and "giant birds"—looked like those of a giant humanoid. This fact was published by Dr. Harkness of San Francisco through the California Academy of Sciences. In his report, the author gave some sketches of said tracks but stated that he had "filled out those areas not clearly shown in the originals." These areas happen very conveniently to go right around the front of the imprints and down their inner sides. As a result they look generally much more human than they would otherwise have done; at the same time just such areas would have cut out any imprints of toes (human or otherwise). Dr. LeConte, of California, agreed in print.

The result of these communications was an immediate response [as was almost invariable in that decade] from (Professors) O. K. Marsh and E. D. Cope. Marsh, of Yale, blasted the idea in his own inimitable style: he did not even bother to refute the matter; he simply stated that the tracks were those of a ground-sloth—either *Mylodon* or what he called *Morotherium* [sic]. No animal has received such a name, but there is, of course, the really giant Ground-Sloth (*Megatherium*); a detail of such a trifling nature would of course not hinder this paleontological free-booter. The most interesting part of this pat pontification is that he appears to have accepted Harkness' quite illegitimate touching up of the tracks, and then to have stated that they were manifestly those of a ground-sloth. Thus, he was, *ipso facto*, wrong in that, as touched up, they were *not* those of such an animal. He seems to have completely missed the further point that *before* touching up they could well have been so. However, he pulled still another boner.

Ground-Sloths—which were actually enormous kinds of shaggy, short-tailed, neotropical anteaters more closely related to the Giant Anteater (*Myrmecophaga*) than to the living Tree-Sloths—could apparently stand on their hind legs but they used their immensely thick short tails as a third prong of a tripod to do so. If they waddled along on their hind legs, their tails must have gouged a deep channel between the tracks of their feet. There were no such channels in the Carson City tracks. Marsh seems to have appreciated this fact so he conjured up some "smaller imprints, obviously those of front feet somewhat outside the main tracks." No such tracks were ever recorded, or sketched. [Cope, Marsh's most implacable foe, simply implied that Marsh was wrong, and that anyone else (LeConte and Harkness) was more reliable.] They probably never existed! This is the way awkward "scientific" discoveries are handled: if they don't fit into the already approved scheme, you make them do so—and by any lie your reputation can get away with. For all one knows, the original tracks may even have been those of a good old *Oh-Mah*.

There is an official wind-up to this but it is almost as extraordinary and inexplicable as the facts themselves. For some reason serious-minded scientists—by which I mean those who still have open minds—*have* concluded that these beastly things are the tracks of one of the Giant Ground-Sloths. We have radiocarbon dated bones of some of these creatures killed and eaten by men of a pretty advanced culture in the Southwest—but I cannot understand how any paleontologist, let alone zoologist, could ever conceive of any form of such a creature [of which we *have* a foot] having ever either walked on its hind legs alone or left a footprint anything like those of Carson City jail. This identification, however, led to all the other large humanoid tracks being dismissed as "just those of fossil animals."

Would that we might pursue the matter of giant skeletons but at this juncture it would be inappropriate. It is (as of now) really a separate subject, and until we obtain a fresh skeleton of one of the large or giant ABSMs, or some competent, trustworthy, and really informed physical anthropologist happens to stumble upon one in a museum, it must be left dangling. So, again, we jump back onto the tracks.

This would be in the year 1890 at a place on the Chetco River about the border between California and Oregon. It appears that about that date and thereabouts, the citizenry had been bothered for some time by really gigantic foottracks that looked, according to the contemporary records, just like those that would be left by enormous naked human feet, which passed back and forth from the forests to the seashore. Then things began to happen at a mining camp some 50 miles inland. Large objects were moved at night and there were unpleasant noises, all naturally attributed to marauding bears, until one fine morning when, after a particularly rambunctious night, during which somebody fired at something, two sets of large human-type tracks showed up all around the camp. A posse was organized and tracked these for a long way into the forest but eventually gave up. A short time later, however, a man was chased into camp by something very large, the looks of which he did not wait to investigate. A watch was set; two men at a time, for a few hours each. Then it happened.

One couple going to relieve a watch found their two companions dead and really grossly mutilated. They had in fact been literally smashed and apparently by being picked up and slammed repeatedly onto the ground so that they looked as if they had fallen off a high cliff onto rocks. The account particularly specifies that there was nothing anywhere near off which they could have fallen. The wretched men had emptied their rifles and there was both spoor and a large blood-trail leading off into the bush. This the whole camp personnel followed with the aid of Amerindian trackers. It led into the Siskiyou complex of mountains to a point far beyond that which any even of the Amerinds had previously penetrated. There, the men are said to have come upon a fresh lava flow. This is an astonishing item. There is volcanicity in the area and there had been an account 14 years before of a great quake and the sound of a far distant explosion, heard on the coast of Oregon, and of a dull glow said to have been seen in the sky for two nights but nothing definite about a volcanic eruption was even suggested at the time. What is more, this report by a party of ABSM hunters was also ignored and it was not till after World War II that lava beds, now re-vegetated but nonetheless of very recent vintage, were located in the area, although they had shown up on aerial maps as irregular patches of some unique form of ground-cover.

This rather gruesome incident appears to have satisfied even local curiosity for some time as nothing much is reported for quite a few years, though a very old Amerindian patriot told me with a whimsical smile of one Chester Johnny of his tribal group who in the year 1905 spent a happy hour watching a large papa *Toké-Mussi* (as the local giant ABSM or *Sasquatch* is called) trying to teach his two youngsters to swim in a river, and to spike fish with sticks. The records I have—though they are, of course, very very far from complete—are almost blank until 1924 when a bunch of hard-boiled loggers came literally roaring into the small town of Kelso, Washington, from their advanced camp in the Cascades, and absolutely refused to go back nor anywhere else in the region out of sight of a highway. They said that their camp had been attacked by a number of enormous hairy wild men who had pelted them with stones and other debris. In view of their tough characters and stubborn attitude, a posse was formed and, well armed, went to investigate. No ABSMs were seen but they had left their enormous tracks everywhere and the cabin of the loggers was not just a shambles but in great part destroyed. That year there had been terrible man-induced forest fires in the region for the first time. I have often wondered if the ABSMs decided to give little "human" men a peremptory lesson in conservation—the best and only really satisfactory approach to which is the total eradication of said little men from the entire locality.

Now, here we are back again at the date 1924. Wherever we go, it seems, and it will recur, there was a marked turning point in ABSM history in the demi-decade 1920 to 1925. I think there must have been a great world-wide historical break at that time which perhaps will not become apparent to historians for many generations. What it was I can only conjecture but more and more I am coming to think that this was the real time of change-over from all that went before to what we call modern times, or the new world. Most of the things that have really affected the outlook of humanity, like the invention of the typewriter, electrical power, and especially light, radio, internal combustion engines, flying machines, and so forth, had taken place before this, but then came the social upheavals of the postwar era. Not even these things had really taken hold before the 1920's, and they took a few years to do so even after that. Man's outlook on life then changed radically everywhere, and he also took a completely new look at his environment. A great number of the shibboleths he had previously held most precious just collapsed, while a whole lot that he had previously considered worthless or redundant suddenly acquired entirely new status. The change was technologically induced but it did not greatly affect basic science—funnily enough—but rather sociological attitudes. It was not that new things began to happen all at once so much as that people began to treat the old ones in an entirely different light. On the one hand, *real* exploration began: on the other, just plain, go-have-a-look exploration came to an end.

Things like abominable foot-tracks went into limbo. The age of "the Curiosity" was over and people were no longer curious. They demanded the facts and in some respects they got them. In others they got falsehood or nothing at all. ABSMs became definitely *de trop*, and gay souls like Prof. Khakhlov, Mr. Tombazi, and others, no longer went barging about the world recording curiosities and writing about them. This initiated the age of skepticism par excellence.

During the 35-year period subsequent to this strange historical turning point, a lot went on nonetheless, and this, due perhaps to its suppression, gradually built up a veritable explosion at the end of the 1950's. It is hard enough to suppress anything, but suppressing truth (i.e. facts) entails its own special hazards. People are more suspicious of truth than they are of falsehood and they almost invariably downgrade it if it clashes with belief or faiths. But "truth will out" seems not to be an altogether valueless cliché. Then again, both false facts, and the suppression of unpleasant facts is apt to be extremely costly; and, if you really want to get at the truth of anything, anywhere, reach first for a pocketbook. I seem to be full of clichés, but it is also perfectly true that if you hit anybody through his pocketbook you are more likely to loosen him up than by hitting him anywhere else. And, as I am in this rut, I might as well add that, while love of money may be the root of all evil, it is still by far the best invention yet for getting at the facts. The moment anything, however curious (or unpleasant) it is considered generally, develops a value, it automatically develops a potential, and when it is founded on fact, it cannot forever be suppressed.

In our case, also, the facts have several special connotations. They impinge directly upon our most basic precepts, such as religion, ethics, politics, and science. A live ABSM would be the greatest propaganda weapon possible; at one fell swoop it would prove Darwinism, and set at nought a great part of religious belief and dogma, while it would also confound a great deal of that which science has written into *its* dogma. Quite apart from all of these high and mighty matters, plain ordinary people have finally become fed up with being called fools, liars, and idiots. The world is full of crackpots but it is rapidly becoming manifest that most of these make a specialty of pursuing beliefs, prejudices, and faiths rather than the facts of everyday life. If you walk into a truck and stagger home bleeding to death, it is quite useless anybody telling you that you are imagining things, that there is no such thing as a truck, or that you ought to be confined. It doesn't help your feelings (or matters as a whole) if somebody suggests that it was undoubtedly a bus and not a truck, or perhaps even a motorcycle. Either the damned thing was there or it wasn't, skeptics notwithstanding; and the average citizen becomes peeved when he is told that he, who saw the thing, is lying, especially by a person who was not present. During the last 40 years, plain, solid citizens have been getting pretty peeved.

During this period ABSMery in the United States contracted in upon itself and became concentrated in and around this Klamath district in northern California, which I described as an example of a virgin montane forest area. It now transpires that somebody has reported something about the matter every year since 1938 in this area while, of course, the Amerinds thereabouts just went steadily and stoically on living with the business and keeping their mouths shut. I won't go into all of these items because they are so exactly alike, and they are all just like the descriptions given of the *Sasquatches*. Hereabouts they are called, as I have said, *Toké-Mussis* by one Amerindian group, the Yurok or Yurock, and *Oh-Mahs* by

[119]

the Hŭppas; there are endless other names for the big ones in accordance with the number of tribes, sub-tribes, and familial dialects of the Amerinds. The name *Oh-Mah*, which is rapidly coming into universal acceptance, actually means something very close to "Devil" as used by our ancestors—a sort of large chap with nasty habits who is dangerous, bestial, potentially carnivorous, and smelly but definitely rather human. *Non*–Amerinds in the area have come to call them "Bigfoot" having the usually mistaken idea that there is just *one* giant of some kind loose in the countryside—just as people speak of *The* Loch Ness Monster, as if it were a lone individual that has been paddling about therein since Cretaceous times, mateless and possibly even parentless. But there is a complication here.

I would have had to come to this sooner or later in any case so I might as well introduce it now, even if it is not the place to go into it in full. To jump ahead, let me say that there are now some hundred separate and isolated areas in the world where or from which ABSMs have been reported—and this is apart from Myth, Legend, and Folklore. The creatures described vary considerably but, with a few notable exceptions, they appear to fall very clearly into four main types—a large (or giant, to us), a medium or man-sized, a small or pigmy, and an excessively bestial creature known as the *Meh-Teh*. These types are not set or patterned, and there is considerable variety in the actual sizes of each as reported. However, they would each seem to form a fairly well defined animal form, having certain particulars, characters, characteristics, and other perquisites all their own.

The giant ones are inhabitants of higher elevations and do actually go around in snow when needs be. They seem to be more carnivorous, at least in winter, like many Primates; they have very human-type feet; and they are clothed in short, thick, hairy fur. The medium-sized are very manlike but clothed in longer, darker hair, have very pointed heads, and very short, broad feet with large toes, the first being extra large and widely separated. They are vegetarian or omnivorous and live in upper montane forests but seldom go up above the tree-line. The little pigmies are also forest dwellers, but in the valleys and bottoms, or at lower levels, and where it is much warmer. They have little manlike feet but with very pointed heels, are clothed in fur but have much longer head-hair that forms a mane down the midback. They are excellent tree-climbers and will take to water. They go about in small family parties and have a sort of primitive language and they may even carry palm leaves or bits of anything that will afford shade from sunlight. They are just about as nocturnal as chimps which move about and feed at night in fine moonlight weather. The giants seem to be almost wholly nocturnal; the medium jobs more diurnal or crepuscular. The *Meh-Tehs* are quite another matter (see Chapter 15).

Again and again and again, these four types will crop up. In Canada I have so far heard only of the giants, and I thought that the same went for the Puget Sound to California area, but I am afraid that I have now to bring up the unruly suggestion that some reports from this area seem definitely to be, or try to be, speaking of both the man-sized and pigmy types. This, you may well think, is going a bit too far, in that it is bad enough to be asked to stomach the possibility of a bunch of giant "ape-men" running around half a dozen of our most up-to-date and worthy states, without being asked to accept also Neanderthalers and "Little People." I would have preferred, as I say, not to have brought this up just yet but, as a reporter, what can I do? The very definite footprints left near Roseburg, Oregon, during the night of October 23, 1959, were definitely of the man-sized type, while literally thousands of the little pigmy type are alleged to have started turning up along—perhaps appropriately—the Mad River Valley about 1950. Thus, as we go along, you must brace yourself for casual asides to the effect that such little ones were seen hither and yon. The Roseburg case is happily so far unique, so that we won't be bothered herein with others of its ilk and so, when I speak of ABSMs hereabouts from now on, it will be of the giant *Oh-Mahs* unless I clearly specify otherwise.

The outburst came in August, 1958.

There was, as usual, an unreported and steady build-up before the event, and there was the usual red herring almost at the outset. This latter was such a bizarre report and was given such wide publicity that it has both diverted public attention and caused many, who might otherwise have investigated the main stream of events with diligence and an open mind, simply to throw up their hands in horror at anything so impertinent. The case is very peculiar, has no precedent and no conclusion. It occurred 3 months after the outbreak of true ABSMery in northern California in August, and it took place 600 miles away from that area, near Riverside in San Bernardino Valley of southern California. Nothing of a similar nature has ever been recorded from anywhere near this place, while all the mountains from the Sierra Nevada south into Baja California may really be said now to have been explored and combed. [At the same time, we might note the proximity of Hollywood and several large mental institutions.] The following is an account from the *Los Angeles Examiner* which speaks for itself, though very facetiously and says everything that there is to be said about the business.

MONSTROUS! Driver Tells of "Thing" that Clawed at His Car

Riverside, Nov. 9 (UP). Funny thing happened to Charlie Wetzel on the way home last night.

A Monster jumped out at him. That's what he told authorities who planned to continue an investigation of the incredible story today.

Wetzel, 24, a resident of nearby Bloomington, reported soberly that he was driving on a street near Riverside when a frightening creature jumped in front of his car.

"It had a round, scare-crowish head," he said, "like something out of Halloween.

"It wasn't human. It had a longer arm than anything I'd ever seen. When it saw me in the car it reached all the way back to the windshield and began clawing at me.

"It didn't have any ears. The face was all round. The eyes were shining like something fluorescent and it had a protuberant mouth. It was scaly, like leaves."

Wetzel said he became terrified when the creature reached over the hood of his car and began clawing at the windshield. He said he reached for a .22 pistol he had in the car.

"I held that pistol and stomped on the gas," he said. "The thing fell back from the car and it gurgled.

"The noise it made didn't sound human. I think I hit it. I heard something hit the pan under the car."

Sheriff's officers said Wetzel pointed at some thin, sweeping marks he said the creature made on his windshield. They went to the scene of the claimed apparition but said they could find nothing to prove or disprove Wetzel's story.

The scene is at a point where North Main Street dips and crosses the Santa Ana River bed, which is usually almost dry. Wetzel said he told the story to his wife and she induced him to phone authorities.

"I kept saying no one would believe a story like this," he said.

Sheriff's Sgt. E. R. Holmes said he thought perhaps a large vulture might have flopped on the hood of Wetzel's car—"sometimes cars hit them when they're in the road eating rabbits cars have killed," he said.

So he searched the area himself today. "But," said Holmes, "I didn't even find a feather."

The build-up to the really valid events may be left till later, for it consists once again of accounts of all the same old things, though, withal, highly confirmatory, and showing that what happened at a place called Bluff Creek in August of that year was neither an isolated case nor anything novel. I will mention these more fully when I come to tell of the aftermath of the Bluff Creek affair.

Before giving the facts of this business I must just hark back for a minute to my description of this country. On Map IV you will see the main roads marked by their route numbers. Apart from the four that surround the area—Nos. 101, 299, 99, and the east-west route over Grant's Pass, there is really only one road through this block of territory. This runs from Willow Creek, diagonally northeast to join Route 99, via Happy Camp. Immediately north of Willow Creek it follows the Hoopa (Hŭppa) Valley and then forks, one small road going back to 101 at the coast, the other major route going into the hills. About 10 miles along this route a new road is being pushed north up a tight valley named Bluff Creek. This road was begun in 1957. I visited the road end in 1959 and it had only gotten in 23 miles, so rough is the country. From this area, the following matters came to light.

In August of 1958—on the morning of the 27th to be precise—a very sane and sober citizen by the name of Mr. Jerald Crew, of Salyer township, Humboldt County, northwest California, an active member of the Baptist Church, a teetotaler, and a man with a reputation in his community that can only be described as heroic in face of certain almost unique personal tragedy, went to his work with heavy-duty equipment at the head of this new lumber access road being pushed into uninhabited and only roughly surveyed territory near the borders of Humboldt and Del Norte Counties. This huge block of territory is crossed kitty-corner from the south at Willow Creek to the northeast by a winding blacktop road, and from east to west by only four other roads of lower grade. Logging trails and some "jeep-roads" now finger into it from these roads and from the main arteries that enclose it to north, west, south, and east, but these are of very limited extent and are hardly used at all. "Jerry" Crew's crawler-tractor had been left overnight at the head of the new road, about 20 miles north of its digression from the narrow blacktop that runs north through the Hoopa [as it is on maps] Amerindian Reservation from Willow Creek to a place with the delightful name of Happy Camp up near the Oregon border.

Jerry was an older member of a crew bulldozing this new road into virtually unexplored territory for one Mr. Ray Wallace, subcontractor to Messrs. Block and Company who had, in turn, contracted with the National Parks Service to carry out the work. He is a local man. His fellow workers were for the most part also local men and included a nephew, James Crew, a very level-headed young chap, others whom I shall mention by name in a minute, and two experienced loggers of Hŭppa Indian origin. The crew had considerable heavy equipment at the scene of operations and had started work in late May as soon as what little snow there is in this area had melted and the much more deadly mud had firmed up. The road had been under construction for two seasons already. The country is mountainous; though this is the understatement of the year, being to most intents and everywhere almost vertical so that you can only go up on all fours or down on your bottom. Unless you make an exaggerated and exhausting climb you cannot see more than about four square miles of the country because you are always on the side of something going either straight up or almost straight down and unless a tree has fallen or been cut out, you can't see *anything* because bare rock is confined to the uppermost summits of the peaks and ridges. The road crawls laboriously up the face of the western wall that encloses a stream known as Bluff Creek. It is still unsurfaced and when I visited it in 1959 was ankle-deep in ultra-fine dust that surpasses anything the deserts of Arizona can produce at their damnedest. All along this mountainous trail there are the stumps of vast trees cut and hauled out, and great slides of friable shales, gray, brown, blue, or even green that have been sliced out of the sheer valley side. The great dozers and crawlers clank and roar in the hot summer sunlight as they gnaw their relentless way into this timeless land. The great trees seem to recoil a little from their mechanical jangling and screeching, but day by day these bright yellow and red monsters munch away ever deeper into one of the last of America's real wildernesses.

Those employed on this work lived during the work-week in camp near this road-head. They had trailers or tents or prefabricated houses and some of them had their families with them and stayed there all week. Others with families resident in nearby communities normally went home on Friday night and returned on the following Monday morning. The younger fellows usually did likewise, for the drive to Willow Creek took only about 2 hours for those who knew the road. Jerry Crew's practice was to return to his family over the week end, leaving his machine parked at the scene of current operations. He had been on this job for 3 months that year before the eventful morning which blew up the storm that literally rocked Humboldt County, California and made the pages of the world press but which then sort of folded in upon itself and was heard of no more for a year.

What Jerry Crew discovered when he went to start up his "cat" was that somebody had inspected it rather thoroughly during the previous night, as could be plainly seen by a series of footprints that formed a track to, all around, and then away from the machine. Such tracks would not have aroused his curiosity under normal circumstances because there were three dozen men at that road-head and the newly scraped roadbed was covered with soft mud areas alternating with patches of loose shale. What did startle him was that these footprints were of a shoeless or naked foot of distinctly human shape and proportions but by actual measurement just 17 inches long!

Of these, Jerry Crew took an extremely dim view. He had heard tell of similar tracks having been seen by another road gang working 8 miles north of a place called Korbel on the Mad River earlier that year and his nephew, Jim Crew, had also mentioned having come across something similar in this area. Being a pragmatic family man he felt, he told me, some considerable annoyance that some "outsider" should try to pull such a silly stunt on him. He at first stressed an outsider because, although his fellow workers liked a harmless joke as much as any man, he knew they were far too tired to go clomping around in the dark after the sort of working day they put in on that job, making silly footprints around the equipment. Then, he tells me, he got to thinking about this outsider and wondered just how he had got there without passing the camps farther down the road and being spotted, and how he had gotten out again, or where he had gone over these precipitous mountains clothed in tangled undergrowth. He followed the tracks up. And that is where he got his second shock.

Going backward he found that they came almost straight down an incline of about 75 degrees on to the road ahead of the parked "cat," then proceeded down the road on one side, circled the machine, and then went on down the road toward

[126]

the camp. Before getting there, however, they cut across the road and went straight down an even steeper incline and continued into the forest with measured stride varying only when an obstacle had to be stepped over or the bank was so steep purchase could be obtained by digging in the heels. The stride was enormous and proved on measurement to be from 46 to 60 inches and to average about 50 inches or almost twice that of his own. Jerald Crew was not only mystified; he was considerably peeved. He went to fetch some of his colleagues. Then he received his third shock that morning.

The majority of them, stout fellows and good friends that they were, refused to even go and look at this preposterous phenomenon that he told them he had found and he had a hard time persuading any of them that even the tracks were there. Eventually, some of the men, who had in any case to go that way to their work, agreed to go along with him and take a look. Then they got their shocks and, Jerry told me, some of them "looked at me real queer." But there were others who reacted differently, and it then transpired that all of them had either seen something similar thereabouts or elsewhere, or had heard of them from friends and acquaintances whom they regarded as totally reliable. The only Amerinds present said nothing at that time. Then they all went back to work.

Nothing further happened for almost a month, then once again these monstrous Bigfeet appeared again overnight around the equipment and farther down the road toward the valley, notably around a spring. About that time, Mr. Ray Wallace, the contractor, returned from a business trip. He had heard rumors on his way in that either his men were pulling some kind of stunt up in the hills or that some "outsider" was pulling one on them. He paid little attention to these reports but he was, he told me, somewhat apprehensive because the job was a tough one, skilled and reliable workers were not plentiful, and the location was not conducive to the staying power of anyone. When he reached the camp and heard the details of the Bigfeet he was more than just skeptical. He was downright angry. Moreover, all he encountered was more talk which he at that time suspected was some sort of prank but just possibly one prompted by more than mere high spirits or boredom.

The matter was until then and for a further 3 weeks a purely local affair known only to the men working on the road, and their immediate families for they did not care to speak about it to casual acquaintances or even friends. Then in the middle of September a Mrs. Jess Bemis, wife of one of the men working on the road and one of the skeptics among the crew, wrote a letter to the leading local newspaper, the *Humboldt Times* of Eureka, which said in part "A rumor started among the men, at once, of the existence of a Wild Man. We regarded it as a joke. It was only yesterday that my husband became convinced that the existence of such a person (?) is a fact. Have you heard of this wild man?" Mr. Andrew Genzoli of that paper told me that he regarded this letter with a thoroughly jaundiced eye but that the longer he saw it about his desk the brighter grew the clear blue light of his built-in news-sense, until he could restrain himself no longer and ran the letter in a daily column that he writes.

There was little response where he had expected a near storm of derision; instead a trickle of tentatively confirmatory correspondence began to come in from the Willow Creek area. This was continuing sub rosa when, on October 2, the maker of the tracks appeared again on his apparently rather regular round leaving tracks for 3 nights in succession and then vanishing again for about 5 days. This time Jerry Crew had prepared for his advent with a supply of plaster of Paris and made a series of casts of both right and left feet early one morning. Two days later he took a couple of days off to drive to Eureka on personal business and carried the casts along with him to show to a friend. While there somebody mentioned to Andrew Genzoli that a man was in town who had made casts of the prints and he was persuaded to go and fetch Jerry. Andrew Genzoli is an old newshand but of the new school; he can sense a good story as fast as any man but he is properly averse to too good a story. When he met Jerry Crew and saw his trophies he realized he had some real live news, not just a "story," on his hands, and he ran a front-pager on it with photographs the next day. Then the balloon went up.

The wire services picked it up and almost every paper in the country printed it while cables of inquiry flooded in from abroad. The first I heard of it was a cable from a friend in London: he seemed to be slightly hysterical. I get a lot of esoteric cables during the year about sea monsters, two-headed calves, reincarnated Indian girls, and so forth, the majority of which I am constrained to do something about because the world is, after all, a large place and we don't know much about a lot of it as yet, but this one I frankly refused to accept mostly because I rather naturally assumed that the location as given (California) must be a complete error or a misquote. I wracked my brains for any place name in Eurasia or Africa that might have nine letters, begin with "K" and end in "ia." The best we could come up with was Corinthia but this was even more unlikely. Then somebody suggested Carpathia, the country of Dracula and other humanoid unpleasantnesses, and we actually spent 6 dollars on a follow-up. There are few people interested enough in such abstruse matters as to spend that sum in pursuit of truth but I fancy there were many on the morning of October 6, 1958 who doubted what they read in their morning papers just as fervently as I did this cable.

The point I want to make is that this whole bit did sound quite absurd even to us, who became immune to such shocks years ago. It is all very well for abominable creatures to be pounding over snow-covered passes in Nepal and Tibet; after all giant pandas and yaks, and an antelope with a nose like Jimmy Durante, and other unlikely things come from thereabouts; and it is even conceivable that there might be little hairy men in the vast forests of Mozambique in view of the almost equally unlikely more or less hairless pigmies of the eastern Congo which are there for all tourists to see, but a wild man with a 17-inch foot and a 50-inch stride tromping around California was then a little too much to ask even us to stomach, especially as we had not yet got the news-stories.

The amazing thing in this case was that the world press took it seriously enough even to carry it as a news-item.

Not so the rest of humanity. One and all, apart from a few ardent mystics and professional crackpots, and including even the citizens of Humboldt County itself rose up in one concerted howl of righteous indignation. Everybody connected with the business, and notably poor Mr. Genzoli, was immediately almost smothered in brickbats. In the meantime, however, a number of other things had happened. Most notable among these was the reappearance of "Bigfoot" as he was called one night before Ray Wallace returned to his operations. Now it so happened that a brother of the contractor, Wilbur Wallace, was working on this job and he, besides seeing the foot-tracks many times, witnessed three other annoying and to him most startling occurrences, which he had reported to his brother. I will repeat these roughly in his own words which appeared to me not only straightforward but most convincing.

First, it was reported to him by one of his men that a nearly full 55-gallon drum of diesel fuel which had been left standing beside the road was missing and that Bigfoot tracks led down the road from a steep bank to this spot where it had stood, then crossed the road, continued on down the hill and finally went over the lower bank and away into the bush. Wilbur Wallace went to inspect and found the tracks exactly as the men had stated. He also found the oil drum at the bottom of a steep bank about 175 feet from the road. It had rolled down this bank and had apparently been thrown from the top. What is more, it had been lifted from its original resting place and apparently carried to this point, for there were no marks in the soft mud of its having been either rolled or dragged all that distance, Second, a length of 18-inch galvanized steel culvert disappeared from a dump overnight and was found at the bottom of another bank some distance away. Third, he reported a wheel with tire for a "carry-all" earth-mover, weighing over 700 pounds, had likewise been in part lifted and in part rolled a quarter of a mile down the road and hurled into a deep ravine. Ray Wallace, however, still remained skeptical even after hearing this from his own brother. However, on his first morning at the location he stopped for a drink at a spring on the way down the hill and stepped right into a mass of the big prints in the soft mud around the outflow. Then, I gather from him, though he is a man with a wonderfully good humor, he got "good and mad." There was for him no longer any question about the existence of these monstrous human-like tracks but there remained the question as to who was perpetrating them, and why. Ray Wallace is a hard-boiled and pragmatic man and he was already experiencing trouble keeping men on the job. Handpicked as they were not a few had just *had* to leave for one apparently good reason or another. Only later did he learn that almost all of them did so not because they were scared by the Bigfoot, but either because their wives were or because of the ribbing they had to take when they went back to civilization, even for the evening to nearby Willow Creek.

Ray Wallace said he at first thought somebody was deliberately trying to wreck his contract and he was not alone. However, the local representative of the *Humboldt Times*, Mrs. Elizabeth (Betty) Allen, set about to investigate the possibility on her own, and discovered beyond a doubt that neither good nor bad publicity, nor any kind of "scare" actually made any difference to Mr. Wallace's contract. First he was a subcontractor; second he was more than up to schedule; third there was no time set on the job; and fourth, it was basically contracted by Messrs. Block and Company with the Forest Service on a performance, not a time, basis. Ray Wallace got so angry he brought in a man named Ray Kerr, who had read of the matter in the press and asked for a job in order to be able to spend his spare time trying to track the culprit. Kerr brought with him a friend by the name of Bob Breazele, who had hunted professionally in Mexico, owned four good dogs, and a British-made gun of enormous caliber which considerably impressed the locals. Kerr, an experienced equipment operator, did a full daily job: Breazele did not take a job but hunted.

Tracks were seen and followed by them. Then one night in late October, these two were driving down the new road after dark and state that they came upon a gigantic humanoid or human-shaped creature, covered with 6-inch brown fur, squatting by the road. They said it sprang up in their headlights and crossed the road in two strides to vanish into the undergrowth. They went after it with a flashlight but the underbrush was too thick to see anything. They measured the road and found it to be exactly 20 feet wide from the place where the creature had squatted to the little ditch where it had landed after those two strides. Spurred by this encounter they redoubled their hunting forays but their dogs disappeared a few days later when they were following Bigfoot's tracks some distance from the road-head. They were never seen again though a story was told—but later denied by its teller—that their skins and bones were found spattered about some trees. Though this story was denied, there is as much reason to believe that this was done to obviate ridicule as to clear a conscience.

All this was, of course, taken with hoots of derision by everybody even in Willow Creek who had not seen any tracks—but with one notable exception. This was Andrew Genzoli and he sent his newspaper's senior staff photographer to Bluff Creek. The party saw fresh tracks at night and photographed them. They also found something else; as did Ray Wallace later. [I have this first hand from these professional skeptics.] At first, the photographer told me, he was more than just skeptical but when he found the tracks and inspected them he not only was convinced that they were not a hoax or a publicity stunt but, as he put it, "I got the most awful feeling that I can't really describe, but it was nearer fright than anything I ever felt when in service." But worse was in store for the newsmen for, in following the tracks down the road, they came across a pile of faeces of typically human form but, as they put it, "of absolutely monumental proportions." He then added, "I can only describe it as a 2-ton bear with chronic constipation." They contemplated going to fetch a shovel and some container and taking this back to Eureka for analysis but it was a very hot night and a 5-hour drive over a dangerous road and also, as they readily admitted, that strange laziness that so often intervenes in offbeat and rather alarming cases of this nature,

[132]

took over and cast the die. Press coverage had gone far enough, and they were not ecologists. Later, Ray Wallace stumbled upon a similar enormous mass of human-shaped droppings. He shoveled them into a can and found that they occupied exactly the same volume as a single evacuation of a 1200-pound horse.

Further foot-tracks and other incidents continued all that fall and throughout the winter until the spring of 1959 ending in February. However, later in the spring, two fliers, a husband and wife in a private plane, were flying over the Bluff Creek area. It was April and there was still snow on the mountain-tops some of which are bare of trees. It is alleged that they spotted great tracks in the snow and that on following them up they sighted the creature that had made them. It was enormous, humanoid, and covered with brown fur, according to secondhand accounts. I tried, and am still trying to locate this couple, with the co-operation of local fliers, several of them having heard of the report, and despite the praise-worthy clannishness of fliers and their willing offers to help, I have not at the time of writing been able to identify this couple. The story may be a rumor or wishful thinking. So also may, three other recent and a whole host of past, old, and even ancient reports of actual meetings with one or more Big-feet in this area.

Among these are alleged statements by two doctors of hav-ing met one on Route 299 earlier in 1958; and of a lady of much probity who with her daughter saw two, one smaller by far than the other, feeding on a hillside above the Hoopa Valley. This lady, to whom a partner of mine talked but who does not wish her name publicized, also stated that when she was a young girl, people used to see these creatures from time to time when they went fishing up certain creeks, and she once saw one swimming Bluff Creek when it was in flood. She also stated that in the olden days people did not go above certain points up the side valleys, due to the presence of these creatures.

More important was a positive flood of further alleged dis-coveries of similar foot-tracks by all manner of local citizenry over a wide area and extending back for many years that came to light as soon as the local press began to take this whole matter seriously. But as these came in, public resentment and ridicule mounted so that the reporters became ever more cagey. Finally, Betty Allen, who as an old-time resident and with experience as an Assistant U.S. Commissioner in Alaska, started talking to the Hŭppa and Yurok Amerinds about these matters and, little by little, an amazing picture emerged.

7. Late North Americans

All possible knowledge has, of course, been right under our noses since the beginning, but we have to dig for it. Often we miss things; sometimes we deliberately ignore them.

You can take the title of this chapter any way you like. Late is a useful word: it has two completely opposite meanings that imply novelty or extinction. There is also a connotation of tardiness about a late-comer. This serves my purpose well.

When Betty Allen started browsing around among her Amerindian friends she brought to light two sets of surprising facts. The first was simply that said friends, one and all, had always known about the *Toké-Mussis* and *Oh-Mahs,* completely accepted them as being quite real, and regarded them as in no wise bizarre. They had, however, and quite rightly, long since decided that they were not a suitable topic for conversation with white men since it seemed to annoy them, while their even mentioning their beliefs about the matter only augmented the general contempt in which all their other ideas were already held. There were those among the Amerinds, even of the older generation who just brushed the business aside or referred to it as folklore. Surprisingly, though, there proved to be not a few among the younger generation who met the white man's skepticism with a deep-rooted scorn of their own, and who affirmed that there was absolutely no doubt that these manlike creatures still exist; and not in too few numbers either, not only all over this territory but over other wide areas. I had the privilege of talking to some of these young people myself and was much impressed—I might almost say startled—not only by their sincerity but also by the matter-of-fact way in which they discussed it, *and* their reasons for not previously discussing it with any outsiders. Though I have the permission of some of these new friends to mention their names, I will refrain from doing so, because they would undoubtedly be subjected to ridicule and an unmerciful ribbing, even in their own community.

I will not report in full what they told me, nor all that Betty Allen learned because it is highly repetitious, is little different from all the other accounts I have given of observations of the creatures, and does not really add any new details. One and all of that category of account of which I speak were firsthand (I have some two dozen on file), alleged encounters with the creatures in and about this block of montane forest which I call the Klamath. The interesting thing is that these reports go back to the 1930's but become increasingly more frequent up till 1958. Since then they have formed a positive flood. My interpretation of this is that, while the age of the tellers naturally showed up statistically, another factor is much more important. This factor is that it is only comparatively recently that roads have been started into these large areas of national forest. The jeep caused the first move in this activity, being one better than a mule in this country, but needing at least a clear path of a certain width that might loosely be called a road. Next, the government decided to open up these national forests to timber-cruising, it having been demonstrated that one of the best ways to conserve timberlands is to cut out the oversized and overage trees which retard new growth. The road-building program for the first time took large numbers of people into areas not previously penetrated, or into which people found it hard to go even to hunt. These are the retreats of the *Oh-Mah*.

The other thing that Betty Allen brought to light was the much more surprising fact that this was not by any means only an Amerindian folk-tale. She began to hear the names of white men and others who, it was said, had also met or seen these creatures. She went after these persons too, and found out in due course that it was so, and that they, in turn, had not been saying anything for fear of ridicule. I withhold their names too, as I do not have the permission of any to publish them and I would no more wish to embarrass them than I would my Amerindian friends. Most of these had also been employed on road construction, but there were others, including two doctors of medicine returning from a mass emergency late one night along Route 299 going east from Willow Creek, who said they had nearly run into one, although they had slowed down, thinking it to be somebody signaling for a lift. They said that it was at least 7 feet tall when it stood up, had straight legs but very long arms, and was clothed in thick lightish brown fur; and who better than (even tired) medical men ought to know? Some of these local stories went back 30 years.

Then, there was the extremely unsavory (to me) interjection of the business of "little people." It is a particularly odd one in this neighborhood for several reasons. First, the Amerinds will not, as far as I have been able to determine, come right out and either assert or deny their existence. Unlike the giants, of which they speak quite factually, they seem to regard these pigmies with a high degree of superstition, and their folk-tales are rife with stories of such little people playing with their children on riverbanks; but, while being visible to youngsters, being invisible to adults. This is a very widespread myth that crops up all over the world about fairies, pixies, and suchlike little folk. However, some white people of higher education, and resident on the outskirts of fully opened-up and settled areas, have told the same story, and perfectly straight, but have also, in several cases, implied that they had assumed, or had definite grounds for supposing that these little hairy ones were the young of the *Oh-Mahs!*

Simultaneously, this dearth of direct claims that these midgets have been *seen* is in marked contrast to reports that their little foot-tracks have actually been found both in snow and mud much more often than those of the giants. I have seen sketches of these drawn to scale, but so far no photographs or plaster casts. Many times they are said to crowd around pools or depressions in snow and to trail into and out of the undergrowth in all directions. They are very funny little imprints, averaging only about 4 inches long and do, for the life of me, look very like those of tiny men but with very pointed heels.

I frankly don't like this: I don't like it one bit: and it also upsets me. All of us almost automatically become annoyed with anything new, and especially when it appears to conflict with our logic and the orderly tenure of our lives. Perhaps you will say that if I can accept the possibility of the presence of giants I ought to be able to take little people in my stride. So I should, but I am afraid that I am a very pragmatic person, and there is something unsubstantial about these little footprints. Perhaps it is that I have not seen them in the fresh state myself? In fact, I find myself performing all the mental gyrations of the most advanced skeptics and debunkers in this case, and I know full well that I am doing my damnedest to explain them away.

The first thing one thinks of—just like the zoologists confronted with the Himalayan *yetis*—is any kind of local animal that might produce these tracks and, by Jove, there certainly is one. This is the large western porcupine. This animal has an astonishingly human-looking hind foot when seen from below, apart from large claws. It has a somewhat pointed heel. But there *is* the problem of its claws; and then there is another objection. The porcupine can waddle along on its hind feet quite well but, like the ground-sloths of old, it has a thick, stubby tail that is directed downward and which forms a tripod with the hind legs when the animal is standing up. However, it can be raised somewhat and could possibly be carried off the ground. At the same time, the claws on the back feet of really large porcupines are actually raised well off the ground so that the swollen pads under the feet can sink into soft substances quite a way before the claws leave imprints. Yet these tracks clearly show 5 toes—not sharply incised claw marks—all of about the same size and arranged almost straight across the front of the feet. In an endeavor to overcome this fact, an ingenious naturalist friend of mine has suggested that the claws of animals sometimes acquire globular encrustations of ice in winter when they are tramping about in wet snow and when a frost is coming on, and that these might produce

[138]

the impression of toes. But what then of the tracks in mud, all over, and by thousands? As I say, I don't like this business; but, I also don't like leaving it up in the air; yet, I have nothing to add to it as of now. Until and unless I can go and find some of the tracks in mud myself, and carry out my own particular kind of investigation thereupon, I shall refrain from further comment. Then there has been another most peculiar business in this area. It transpired that nearby, certain persons who are free, white, family folk, live in rather expensive houses sometimes of the split-level ranch type, on blacktop roads around which school buses parade daily to take their offspring to be educated. In many cases they own houses which stand in several acres of land backed up against solid forest that has not been touched except for logging of large timber a century ago. They had something most unpleasant to report. These people live not more than 30 miles from a large and bustling modern city. They stated, in confidence and off the record, to certain locals for whose veracity I will vouch, that they had for long experienced a problem. This was simply that their kids—i.e. under 7-year-olds—had been found to be playing in the back fields up by the borders of the forest with certain fairly small hairy ones, who, when alarmed by the approach of human adults, allegedly took to the trees.

Said human kids, on reaching the age of reason, turned out not to want to talk about this abomination, while their parents most definitely did not and do not want it talked about. Nonetheless, they have talked a bit, and I pass it on to you for what it is worth. This is the kind of thing that gets people really riled: it also seems to me to slop over into the realm of "Little People" that only kids can see. Let us just suppose for a moment that *Oh-Mah* mothers permit their kids to play with ours (up to about the age of 7) but tell them to cut out the moment one of our adults appears over the fence! Naturally it would be only the kids who see the little hairy ones. There is no better playmate for a child than a 2-year-old chimpanzee.

There are other items connected with ABSMery generally in this area and to the north of it, which I also do not like but which should be presented and also without comment. This comes from, of all places, Albany in Oregon, which is in the Willamette Valley at the foot of Mt. Jefferson, and concerns a certain Lake Conser. A brief notice of this was published in *Fate* Magazine's issue for January, 1961 and read:

Albany, Ore.—The monster of Conser Lake is still on the loose. The creature reportedly stands on two *webbed* [italics mine] * feet, is 7 to 8 feet high [tall], and with its shaggy white hair somewhat resembles a gorilla. It has kept pace with a truck going 35 m.p.h. Never harmed anyone though.

This is a nasty one, but let me give you some further details. These were contained in a letter to a friend of mine, dated October 27, 1960.

Creatures (several) last report, being sighted on farmer's farm. An attempt is being made to contact farmer whom to date wants his name and address held secret. Have made 5 investigation trips and have for evidence a finger print lifted off a house window including a plaster cast of a foot print (right). Have personal taped accounts of this creature plus many interviews, this includes photographs. He is all of 7' tall, 400 lbs., can move at tremendous speeds, jump tremendous distances. No news items concerning this being have been printed in the Portland Papers. He displays extreme cunning, walks and runs erect, appears frustrated, acts as if would like to communicate. He makes extremely high pitched sounds. His hair or fur has a slight glow in the dark and is 3 to 4 inches long. He walks with feet 19 inches long that make a sqeeshy sound. Has been seen in daylight and at night and seen to disappear once into the lake. Will send you complete report as soon as I can.

Creature first sighted several miles north of Albany, Oregon in a dense land area approximately 3 sq. miles. Open land extends all around this area & dotted with farms. Have any ideas how he got there?

Sorry for the delay for there has been new developments. A farmer who wishes to remain anonymous has sighted several on his farm. He is attempting to make friends with them. One is brown and one is white. At times they imitate his voice when he talks to them. Mr. farmer is an animal trainer and at the last report steady progress is being made. Hal Starr was contacted by this farmer and has promised that the location and that his name not be revealed. I would like to investigate further but am handicapped. They are up to 7' height covered with long hair which

* For the significance of the use of the word webbed here, see analysis of the imprints of the *Oh-Mahs* in Appendix B.

hangs over their faces. They walk erect and with all fours. They have taken a shine to the horses but the horses were frightened of them. Lots of foot prints around and are cloven.

Two weeks ago a sheriff of Salem told me that he heard on the radio KBZY that a person had called in saying that he had seen a creature near Hwy. 99. I talked to the announcer in Salem and verified this event. I am busy writing you a complete report. Hope this will suffice for a while.

I am afraid this did not "hold me" even for a little while because it is altogether one of the most shocking reports that has yet come into my hands. I have been pursuing the matter diligently with, however, no result whatsoever.

This remark about going into water on the part of an ABSM is fairly common and causes me to think furiously on two counts. First, it is really a very bizarre thing for anybody who is making a good story out of a series of lies, to think up. Into a cave, or even into a swamp, yes: but into water, per se, just as if it were an aquatic or at least semi-aquatic creature, is very weird. At the same time, one just has to take into account the perfectly astonishing theory put forward by Professor, Sir Alistair Hardy of Oxford early in 1960 and which, utterly bizarre as it at first sounds, has been most seriously considered by scientists and fully accepted as at least possible by many.

This suggests that one branch of the general Anthropoid stock—and, although Sir Alistair calls them "apes," I think we should surely name them Hominids, or at least as already being on the *Man* branch of that stock rather than on the Pongid or ape branch—about a million years ago took to semi-aquatic life and especially along seacoasts. But let this bold savant state his case in his own words:*

"Many apes were driven to hunt in the sea by fierce competition for food in the forests. At first they waded and groped in the water, but gradually learned how to swim. Over a period of several hundred thousand years, the species lost its hair as it carried on its marine life. The only hair left was on the very top of the head to help protect the creature from the sun.

"The sea ape learned to stand upright because water helped support

* Quoted from a story in the *New York Herald Tribune*, of March 7, 1960, from a March 6, verbatim, A.P. report on a conference of "Marine Scientists" at Brighton, England.

the body. It developed longer legs than its land-based brother ape for swimming. Its hands became sensitively shaped to allow it to feel along the sea bed for shellfish and open crabs. It learned to use tools by picking up stones to crack open sea urchins. It would be only a step for man to discover that flints chipped into sharper and more useful tools, knives, and arrows. Then, armed with such equipment and his erect posture, he was all set for the chase. He could now reconquer the continents, running and hunting the animals of the plains. I estimate that apes were driven into shallow sea waters a million years ago. They emerged as men about 500,000 years ago."

He said he had discussed his theory with many other scientists and they had been unable to find a flaw in it. (A.P.)

If in the sea, why not also, or even previously in rivers, lakes, and ponds, more especially as swamps and marshes were much more prevalent in the past than they are now, particularly in the pluvial periods following the ice-advances and retreats of the past million years. Then again, there is another most convincing aspect of this idea, as follows. If at the beginning of the Pleistocene there were a variety of primitive anthropoids of the Hominid branch scattered about the earth, and if all of these were hairy, but did not all become extinct, as we have until now supposed, we have some ready-made characters for our ABSMs.

Let us suppose that several of these started going into water after food, and that one (or perhaps several) types did very well at it; lost their body hair; learned to crack stones and all the rest; and then came back to conquer the land as *Men*, just as Professor Hardy suggests. This still need not presuppose that *all* of them did so. Some of the types that started the practice may never have gotten farther than ducking into inland lakes and, while they did not keep at it fervently enough to lose their body hair, they did develop very long toes with an almost complete web between them. Do not forget that *we* still have two half-webbed toes ourselves—our third and fourth—and please don't fail to flip over to Appendix B and take a look at the California "Bigfeet," in which the second "ball" appears to be an enlarged basal big-toe joint. All the toes of this type must then, be *very* long and be webbed, because the mud or

[142]

snow does not squish up between them but forms (and always forms) a tall angular ridge running at right angles to the direction of travel, just where it would be bunched up if the foot were webbed. We might therefore legitimately conceive of the *Sasquatch–Oh-Mah* type of ABSMs at least being relics of early hominids with semi-aquatic habits. This would explain any failure to have tools!

However, to interject at this point, I recently received a report from a neighboring area which would seem to indicate something of the same nature. This came to me from a young man in our Air Force whose wife is part Amerind. He lived until recently on the Makah Indian Reservation at Neah Bay, Washington. This young man got in touch with me through a magazine publisher, stating that he had some information that might interest me. I wrote him, and in reply received some very charming and highly informative letters, the contents of which I see no reason to question. Among these he wrote:

In my letter to you I mentioned the 18½ inch foot prints that were found out on the beach. I know these weren't made by any man going around with a foot cut out of a piece of wood. This beach is about 8 miles in the back woods and is a very hard spot to get to.

On another occasion last summer one of the fishermen out here was going to bed and heard a lot of splashing going on in a swamp in his back yard. From what he told me, he got a flashlight and went out there to take a look around and seen this huge creature tearing back into the woods after the light hit him. Up to this date there has been nothing more seen of it. Altho many people are waiting for it to come back. The day this person told me of what happened I took a gun and went into the swamp to look around. I actually found huge hunks of hair that must have been pulled loose when he ran back into the woods that night.

I have hunted and killed quite a few bears around here but that hair that I found that day was definitely not hair from a bear. For one thing, there was a couple of hairs that I measured to be close to 14 inches long and these hunks had a very strong odor unlike any bear that I have killed.

There is also one other occasion that makes me think that the Abominable Snowman is up around this neck of the woods. This happened to me some time before I read your articles in the *True* Magazine. One evening I went up this unused logging road to hunt bear. I was some 13 miles up this road and there is not one person living for about 20 miles around. On this occasion I happened to be alone. Well anyway I was sitting on a stump and was sitting there for about an hour when I heard this high pitched scream like a baby but this went on for almost an hour and the more I listened to it the more I decided that it wasn't a mountain lion. Then after a while it stopped and I never heard it again and I left without looking around. Then after I read your article I thought it might have been a Snowman up there. I went up there quite a few times after that but never heard or seen anything.

Once again, I have received nothing more from this source!

It has always been my firm belief, as a reporter, that children don't lie. By this, I mean that, while real kids (say, under seven) live in a world of their own, peopled by many things that are not of our world but which are still most real to them, and while young persons from seven to the age of puberty delight in pulling the legs of their elders with tall tales, all young persons are much more basically honest than grownups. More, important, I do not believe that a young person can carry a lie forced upon him or her by an elder for any length of time; and, especially, under sympathetic questioning. I am therefore always interested in what young people have to say, provided that they know that I am sincerely interested, have an open mind, and am not critical of their age. Young people are also extremely keen observers, perhaps because they take a more nearly worm's-eye view of life and because their senses are more acute. Thus, when somebody tells me something that happened to them when they were young, I like to listen. This then from a young person about an incident when she was still younger:

Dear Mr. Sanderson:

I have just finished reading your story concerning the abominable [sic] man of Northern California. Before I write any further I would like to say that what I am about to say is positively true and I have never told anyone this story before for fear that they would think that I was *half cracked and out of my mind*.

I have seen this man-monster and can give you a detailed description of him. He is far from being pretty and I still wake up nights dreaming of him.

[144]

When: About 9 years ago, at about 10 o'clock in the morning. Where: Near the Eel River above Eureka, California. At the edge of a meadow near the river's edge. Under what circumstances: My family and I were fishing on the Eel River. We had been camped in the vicinity for about 2 weeks and had had poor luck when it came to fishing. I used to go for a short walk before breakfast because there was a very pretty meadow about a mile or two from our camp and I used to love to see the mist rise off the grass. I was only about 10 years old at the time and the world of nature was something which both fascinated and enthralled me. I entered the meadow and proceeded to cross it in order to reach a small knoll at the other side. When I approached the foot of the knoll I heard a sound. It was the sound of someone walking and I thought perhaps my little brother had followed me and was going to jump out and try to scare me. I hollered, "All right, stinker, I know you're there." Needless to say it was not my brother that appeared. Instead it was a creature that I will never forget as long as I live. He stepped out of the bushes and I froze like a statue. He or "it" was about 7½ to 8 feet tall. He was covered with brown stuff that looked more like a soft down than fur. He had small eyes set close together and had a red look about them. His nose was very large and flat against his face. He had a large mouth with the strangest looking fangs that I have ever seen. . . . His form was that of a human and he had hands and feet of enormous size, but very human looking. However, there was one thing that I have not mentioned, the strangest and most frightening thing of all. He had on clothes! Yes, that's right. They were tattered and torn and barely covered him but they were still there. He made a horrible growling sound that I don't think could be imitated by any living thing. Believe me I turned and ran as fast as I could. I reached camp winded and stayed scared all while we were there.

I have often thought that perhaps it was a mutation of some kind. I think this thing is highly dangerous and something should definitely be done about it.

I would be willing to testify to anything I have stated in this letter. I am not a crackpot and am completely sound of mind and body. I just thought you might be interested to know what your man-monster really looks like. Believe me if you saw him he would scare the wits out of you. I know!

Yours truly,

(signed)   (Miss) B. C.*

* Name and address on file, but for release in special circumstances only.—Author.

It is not perhaps quite proper to interject the following comments at this point but, I contend, a reporter has the right to indulge some speculation upon matters that he has investigated firsthand. This may be an infringement upon editorial rights but can be fobbed off as background information. It seems to me that there is something to this whole bit in California, Oregon, and Washington, and that it is pretty fatuous to try and put it all down to any of the standard explanations such as the hoax, the publicity stunt, the Indian folk-tale, mass hypnotization, mass cases of mistaken identity (of known animals), or other suggestions of that nature. We are all pretty odd, but we are not all liars or crackpots. Further, I do not feel it to be either right or justified to dub all Washingtonians, Oregonians, and North Californians as either; just because they say something we don't like, or which does not fit into our orderly pattern of what is or is not supposed to be. At the same time, I don't give a hang what any "expert" actually says. There are enormous areas in those three states about which nobody—not even the majority of their inhabitants—knows anything. I really cannot see why some new things should not turn up in those states.

If you could read all the reports that I have; and, much more; if you could listen to my recordings or have been with me when I interviewed and got to know the good people who had the guts to tell these stories, I think everyone interested would be not just amazed but somewhat shamed. It is so easy to sit back in one's own home, surrounded by all the normal, known things of modern life, and say "Phui"; but, get out in the woods and get hungry. A person will begin to see a lot of things he never saw before, and would never have seen if he had not got lost and run out of food. Thus, when a teen-ager writes to me from the delightfully named Happy Camp at the edge of the Klamath area and says: "Reading your story of America's abominable snowman, I find very interesting. But I think they've only found the baby. Here, in Happy Camp, our cars are turned over and rolled into the river, 6-foot trees uprooted, slides in the mountains, and when it snows 10 feet deep, 1-inch power lines are snapped in two. The daddy must cause this." I do not yell for Paul Bunyan and go into gales of laughter. Maybe there *was* no flood that shifted the cars, and the trees *were* 6 feet tall, not thick.

If things as bizarre can happen, or be alleged to happen, right in our own back yard, we should be doubly careful of criticizing things that are reported to happen beyond our borders. And when these form a logical concomitant to happenings in our own bailiwick, we ought to listen most carefully. Of course, there is the damnable, added frustration in dealing with foreign matters inherent in their very foreignness—one can't often go and look into them firsthand, and if one does, one has language and other difficulties. Moreover, if we doubt our own citizens, how much more so may we not those of other countries? This is all a pity but nonetheless the way things are. From now on, therefore, I won't expect anyone to believe what I report at all. We go first over the border south to our sister republic of Mexico.

8. On the Tracks of . . .

All peoples have always thought all other peoples
to be both stupid and at a lower state of culture.
This is both stupid and uncultured.

The title of this chapter is an acknowledgment of a good
friend and fellow zoologist. He, Dr. Bernard Heuvelmans,
Consultant to the Musée Royal D'Histoire Naturelle de Bel-
gique, but resident in Paris, is the author of the only book
that covers the ABSM problem world-wide. It covers also
many other items of a crypto-zoological nature, and is entitled in its English version, *On the Track of Unknown Animals.** I shall be leaning very heavily upon this work from
now on, with its author's more than generous permission.
Bernard and I have been on these tracks separately for many
years now but, as we have constantly exchanged information
and discoveries, a considerable amount of what we have to
say has similar origins. However, there is much that both of
us have unearthed [either firsthand or by burrowing assidu-
ously through published material], that the other has missed.
Frankly, neither of us knows any longer, in many cases, exactly just which items came from which in the first place;
and, as constant acknowledgments in the text would be irksome, Bernard has given me permission just to barge ahead
and gobble up anything that may seem to me to be pertinent.
However, while we were both once professional zoologists, we
specialized in different aspects of the science. I started out
as and always really remained a field ecologist but have
specialized in the major distribution of animals in accordance

---

* Published in England by Rupert Hart-Davis, 1958; published in the
United States by Hill & Wang, 1959.

with that of vegetational types, and therefore approach most, if not all, matters from that angle. Thus, there may be times when I disagree with my good friend Bernard and, since I have never then failed to say so to him, I shall also be mentioning the fact in this text as I go along, if occasion arises. It was on a framework of phytogeography that I tackled ABSMery in North America. From now on, and especially in the tropics, it becomes the main theme of my story.

At this point I have to revert to type and refer to Map VI. I also have to ask the reader to plunge once again into botanical geography. In addition to being all of the other unpleasant things that we have accused them of, people are very chauvinistic and, from a national point of view, frankly bucolic. This shows up in various ways, like wars and tariffs, but most noticeably on maps. It is almost impossible to buy a map of any country, *in* that country, that shows anything starting immediately beyond the borders of that country. Thus, not only road maps but even our school atlases have a habit of going along splendidly to the Rio Grande and to some arbitrary, somewhat jiggly, and quite nonexistent line from a point about El Paso, Texas, west and just north of the upper end of the Bay of Baja California and thence to the Pacific coast a few miles south of San Diego, California. Beyond that, southward, there is a great white blank. While the United Staters of North America are outstandingly obtuse in this respect, we cannot really exonerate the other United Staters of this same continent—the United States of Mexico—from indulgence in the same idiocy. Their maps customarily run *up* to that same ridiculous line; above which a ghostly "Pais de los Gringos" may be seen—in strong light.*

* Sometimes things get much worse, as when Guatemala published a map of her country which included the whole of British Honduras, because they "claim" it; and then the Mexicans countered with a map of their southern states from the Isthmus of Tehuantepec that showed the northern half of that hapless little independent colony as being a part of their Territory of Quintana Roo. Happily, the Republica de Honduras, being betweeen governments, only issued a pamphlet which claimed all the cays and islands off British Honduras. (There are *five* United States in all America—ours, Mexico, Venezuela, Brazil, and the Argentine.

MAP V. GUATEMALA

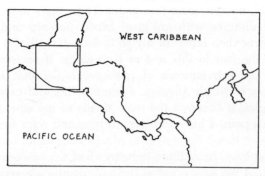

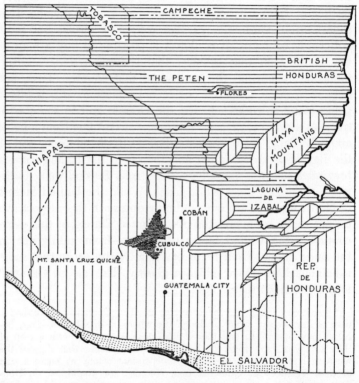

See facing page for legend

[150]

The American Geographical Society of New York has published some most excellent maps that show the whole of North America and especially the overlap between our country and Mexico. These are very revealing in that one learns from them—as one does if one actually travels through that strip of territory—that practically everything goes on just the same right across the border. The Tularosa Basin just flows on south into the great Bolson (basin) of Mapimi; the Rockies pass on through the Sacramento and Guadalupe Mountains, via the Chisos, straight into the Sierra Madre Oriental; and the endless semi-parallel ranges that bestrew southern Arizona go right on to become the Sierra Madre Occidental, while the mountains of southern California march on stolidly to become those of Baja California. Nothing much else changes either: even a parrot (a Conure) and the jaguar get on to the north side, and millions of tourists on to the south. The only things that change abruptly are the brands of beer and the length of the women's skirts—there is a strong European influence even just over the border. [Money is interchangeable for a time but the gasoline does, I must admit, seem to be of quite a different substance on the two sides of this otherwise arbitrary border.]

The really funny thing is that practically nobody knows any-

MAP V. GUATEMALA

The position of the Republic of Guatemala and the over-all area of this map is shown by the square in the box, which, in turn, encompasses what is popularly called Central America. Guatemala is divided into two very distinct parts—the northern, called the Peten, which is a lowland, heavily forested plateau; and the southern which is mountainous and where there are large numbers of volcanoes, both active and idle. To the west, these mountains are contiguous with the eastern rim of the Chiapas in Mexico. The southern coastal plain is arid. In the northeast corner of the country, which reaches the Bight of Honduras in the Caribbean, there is a limited sealevel triangle containing the so-called "Lake Isabel"—actually the Laguna de Izabal. This is really an arm of the sea and is connected to it by a river-like channel. The area from which ABSMs have been reported centers around the peak named Sanché in the Sierra de Chuacus.

[151]

thing about the first great belt just south of our border. There
are four major roads and three major railroads to get south
and that is all; and all of them roar across a variably unpleasant and dreary desert for a long way before they come to anything important. Mexico lies on its side, so that what we call
its west coast is really its south coast. It consists of a central
core of enormous volcanic mountains from which two great
tines of mountain ranges extend toward the United States, like
a pair of giant scissors. Behind, or to the southeast of the core,
there is a narrow neck of lowlands, the Tehuantepec Isthmus,
and then what is really quite another country named Chiapas
that stands up like a large flat salad bowl on a footstool, or
actually more like a flamingo's nest.

To the northeast of this is still another Mexican country
called Yucatan, which is a low plateau formed of limestone
marl, riddled with caves, and separated from Chiapas by a
great swath of swamps more or less at sea level and clothed
in an awful, low, tangled, spiny growth called *akalché*. Yucatan, which includes the state of that name, as well as the
Territory of Quintana Roo, and the states of Campeche, and
Tabasco, is the land of the Mayas. Behind Chiapas, on the
south side, lies Guatemala; an arbitrary hunk of volcanos and
aggressive mountains that really forms part of a much larger
mountain block that extends to the great lakes district of
Nicaragua.

The Sierra Madre Oriental, along with her many associated
ranges, are still not much known, though they are—and have
been for countless centuries—well inhabited. Among them
are some valleys filled with a truly tropical type of vegetation.
The Sierra Madre Occidental, on the other hand, is almost
entirely unknown. There are people living in it but they
don't have anything to do with anybody and, least of all and
if possible, with Mexicans, whom their inhabitants call
"guachés" [which is a slang expression for a very old bus
more or less held together with bits of string]. Among these
people are the Yaquis who played a great part in modern
Mexican history; who still write in hieroglyphs; and who were

alleged to have been scalping on the main Pacific Highway in the early days of World War II. They are very splendid people—everything an Amerind should be, both in fact and in fancy. This southern (i.e. to us, western) block of mountains runs for 800 miles southeast till it hits the comparative lowland break of Guadalajara. It is crossed by only one road, from Durango to Mazatlan; it has a canyon in it that has been estimated to be two hundred times the size of the Grand Canyon when all its measurements are taken into consideration, though you may console yourself about this, because nobody has ever explored it. I have seen one end of it and very impressive it is. Most of its bottom is choked with forest and there are said to be "people" in there—at least my Yaqui Indian friends told me so. These are said never to come out, to be very big, and to be *hairy* all over!

The rest of Mexico down to the northern escarpment of Chiapas is charming and much more civilized than any of us northerners realize or like to think. They had universities down there 200 years before our country was founded, and some of their modern ones are carrying on studies that are so far ahead of anything being prosecuted in ours that it makes us look a little silly. [That may be why we don't hear about them.] The best work that I have seen on vegetational distribution, not excluding Soviet Russia, have been done at, and published recently by, the University of San Luis Potosi. The indigenes—for we can hardly call them natives—of this main, central area are too busy even to turn up any folklore about ABSMs, but they have dug up some awfully funny-looking statuettes. But, this is another subject that I cannot get involved in here.

Now, while the plateau of Chiapas is rather an unpleasant, dusty, cactus-strewn place, and looks not unlike one of our lesser deserts (due to its altitude), it is ringed by well forested mountains with gorges that are filled with real "jungle." Also, it flows back into the uplands of the main Central American block; and it is really part of that block. Were it nearer sea level, it would be properly tropical, and it is in any case only just "North" American. The true dividing line between the two continents of *Erica* and *Columbia* (see Map XV) is a very complicated line that meanders about all over the place on its way from the Pacific to the Caribbean. Plants and animals respect this line mightily. In fact, you are hard put to it to catch one of the party in the first part in the territory of the second part; and vice versa. Possibly certain ABSMs show the same respect for Nature here, too.

There is nothing like the wealth of material on the subject of ABSMs in the tropics, and notably in South America, that there is in North America, in the Himalayas, and in central Eurasia. What is more, what there is, looks extremely spotty and lacks any pattern unless it be *mapped:* and mapped on phytogeographical grounds at that. When this is done, how-ever, it begins to make a great deal of sense. Despite an enor-mous volume of literature on the geography and the distribu-tion of plants and animals in South America, there are still many widely held misconceptions about the constitution and history of that continent—held by profound students of the matter as well as by the general (and not technically inter-ested) public. The general impression of the continent is that it is a vast tropical jungle all over but, while a lot of it is covered with closed-canopy forests—whether you should call them jungles or not is a matter of much controversy in any case—the major part of it is *not;* and, a large portion down at the bottom has a temperate climate tailing off to a sub-polar one. Then, there is the great Andean upland and mountain ridge that occupies its whole western side. Least understood of all, however, is the area which is occupied by Brazil.

Looking at Map VI, you will perceive that, in addition to the two mountain blocks in Central America, and the three arbitrary divisions of the Andean ridge, there are three other upland massifs on this continent. These are the Guianese, the Matto Grosso, and the enormous Caātinga. The last is the most puzzling to foreigners, because one's impression of Brazil has been gained from the periphery of this grim slop-ing plateau, and this periphery is almost everywhere a lush lowland belt of forests and other massed vegetation. The appalling aridity of this still so-called "Terra Incognita" which reaches its climax in the northeastern bulge of the continent, is not generally known. If you want to get a clear picture of it, you should read a book entitled *Tukani* by Helmut Sick, a scientist who accompanied the first official expedition to cut right across this terrible territory to the Amazon Basin. In this, you will very soon see the complete difference between these uplands, their vegetation, climate and fauna, and that of the equatorial forests of the Amazon. The two are abruptly different worlds and, as one approaches the latter from the former, one comes up against an actual *wall* formed by tall evergreen vegetation.

If one raises the subject of animal life in South America, everybody invariably yells "Green Hell," and thinks of the Amazon Basin. It is a funny thing, but there is nothing hellish about any jungle and rather especially about that of the Amazon. It is, like all equatorial forests, never too hot or too cold, singularly free of noxious insects, completely free from disease [provided you keep away from human beings and don't carry any pestilence in with you when you enter], is well supplied with food that is easy to obtain, has plenty of good water, and is not too badly infested with indigenous people who resent one's presence. There are poisonous snakes and jaguars but you really have to look for them, and they are absolutely harmless as long as you look where you are going and don't molest them. [I once persuaded a jaguar to leave the ridgepole of our bush-house in which my wife was sleeping, one night, simply by saying "Boo" at it.] Then there is this Amazon bit.

It so happens that the basin of this name, which contains the greatest river, and river system, in the world, was, until not long ago geologically speaking, an arm of the South Atlantic—a great inland sea. Further, there is evidence that long since it became land it may have been completely flooded again for briefer periods off and on, and some Brazilian scientists claim that they have evidence that the last time this happened was only about the year 1200 B.C. It is indeed today a sort of enormous botanical cum zoological garden but, actually, its flora and fauna in no way compares in diversity with that of all the surrounding areas combined. In fact, it has manifestly been repopulated quite recently by several streams of animals and plants *from* those areas, which must have remained above sea level either as great islands or massive peninsulas attached to the rest of the continent. Moreover, there were jungles and other wet forests on those blocks as well as the vegetation and appropriate wildlife of their drier uplands. Many of those areas are also extremely ancient; meaning, that they have remained above sea level for a particularly long time. The most isolated and perhaps the oldest is the Guiana Massif, but seniority may be claimed for the Colombian Massif. This was certainly there before the Andes were pushed up. The Andes themselves are really comparatively recent, and they might be *very* new. This is not of our story but it is germane to it, in that the age of the montane forests of the Andes has a very important bearing on the recently past history of ABSMs and their possible distribution there.

The point I am trying to make here is that if I were asked to go find an ABSM, or any other as yet uncaught kind of animal, in South America, the last place that I would go would be the Amazon Basin itself. I would tackle the Guianese Massif first, next the Colombian Massif, and then move on to the uplands surrounding the Matto Grosso. After that I would do what I could about the Caātinga, and then Patagonia, and then the Andes, but would leave the Amazon till last. As a matter of fact, I would do a thorough job on the northern Central American Block before even going to South America at all, and this is just what I now propose to do.

The limits of this last block are very clear on Map VI, and are confined between the Isthmus of Tehuantepec on the west and the gutter filled by the great lakes of Nicaragua on the east. The smaller southern block, running from the latter line to the valley of the Atrato River, that cuts the Panamanian isthmus off from the Colombian Massif, will not concern us. There are some exceedingly strange small animals in that block, and there is some odd folklore but I have nothing concrete upon our subject from it. The main or western block is enormously mountainous, and constitutes one of the major areas of volcanicity in the world. The number of volcanos you can count from a point above Guatemala City is variously estimated and often grossly exaggerated but it is none the less quite remarkable. The southern edge of this block drops abruptly to a narrow, cactus-covered, dry, coastal plain; but the northern face steps down through ever-decreasing banks of mountains and hills to a wide forest-covered coastal fringe. Its real border is the valley of the River Usumacinta in Campeche, but north of this there are some ancient low hills in the Peten, and these mount up to the east into what is probably the most remarkable little mountain massif in the whole of Central America. This is called the Maya Mountains and lies in southern British Honduras.

I have been carrying on a very long-distance correspondence with an American lady for long resident in what is really the outer periphery of the Mexican state of Chiapas. She was introduced to me by a man in the publishing field with the very highest reputation and whom I most greatly respect. Were it not for this, I simply could not bring myself to record the following, even in a purely reportorial way. As of going to press I have not received a reply to my written request—and letters have to be paddled up a river to her, taking several days—to enter this information over her name.

However, I heard from her that a form of ABSM is not *quite* but *very* well-known in the forests nearby where she lives. [This, incidentally, is a continuation of those montane forests about which my friend Cal Brown writes (see below).] This she tells me is known locally by various names such as *Salvaje, Cax-vinic,* or simply *fantasma humano.* She then goes on, deadpan, to write: "I have seen this creature on various occasions and heard it frequently—the last time was about a year ago however. Some of the things I know about [it] coincide with your information [from other areas] but I can't reconcile the cry described with mine. I don't think I have ever heard anything so disturbing—not frightening but more dreadful and haunting, and full of threat I couldn't imagine.

[157]

I suspect that from this cry alone men living in this jungle could assume it to be a 'fantasma humano.'" As a friend of mine remarked on reading this, "And I suppose she rides one of the mastodons that the locals use for plowing."

This almost casual letter is somehow quite shocking to me, though knowing what I do of this matter in other much more settled areas, and in view of the fact that it is hard by Cal Brown's pinpointed area for something very similar-sounding, there is really no need to be upset.

As I remarked in a previous passage, Chiapas of Mexico is shaped like a salad bowl held on high. Its eastern rim abuts on to the mountains of Guatemala and these tumble down into the Peten in a tremendous jumble of tall, tight peaks and ridges with deep narrow valleys and gorges in between. The whole is choked with wet tropical forest, is unmapped, unexplored, and just plain not known. I have a group of young associates under the leadership of this Kenneth (Cal) Brown, who have for some years been working in this area collecting scientific specimens for botanical, zoological, and petrological studies, and I once lived for several years in that area myself, flew over almost all of it repeatedly during the war and have walked all about it. Comparing notes (after 20 years of this) Cal and I have come to the conclusion that this is one of the oddest areas on earth, made the more strange, almost eerie in fact, by the presence of many ancient Mayan ruins therein, which one stumbles across everywhere. There is something uncanny about these gigantic artificial hills, with their endless, writhing carvings, courts, passages, mighty flat-roofed halls, now filled only with the chitterings of bats; utterly abandoned in vast uninhabited jungles that just breathe silently in the noonday tropical sun. There are many strange things in these jungles and some of these pertain to our quest.

Cal Brown has pinpointed for me a valley to which his party once attained and where some of those odd incidents occurred that so often crop up when actually exploring. You can't really put your finger on them, and often one misses even recording them. It may be plants freshly broken in a way that is just not right; or very strange calls; or a certain reluctance by any native people around to go any farther or even to talk much. So powerful was this atmosphere at this place that one of Cal's partners—Wendell Skousen, a geologist, and one of the most pragmatic men I have ever met—corralled the locals almost by force and demanded to know what was going on. Then it came out. The locals explained:

There live in the mountain forests very big, wild men, completely clothed in short, thick, brown, hairy fur, with no necks, small eyes, long arms and huge hands. They leave footprints twice the length of a man's.

The area in question was in Baja Verapaz, around the town of Cubulco. Cubulco is the last vestige of civilization, the road ends there, and for all intents and purposes so does everything. The range of mountains in question is the Sierra de Chuacus, whose greatest peak is Mt. (Cerro) Sanché, 8500 feet elevation. Depending on which direction you're coming from, there are between 5 and 7 ridges from the floor of the Cubulco Valley [Rio Cubulco, which eventually joins the Rio Negro to the north roughly 20 kilometers] to C. Sanché. Further than this, I would not want to speculate as to range of this alleged creature. I have coloured in a patch on the enclosed map which depicts the approximate range according to what the natives told me, which means it would range into the departmento of El Quiché. (See Map V.)

Cubulco itself, at about 4200 feet, is really "tierra templada," and the area in question ranges up to "tierra fria." The vegetation is open pine and oak forests on the slopes, and many high plateau areas are covered with grass, as is the Cubulco environ. Along the margins of the highlands where rainfall is greatest, the oak and pine forest merges with the rain forest. Temperature ranges from 30°F to 90°F, and while I have no good figures on rainfall, it is considerably less than, say, Coban.

Now, as to "what the natives said." They referred to a large, hairy creature, which sometimes walked on two legs, and apparently ran on all fours. I considered bear first of all, and queried them regarding size, shape, appearance, etc. The answer was that it looked like a bear, but it wasn't from the description they gave—no conspicuous ears, no "snout"—it was somewhat taller than a man, and considerably broader, covered with darkish hair, and the locals live in mortal dread of disturbing it. Occasionally, one or two of the natives who got drunk or particularly boastful would go half way up the ridge and make a big show of "hunting" it, but no one has ever killed one that I learned. Several persons reported they were chased by it down the mountain, although with the fear they have of whatever it is, they probably just caught a glimpse of it and ran all the way down the mountain at top speed. No one seemed very anxious to guide us to the spot, or spots, but one of the braver souls agreed to do so finally. Unfortunately, we never got to it, for which you will curse, no doubt. I have no way of determining from their descriptions whether it was a bear or a *Sisemite* or something else, but it would seem reasonable that something is back there. You will be somewhat interested in the fact that the natives reported to me that this thing "calls" every so often, and they hear it from time to time when they are travelling about the ridges.

One cannot lay any store by "calls," for the tiny Dourou-couli, or Night-Monkey of South America (*Aötes*), can almost blast you out of bed when it really gets going, and the Howler Monkey (*Alouatta*), can individually make a series of noises that sound just like a dozen jaguars fighting in a thunderstorm. My point here is that I know Cal Brown and Wendell Skousen and the others very well indeed and have done so for many years. They are the hardest-boiled collection of skeptics I have ever met; yet, they were more than just impressed—they were astonished.

What they have told me, moreover, acquires a certain added interest when one reads in *The Museum Journal* (Vol. VI, No. 3, September, 1915), published quarterly by the University Museum of the University of Pennsylvania in Philadelphia, the following excerpts on what are therein described as Guatemaltecan mythology [sic]:

There is a monster that lives in the forest. He is taller than the tallest man and in appearance he is between a man and a monkey. His body is so well protected by a mass of matted hair that a bullet cannot harm him. His tracks have been seen on the mountains, but it is impossible to follow his trail because he can reverse his feet and thus baffle the most successful hunter. His great ambition, which he has never been able to achieve, is to make fire. When the hunters have left their camp fires he comes and sits by the embers until they are cold, when he greedily devours the charcoal and ashes. Occasionally the hunters see in the forest little piles of twigs which have been brought together by El Sisemite [also called Sisimici] in an unsuccessful effort to make fire in imitation of men. His strength is so great that he can break down the biggest trees in the forest. If a woman sees a Sisemite, her life is infinitely prolonged, but a

[160]

man never lives more than a month after he has looked into the eyes of the monster. If a Sisemite captures a man he rends the body and crushes the bones between his teeth in great enjoyment of the flesh and blood. If he captures a woman, she is carried to his cave, where she is kept a prisoner.

Besides his wish to make fire the Sisemite has another ambition. He sometimes steals children in the belief that from these he may acquire the gift of human speech. When a person is captured by a Sisemite the fact becomes known to his near relations and friends, who at the moment are seized with a fit of shivering. Numerous tales are told of people who have been captured by the Sisemite. The following incident is related by a woman who had it from her grandmother:

A young couple, recently married, went to live in a hut in the woods on the edge of their milpa * in order that they might harvest the maize. On the road Rosalia stepped on a thorn and next morning her foot was so sore that she was unable to help Felipe with the harvesting, so he went out alone, leaving one of their two dogs with her. He had not been working long when the dreaded feeling, which he recognized as Sisemite shivers, took hold of him and he hastily returned to the hut to find his wife gone and the dog in a great fright. He immediately set out for the village, but met on the road the girl's parents, who exclaimed, "You have let the Sisemite steal our child, our feelings have told us so." He answered, "It is as you say."

The case was taken up by the authorities and investigated. The boy was cross-examined, but always answered, "The Sisemite took her, no more than that I know." He was, in spite of the girl's parents' protests, suspected of having murdered his young wife, and was thrown into jail, where he remained many years.

At last a party of hunters reported having seen on Mount Kacharul a curious being with hairy body and flowing locks that fled at the sight of them. A party was organized which went out with the object of trying to capture this creature at any cost. Some days later this party returned with what seemed to be a wild woman, of whom the leader reported as follows. "On Mount Kacharul we hid in the bushes. For 2 days we saw nothing, but on the third day about noon this creature came to the brook to drink and we captured her, though she struggled violently. As we were crossing the brook with her, a Sisemite appeared on the hillside, waving his arms and yelling. On his back was a child or monkey child which he took in his hands and held aloft as if to show it to the woman, who renewed her struggle to be free. The Sisemite came far down the

* Cornfield.

[161]

hill almost to the brook; he dropped the child and tore off great branches from big trees which he threw at us."

The young man was brought from his cell into the presence of this wild creature and asked if he recognized her. He replied, "My wife was young and beautiful; the woman I see is old and ugly." The woman never spoke a word and from that time on made no sound. She refused to eat and a few days after her capture she died.

Felipe lived to be an old man, and the grandmother of the woman who told this story remembered him as the man whose wife had been carried away by the Sisemite.

This account would have been relegated to "Myth, Legend, and Folklore," had not an almost identical story, in the form of a complaint on a police-blotter, turned up in Coban, in the same region in the early 1940's. This was made by one Miguel Huzul and was to the effect that his son-in-law was delinquent in having permitted his daughter to be seized by a creature of the mountains to which he gave a name that was apparently too much for the recording officer and which he therefore put down as "a sort of gorilla or man" as far as it could be deciphered and transliterated. I had a copy of this document once, with a tracing of this passage, made for me by a Puerto Rican American who was baffled by the local Spanish and did not know any Mayan. Unfortunately my original went up under a wartime bomb, but we are searching for the records from which it came. All I can add is from memory, but this is pretty vivid in this case as you can imagine, for it was "in my district" at the time, I then being engaged in collecting in the area. It related, in substance, that the *Sisemite* had entered the young man's house and in the presence of other witnesses gathered up his young wife and carried her off while he, the husband, just sat there shivering. No action was taken because the father was disbelieved, while it was rather nicely pointed out that if all that is said about the *Sisemite* is true, the young man could not be accused of cowardice and/or delinquency. I presume there is no precise law covering the matter!

Even then, I would still relegate both stories to Chapter 17, were it not for my own personal observations, very close by in British Honduras. While there, my wife and I penetrated some distance into these Maya Mountains, not an easy task in the absence of any paths or people, their almost straight up and down topography, and the virtual nonexistence of people willing to carry things in all surrounding areas. While camped up there, the Senior Forestry Officer of the colony—one, Mr. Neil Stevenson—visited us, and we took a day's exploratory and collecting trip up to the top of the next ridge into the magnificent montane palm forest which is sufficiently "open" to be able to permit a view. On the ridge beyond that, then and still now totally unexplored and never even yet attained, there were rectangular areas of forest of distinctly different color, showing that they had once been cleared for cultivation. Later, we saw smoke rising from those forests, and Mr. Stevenson heard cocks crowing therein in the clear mountain air at dawn. When the Shell Oil Company later made a detailed survey of that whole mountain block by aerial, stereoscopic, photography, they brought to light further evidence that there were people living there. Yet, this mountain block stands up like an island in a sea of lowlands which have been crisscrossed for generations by mahogany workers and chicle collectors. Not one single human being has ever been known to come out of it.

Who are these people? Some Mayas left over since precolonial days; *pre*-Mayan people; or whom? Whoever they may be, they must be getting a strange education, for their home lies under one of the main commercial airline routes [from Florida, New Orleans, and Merida, Yucatan, to Guatemala City], while we ourselves once sat up on the lower slopes and watched the *Queen Mary* glide majestically by below, down the Gulf of Honduras on her way to Puerto Barrios, on a cruise! This is only a couple of hours flight from Miami, and yet there are apparently *people* living there who have never contacted other people since the time of Columbus.

Now, I am not suggesting that these tree-clearing, chicken-raising chaps, whoever they may be, are ABSMs; but, what I am suggesting, is that if such people can continue to live in magnificent isolation for 450 years, in a tiny country such as this, not more than 50 miles from a number of settled communities [in all directions, as a crow is alleged to fly], there could perfectly well be all sorts of other types living nearby too. And this is just what the people who live *around* the area affirm.

These people are of two major types—Amerinds, and sundry settlers of mixed Amerindian stock in Punta Gorda on the south, and related kinds of people to the north, plus what are called the *Caribs,* along the coast. These latter are not in any way the Amerindian Caribs who gave their name to the Caribbean, but are a group of West Africans of Sudanese Negro stock, who obtained their freedom on the Lesser Antilles in early days, and then sailed their own ships to the mainland coast. They are very strange people with their own language, customs, and religion; great boating people; fearless, and rather fearful. They don't trust anybody and they don't seem to like anybody, and whatever they say they should not be trusted—not because they are untrustworthy at all but because they have learned long ago never again to trust any white.

Both these peoples—the regular British Hondurans or Belizians, and the Coast Caribs—assert that there dwell in the tall, wet forests of the southern half of their country certain small semi-human creatures which they call *Dwendis,* a form of *Duende,* Spanish for goblin. To the very well-educated Belizians, these are regarded more as we regard fairies than as real entities—*unless* they have lived or worked in the southern forested area. Then they, like the Caribs, take quite another view of the matter. I lived in that country off and on for years while we traveled Central America and the West Indies, and I talked to innumerable people there about them. Dozens told me of having seen them, and these were mostly men of substance who had worked for responsible organizations like the Forestry Department and who had, in several cases, been schooled or trained either in Europe or the United States. One, a junior forestry officer born locally, described in great detail two of these little creatures that he had suddenly noticed quietly watching him on several occasions at the edge of the forestry reserve near the foot of the Maya Mountains when he was "cruising" and marking young mahogany trees. His description of them coincided with that of all the others who were serious.

These little folk were described as being between three foot six and four foot six, well proportioned but with very heavy shoulders and rather long arms; clothed in thick, tight, close, brown hair looking like that of a short-coated dog; having very flat yellowish faces but head-hair no longer than the body hair except down the back of the neck and midback. Everybody said that these *Dwendis* have very pronounced calves but that the most outstanding thing of all about them is that they almost always held either a piece of dried palm leaf or something looking like a large Mexican-type hat over their heads. This at first sounds like the silliest thing, but when one has heard it from highly educated men as well as from simple peasants, and of half a dozen nationalities and in three languages, and all over an area as great as that from the Peten to Nicaragua, one begins to wonder. Then, one day, I came across a lone chimpanzee in West Africa in an open patch of forest and on the ground; and, *by jingo*, it was solemnly holding a large section of dead palm frond over its head, just like an umbrella and looking exactly like a large Mexican straw hat!

Dwendis are said to appear suddenly in the forest both by day and night and to watch you from a discreet distance. They are silent but seem to be very curious. I heard of no case of their ever making any threatening move, but I was time and time again told of them chasing, sometimes catching, and carrying off dogs. They are said to leave very deep little footprints, *that have pointed heels.*

One does not really know quite what to make of all this. If you go to Belize—and a more delightful spot there can hardly be on earth for a vacation or just to live—and ask around about these things you will be met with gay smiles and probably a healthy quote from some classic such as *The Water Babies* but if you persist you will quite soon find some man who has timber-cruised, or been in the bush farming, and he will surely come out with some details about these mysterious little imps of the forests.

Perspective is a hard thing to evaluate on ancient carvings since captives bearing gifts to an important potentate may be made very small, compared to the monarch. Nevertheless, there are many Mayan bas-reliefs that show pairs of tiny little men with big hats but no clothes, standing among trees and amid the vast legs of demi-gods, priests, and warriors. They are also much smaller than the peasants bearing gifts to the temples!

As we have gotten on to the Pigmies again we might as well follow them. I have a letter from a well-known animal dealer of Guayaquil, Ecuador—Herr Claus U. Oheim—who knows his zoology, and who has a very long and intimate experience of the forests of his country and those of Colombia on the Pacific slopes of the Andes. In this he says:

The so-called *Shiru*, I have heard of from the Indians and a few white hunters on both sides of the Andes, but decidedly more so on the eastern slopes, where vast mountainous areas are still quite unexplored, and rarely if ever visited. All reports describe the Shiru as a small [4–5 feet] creature, decidedly hominid, but fully covered with short, dark brown fur. All agreed that the Shiru was very shy, with the exception of one Indian, who claimed having been charged after having missed with his one and only shot from a muzzle loading shotgun, a weapon still used by the majority of Indians, along with the blowgun. These reports were rather sober and objective, and in no way tinged with the colorful imagination, into which Latin-Americans are prone to lapse.

This business of the "eastern slopes" is going to get us into unwarranted difficulties unless we once again resort to a map. I think the best way to contemplate South America is as if it were made up of a number of large islands comprising those blocks of territory today enclosed within the 500-meter contours. This gives us a picture like that shown on p. 168 (see Map VI) on which both the 200-meter and 500-meter [1500-foot] contours are shown, and from which it may be seen that the uplands consist of the continuous line of the Andes; the Guiana Massif; and the Brazilian Uplands (composed of those surrounding the Matto Grosso, and the great Caātinga).

The 200-meter contour shows how these would be connected if there was any slight lowering of the land or an uprise of the sea. The Caãtinga would still be joined to the Matto Grosso, and then both by a narrow land-bridge to the Andean chain. The Guiana Massif, which is the most isolated, would in turn be joined to the Colombian Massif by a lowland bridge.

The "spine" of the Andes runs just about down the middle of that colossal range. The important fact to grasp is that this forms a complete break between the forests of the Amazon and the eastern part of the continent on the one side, and the small patch to the west, on the Pacific slopes on the other. This latter small area, has a noticeably different fauna and flora from that of the east and the Amazon. It is terminated to the south on the Pacific coast by the southern deserts. ABSMs in South America are reported from both sides of the Colombian Massif, from the Guiana Massif, and from the Matto Grosso. [The Patagonian affair is, I believe, something quite else.] I have some extremely funny reports from the Pacific side of the Colombian block but, while the strangest things have recently been found there* and monstrous foot-tracks have been reported in the same area, there has not been any suggestion that any of the latter were humanoid. Colombian scientists have taken the matter of what they call "an ape" fairly seriously but all the talk has concentrated on the forests of the eastern slopes. It was once thought that Pigmies, or ABSMs of the little *Orang Pendek* type had cropped up again in the Motilone territory in that area but, as Heuvelmans points out, a perfectly good Amerindian tribal grouping named the Marakshitos, averaging only about 5 feet in stature (like the central Mayas, incidentally), have been fully studied by the Marquis de Wavrin, while surrounding peoples admit that

* I have for some years been interested in the reported existence of giant Earthworms in this area, based upon some correspondence and some extraordinary bas-reliefs on ancient pottery from that country. In 1956 and again in 1957, Mrs. William (Marté) Latham made trips to the Pacific slopes of the Andes and obtained numbers of these—5 feet long when contracted, and over 2 inches in diameter. Preserved materials of them is lodged with the Smithsonian Institution but the animals do not as yet even have a generic name.

MAP VI. CENTRAL AND SOUTH AMERICA

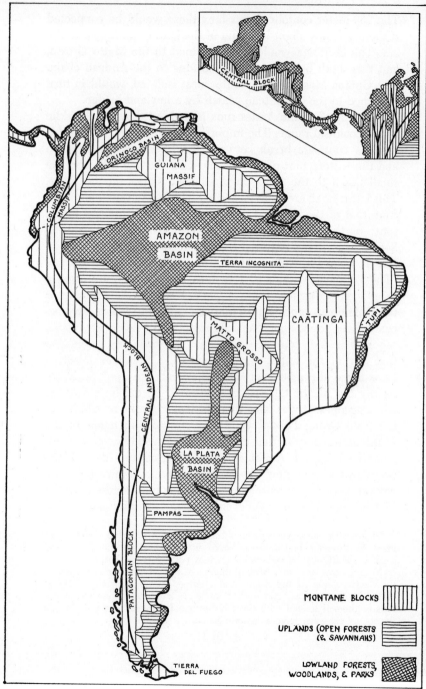

ORINOCO BASIN

GUIANA MASSIF

COLUMBIAN MASSIF

AMAZON BASIN

TERRA INCOGNITA

CAĀTINGA

TUPI

MATTO GROSSO

CENTRAL ANDEAN BLOCK

LA PLATA BASIN

PAMPAS

PATAGONIAN BLOCK

TIERRA DEL FUEGO

CENTRAL BLOCK

MONTANE BLOCKS

UPLANDS (OPEN FORESTS & SAVANNAHS)

LOWLAND FORESTS, WOODLANDS, & PARKS

See facing page for legend

these are the "creatures" that they call *Guayazis* and which they regard as bestial. In the Colombian block, a man-sized kind has been rumored. This has been very greatly muddled and muddied by a most preposterous business about a photograph of a Spider-Monkey (*Ateles sp.*) for which the most extravagant claims have been made, and for which a number of serious-minded and otherwise highly critical people seem to have fallen.

As this matter has played such a prominent and, in my opinion, harmful and misleading part in ABSMery, I would like to try and dispose of it once and for all—or, at least, once again; for this has really been done several times already.

First, this picture produced by one Dr. François de Loys is obviously that of a Spider-Monkey which is a very distinct type of South American primate that may be seen in any zoo. It displays all the characteristics of that genus—narrow shoulders and pinched chest; comparative lengths of upper and lower

---

MAP VI. CENTRAL AND SOUTH AMERICA

This continent is most notable for its lack of associated islands. It is today composed of three subcontinents joined by extensive lowlands. The former are: first, the Andean chain of mountains and their contained Alto Planos; the ancient Guiana Massif; and, the eastern uplands. The last is divided into two parts—the mountains around the Matto Grosso swamps and the vast arid Caātingas. Between these three major upland blocks there are three enormous drainage basins—those of the Orinoco, Amazon, and La Plata. All these are multiple river systems with innumerable tributaries that meander through extensive forested lowlands. Surrounding the upland massifs and bordering these river basins are intermediate plateaus. These are mostly clothed in savannahs. In the southern tip of the continent, south of the La Plata, these intermediate lowlands are covered with the tall grass Pampas and farther south with scrub. In the extreme northwest there is a block of equatorial forest on the Pacific side of the Andes cut off abruptly to the south by the excessively arid western coastal fringe. ABSMs have been reported from this Colombian area; from the Guiana Massif; from the mountains around the Matto Grosso; and from a few points in the central Andean highlands. "Bigfeet" have long been rumored from the Patagonian region, but the matter is there muddled with the Ground-Sloth business.

arms and legs; the hands and feet in detail; and the enlarged clitoris of a female. In fact, it is a pretty clear picture of one of these animals—dead. However, of much more importance is the box on which it is perched. Anybody who has ever been outside a tourist hotel in the tropics will have run into the fuel problem. Since the discovery of petroleum oils, they—including gasoline and kerosene—have been shipped all over the world in pairs of 5-gallon cans, or rather light tins, fitted into cheap wooden cases, measuring exactly 20½ inches long, by 10½ inches from front to back, and 15½ inches high. The better grade boxes are bound with metal tape around the two ends. The case shown in de Loys picture is such an object, and stenciled lettering may be seen on it under the monkey's right leg. Such lettering is also standard and is usually stamped over two of the four 4-inch bits of board of which the sides are invariably constructed. Thus the animal, with its head poked up to an unnatural degree by a stick, measures about 27 inches [it measuring 10x:6x as against the box]. This is a fair-sized Spider-Monkey but not even a large one.

The original photograph is not just a case of mistaken identity; it is an outright hoax, and an obnoxious one at that, being a deliberate deception. I would have thought that anybody might have suspected this, even without seeing the picture, from the originator's story. According to this, he was threatened by this creature and its mate on the ground when in company of one of his assistants; shot it; photographed it; counted its teeth; and then—despite the fact that he was a man of scientific training, and considered his specimen so odd (though out of his field), as to warrant all this trouble—solemnly gave the head to his cook to boil, and permitted that worthy to employ the cranium as a salt container, which "dried up and was lost bit by bit." But worse than even this is a lot of mumbo-jumbo about having other photographs that were lost in a river during a flood. This is the kind of nonsense that has done more harm to the cause of any serious search for ABSMs, and other creatures as yet unknown, than anything I can name, and it is to be most utterly deplored.

Quite apart from anything, the picture alone, if analyzed, displays the creature shown, to be a maximum of 48 inches from crown to heel. This is indeed large for a female *Ateles* but is really substandard for large females of the northern *A. beelzebub* group. Then again, gigantism is not uncommon among all the South American *Cebidae*. Finally, I may add, de Loys' photograph shows an animal that I would say had started to decompose and was well on the way to being "blown," a condition common in the tropics in daytime in a few hours, in which not just the body cavities but the whole body becomes puffy and bloated. Even if this should be a very large specimen of an as yet unknown *species* of Spider-Monkey (and even if, by some accident or deformity it happened *not* to have had a tail, which I very much doubt), there is no justification whatsoever for giving it a technical name on the strength of a single photograph, and especially one so grandiose, so misleading, and unscientific as *Ameranthropoides loysi* (Montandon) which means, literally "Mr. Loys' Ape-like American."

The harm done by this obnoxious effort has been widespread. Above all it has put the whole of ABSMery, in this area, into eclipse. No serious-minded person, zoologist or otherwise, seeing this ridiculous picture and having heard the equally ridiculous claims made by some for it, can be expected either to lend any credence to or even listen to the accounts of others who state that they have met unknown creatures of a Hominid form in this country. Yet, there have been some vague accounts thereabouts.

The earliest is that of the Baron Alexander von Humboldt, being a careful record of the local Amerindians' descriptions of a creature they called the *Vasitri* which, they said, constructed primitive huts, was carnivorous, and would eat men but carried off women for breeding purposes. There is nothing outrageous about this, for many ABSMs have now been reported to be carnivorous (at least at times), and their carrying off of women for reproduction is almost standard. [Something, incidentally, that all Africans that I have met who know and live among gorillas and chimps absolutely deny that those apes ever do.] Several other early writers are said to have mentioned the same creatures in this area.

Bernard Heuvelmans discusses an alleged encounter with an ABSM in this area by a Mr. Roger Courteville but shows that we cannot place any reliance upon it. The raconteur's description does include some odd items that are not otherwise to be noted in accounts from South America but which concur with, of all people, Mr. Ostman's description from *British* Columbia. These are: a tuft of thick hair running across the forehead; a powerful neck towering from a V-shaped torso; and long body-hair. However, the rest of the description, and particularly the "darting" blue-gray eyes, leave one in the gravest doubts. The only other definite information I have ever seen from this whole area is derived fourth hand from the Motilone Indians who are said to state that there is an "apelike" terrestrial creature in the Sierra de Perijaá, the scene of de Loys' exploit, and which is quite common. Thus, apart from the little *Shiru* and the possibility that von Humboldt left us a record of something real, there is actually *no evidence* whatsoever for any ABSM in this whole area. Apart from one locality—the somewhat mysterious Guiana Massif, there is not, as a matter of fact much if any ABSMery in the whole of South America.

There is, however, the strange matter of "giant footprints" in Patagonia but I do not know of any proper investigation of these, either firsthand in the field or even bibliographic, ever having been made. From what I have been able to unearth it would seem that the imprints mostly refer to those of ground-sloths and in some cases those of the Giant Ground-Sloth (*Megatherium*) itself, which are altogether bizarre, since it walked on the outsides of its enormous feet. There has been a terrific rumpus about ground-sloths in the Argentine that has been going on for decades. A dried skin of one, found hanging over a fence on an estancia in 1898, led certain persons to prosecute a hunt for the animal, believing it still to be alive. This led to a cave in which strange stone corrals were found deeply piled within with the dung of these huge beasts, while other evidence seemed clearly to indicate that they had been penned therein by men. After considerable excitement promoted by the notion that some of the smaller forms at least might still be found alive, and after the discovery of early records by the Spanish colonizers to the effect that the local natives caught huge shaggy animals in pits and killed them by building fires on top of the hapless beasts [because their skins were so thick and filled with little bones that they could not be pierced with their primitive stone-headed weapons], the whole thing died down.

However, mixed up in all this uproar there were reports of giant footprints of a very humanoid form being seen all over Patagonia. There was a period during which there was much speculation upon the possibility of a giant race of Amerinds living in that region but this later became a somewhat debilitated notion—to wit, that some large indigenous Patagonians had large feet. It is true that some now almost extinct southern Amerinds were among the tallest races of men ever on record, and they seem to have been large all over.

Today, most of Patagonia is sheep country. It was cleared of its indigenous human population over wide areas by the simple and ingenious procedure of poisoning all the available wells and other available water supplies. It is now a vacation land for the more rugged "sportsmen" and it must be admitted that the best trout fishing in the world is there available. However, there still are some enormous areas of a kind of desiccated tangle of large bushes that somehow manage to grow in endless blankets upon utterly dry ground for mile after mile. In these it is quite possible that some smaller types of ground-sloth, such as that called by the aborigines the *Ellengassen*, might still exist; but of ABSMs there is no trace —reports of giant humanoid footprints notwithstanding.

Almost the same may be said of the Caātinga. Herr Sick, the author of the book mentioned above, makes some casual remarks about unknown animals possibly still remaining to be found in that desolation; but he also makes some very dubious remarks, such as that "desiccated *Hyaena* droppings may be found" there. Not even the ebullient Argentine Professor Ameghino suggested the presence of *that* group of animals in South America; so one must take all these statements with more than just the average grain of salt. But, when we come to the Matto Grosso, matters are rather different.

Here we hit something much more persistent and much more concrete. Not only are there endless accounts of giant human-type footprints and tracks, usually given as being some 20 inches in length, but there is the matter of the mass slaughter of cattle for months on end from time to time, by the extraordinary device of ripping their tongues out. These inexplicable excesses are reported to be accompanied by roarings so terrible that even the locals—who are profoundly Amerindian, be it noted—become very nearly hysterical. The perpetrators of these dastardly acts are, the locals assert, to be called *Mapinguarys*, and to them the locals attribute all manner of appalling qualities. In fact, we have here for the first time on our trip to contend with some real imaginative and traditional frills and furbelows. There is obviously a gross clash here between the perfectly prosaic Brazilian *estancieros*, with their modern herdbooks and statistics, on the one hand, and a local population of truly "superstitious natives" on the other. This clash has not been resolved to anybody's satisfaction, least of all the herd owners who are periodically rendered clean out of pocket by some hundreds of head of good cattle. The "natives" seem, every time that this has happened, to have adopted a sort of "We told you so" attitude. This is not very helpful, but the Brazilian Government apparently has had no better ideas.

The only known animal that can kill cattle in that part of South America is the jaguar, but these large cats don't, and cannot, go around tearing the tongues out of steers. They jump on their backs and break their necks by pulling their noses around with a forepaw—when they attack cattle at all. Try pulling the tongue out of, say, a dead rabbit sometime. You will find that despite one's enormous size compared to the rabbit, plus inborn finger dexterity, you will have one heck of a hard time. Pulling tongues out of oxen calls for both extraordinary hand dexterity and positively phenomenal strength. What on earth may have such strength? The locals say *"Mapinguary,"* and point to giant humanoid foot-tracks on sandbars. Cattle owners just fume and say "rubbish." This is obviously not getting the latter anywhere since this sort of thing seems to crop up every few years.

Nobody seems ever to have seen this ox-tongue-puller but there is one story that Bernard Heuvelmans has dredged up from what can only be described as unimpeachable sources. With his permission I reproduce this in its entirety from his book, *On the Track of Unknown Animals.* This account was given to Dr. Heuvelmans by a correspondent, Senhora Anna Isabel de Sa Leitao Texeira, who obtained it from Dom Paulo Saldanha Sobrino, a much respected Brazilian writer with a very wide knowledge of his country. His informant was in turn the principal in the account; one known simply as Inocêncio. Heuvelmans writes:

In 1930 he went on an expedition of 10 men led by one Santanna. They went up the Uatumã towards the sources of the Urubú. When their boat came to an impassable waterfall they cut out across the jungle to reach the Urubú watershed. After 2 days they reached a stream which the leader decided to follow. Inocêncio was in the party going upstream, but after 2 hours' march he was led astray by a troop of black monkeys which he followed in the hopes of shooting one. When he realized that it would take him some time to reach the stream again, it was already too late. He shouted and fired his gun, but there was no reply except the chatter of monkeys and squawks of angry birds. So he began to walk almost blindly, feeling he must do something in such a critical situation until night fell, when he climbed into a large tree and settled himself in a fork between the branches. As it grew dark the night was filled with jungle noises, and Inocêncio rested happily enough until suddenly there was a cry which at first he thought was a man calling, but he realized at once that no one would look for him in the middle of the night. Then he heard the cry nearer at hand and more clearly. It was a wild and dismal sound. Inocêncio, very frightened, settled himself more firmly into the tree and loaded his gun. Then the cry rang out a third time and now that it was so close it sounded horrible, deafening and inhuman.

Some 40 yards away was a small clearing where a *samaumeira* had fallen and its branches had brought down other smaller trees. This was where the last cry had come from. Immediately afterwards there was a loud noise of footsteps, as if a large animal was coming towards me at top speed. When it reached the fallen tree it gave a grunt and stopped. . . . Finally a silhouette the size of a man of middle height appeared in the clearing.

The night was clear. There was no moon, but the starry sky gave a pale light which somehow filtered through the tangled vegetation. In this half-light Inocêncio saw a thick-set black figure "which stood upright like a man."

It remained where it stood, looking perhaps suspiciously at the place where I was. Then it roared again as before. I could wait no longer and fired without even troubling to take proper aim. There was a savage roar and then a noise of crashing bushes. I was alarmed to see the animal rush growling towards me and I fired a second bullet. The terrifying creature was hit and gave an incredibly swift leap and hid near the old *samaumeira*. From behind this barricade it gave threatening growls so fiercely that the tree to which I was clinging seemed to shake. I had previously been on jaguar-hunts and taken an active part in them, and I know how savage this cat is when it is run down and at bay. But the roars of the animal that attacked me that night were more terrible and deafening than a jaguar's.

I loaded my gun again and fearing another attack, fired in the direction of the roaring. The black shape roared again more loudly, but retreated and disappeared into the depths of the forest. From time to time I could still hear its growl of pain until at last it ceased. Dawn was just breaking.

Not until the sun was well up did Inocêncio dare to come from his perch. In the clearing he found blood, broken boughs of bushes and smashed shrubs. Everywhere there was a sour penetrating smell. Naturally he did not dare to follow the trail of blood for fear of meeting a creature which would be even more dangerous now that it was wounded. Taking a bearing on the sun, he at last reached a stream and rejoined his companions, who fired shots so that he should know where they were.

I maintain I have seen the *mapinguary* [Inocêncio said to Paulo Saldanha]. It is not armoured as people would have you believe. They say that to wound it fatally you must hit the one vulnerable spot: the middle of the belly. I can't say where it was wounded by my bullet, but I know it was hit, for there was blood everywhere.

I have heard many stories like this but, like Bernard Heuvelmans, I feel there is something sincere about this one. No; not just sincere; but factual. I have lived through some much lesser experiences myself in the tropical rain-forests that I

[176]

could never have reported so pragmatically; and there are junctures in the telling of this one that are so frightfully "right." If the teller had wanted even to exaggerate he could so very easily have done so but he did not. And yet, of course, it is ridiculous. But is it? There are still those tongue-twisters to be accounted for, and their little efforts are on the books. Do we therefore have a rather rough race of the otherwise bland and retiring *Sasquatch–Oh-Mah* type ABSM tucked away here in the soggy wilderness of the Matto Grosso who somehow got cut off, sometime, by a mass flooding of the continent that they had strayed into? If puny little Amerindian Man came over the Bering Straits and got right down to Tierra del Fuego, millennia ago, there is no conceivable reason why some more lowly type of Hominid may not also have done so. Perhaps he got there before "the flood" as it were.

The Matto Grosso uplands seem to have been above water for quite a long time but, according to their flora and fauna today, which is not particularly odd, they do not seem to have been so privileged as another area. This is the great Guianese Massif. Here, if anywhere, is the place where really ancient relics should have been able to linger; and there are some real lulus that have done so there. It is notable that the representatives of almost all the great groups of mammals, birds, reptiles, amphibians, and especially of fishes and insects, found in South America turn up there in strange and sometimes fabulous guises. There are great numbers of living fossils in this area; creatures like the Hoatzin or "Stinking Pheasant," a bird that, when young, has a clawed finger on its wing, like an Archaeopteryx. This block of ancient mountains seems, indeed, to have been a refuge from flooding throughout geological ages—a sort of last retreat for wave after wave of creatures throughout time, driven out of their previous habitats by shifts or submergences of the earth's crust. This is where we would most expect to come across ABSMs if there are any, or have ever been any, on this continent. And it is indeed from there that the most reports, and the most definite ones, have come.

In the Guianas—Venezuelan, British, Dutch (Surinam), French, and Brazilian—the name for these creatures is everywhere something like *Deedee* or *Didi*, with sundry prefixes and suffixes like *"Dru-di-di"* or *"Didi-aguiri,"* most of which mean something simple, such as "nasty" or "of the water." The whole concept is, however, as far as I was able to find out, very muddled in the native mind. This is probably because most of the current "natives" are not indigenous or in any way *native* to the country. It is only when you go among the now rapidly disappearing Amerinds—Caribs, Arawaks, and such—that you get any clear picture of this creature. Conversation with these folk is almost impossible as their languages are not known and are extremely difficult to apprehend. Also, they are naturally very cagey.

I first stumbled across this business when on my constant quest for animals which entailed endless patience in asking anybody and everybody about *all* the kinds of animals they had ever heard of. It was with the Primates—or monkey kingdom—that I kept getting information about more and ever more kinds that I had not yet seen. This started in British Guiana and went on in Surinam [then Dutch Guiana]. It seemed that there was no end to the kinds available and, to my great surprise, the locals were as good as their tales, for more and still more kinds were brought to us—or we were taken to them. I saw monkeys alive—and in captivity—in that country to which I could not and still cannot give even a familial name. And from quite early on we kept being told about these *Didis*. They lived way back in the hills, and they were pretty smart "Kwasi," which is the generalized name for all Primates in that area. Also, they had no tails, lived on the ground, had thumbs like men, and built crude bush-houses of palm leaves. They usually ran away but if a large party of men should penetrate into those completely uninhabited mountains they would come, a lot together, and throw sticks and mud at your canoe. So went the stories.

I never saw a *Didi* but then we never got really far into the uninhabited territory but I did come across some extremely large human footprints in the mud of a tiny side creek off a main river right up by the first cataract and 40 miles above the last known village. I put them down to visiting Jukas upriver to hunt, or to a band of roving Amerinds —for there were still some in the district though nobody had seen them for over a decade—but I was mystified. I did not connect them with the *Didis;* but I have since often wondered. It is so easy to find plausible explanations of odd facts. Besides, some other dashed rum things happened at that camp.

The story of the *Didis* goes back to the first days of European exploration of the Guianas. Sir Walter Raleigh's chroniclers mentioned them; the early Spaniards said that the natives spoke of them; and in 1769 Edward Bancroft* wrote of them, saying that the Indians said they were about 5 feet tall, erect, and clothed in black hair. Once again also, the redoubtable Bernard Heuvelmans has brought to light some specific statements on these elusive creatures. These he gives us as follows:

In 1868, a century after Dr. Bancroft, Charles Barrington Brown, who was then Government Surveyor in British Guiana, heard new rumours on the Upper Mazaruni on the Venezuelan frontier that a sort of hairy men lived there. Oddly enough, it was after hearing the "plaintive moan or howl" which Cieza de Leon also alleged these ape-men made.

The first night after leaving Peaimah we heard a long, and most melancholy whistle, proceeding from the direction of the depths of the forest, at which some of the men exclaimed, in an awed tone of voice, "The Didi." Two or three times the whistle was repeated, sounding like that made by a human being, beginning in a high key and dying slowly and gradually away in a low one. . . .

The "Didi" is said by the Indians to be a short, thick set, and powerful wild man, whose body is covered with hair, and who lives in the forest. A belief in the existence of this fabulous creature is universal over the whole of British, Venezuelan and Brazilian Guiana. On the Demerara river, some years after this, I met a half-breed woodcutter, who related an

* Called himself Jacobus Van Zandt, spied on Benjamin Franklin in Paris, and almost kept France from joining our War of Independence. He was a well-known botanist and naturalist, as well as a doctor!

encounter that he had with two Didi—a male and a female—in which he successfully resisted their attacks with his axe. In the fray, he stated, he was a good deal scratched.

In 1931 Professor Nello Beccari, an Italian anthropologist, Dr. Renzo Giglioli and Ugo Ignesti, made an expedition to British Guiana, where one of their secondary objects was to attack the problem of Loys' ape. For in this area the fauna, flora, climate and indeed the whole ecological pattern—what is now called the "biotope"—are the same as in the Sierra de Perijaá; * and Beccari had read in Elisée Reclus's geographical encyclopedia that according to Indian legend the forests in British Guiana were haunted by fabulous hairy men called di-di, which all the Indians fear, although they have never seen them. But it was not until he was just about to return to Italy that he heard any definite information about where this beast lived.

On his return from several months in the interior, he met the British Resident Magistrate, Mr. Haines, who was then living on the Rupununi. Haines told him that he had come upon a couple of di-di many years before when he was prospecting for gold. In 1910 he was going through the forest along the Konawaruk, a tributary which joins the Essequibo just above its junction with the Potaro, when he suddenly came upon two strange creatures, which stood up on their hind-feet when they saw him. They had human features but were entirely covered with reddish brown fur. Haines was unarmed and did not know what he could do if the encounter took a turn for the worse, but the two creatures retreated slowly and disappeared into the forest without once taking their eyes off him. When he had recovered from his surprise he realized that they were unknown apes and recalled the legend of the di-di which he had been told by the Indians with whom he had lived for many years.

When Miegam, the guide of the Italian expedition, heard this story he remembered that he had had a similar adventure in 1918. He was going up the Berbice with three men, Orella, Gibbs and an American whose name he had forgotten. A little beyond Mambaca they saw on a sandy beach on the river-bank two creatures which from a distance they took for men, and hailed them to ask if the fishing was good. The unknown

* This statement is not strictly true. There are most marked botanical and zoological differences betwen these two areas, while their forests are completely separated by a wide belt of orchard-bush and open savannahs which form a barrier just as complete as if a sea. The Guiana Massif, moreover, is the more isolated in technical parlance, and has been so much longer and more often in the past. Sub-hominids and submen would be just about the only animals that could cross the open country but even modern forest peoples prefer not to do so.

creatures did not reply but merely slunk away into the forest. The four men were puzzled and landed on the beach, where they were staggered to find that the footprints were apes', not men's. Miegam could not say whether the creatures had a tail, but it could hardly have been conspicuous, or he would not have mistaken them for men. He did, however, say that two other settlers called Melville and Klawstky had similar adventures in other places.

Professor Beccari obtained further information about the *di-di* from an old negro at Mackenzie, famed for his wisdom, learning and experience. Everyone on the banks of the Demerara called him "Oncle Brun"—presumably he had come from French Guiana or the French West Indies—but the few Indians that survived in the neighborhood respected him so much they named him "The Governor." "Oncle Brun" had been told by the Indians that the *di-di* lived in pairs and that it was extremely dangerous to kill one of them, for the other would inevitably revenge its mate by coming at night and strangling its murderer in his hammock. Beccari did not trust the more fanciful part of this story, but felt that it must have a kernel of truth. Loys, like Haines and Miegam, had also met a pair, and so had Barrington Brown's woodcutter. Most South American monkeys live in largish troops, and this habit alone suggests that this is a very peculiar species.

The most significant single fact about these reports from Guiana is that never once has any local person—nor any person reporting what a local person says—so much as indicated that these creatures are just "monkeys." In all cases they have specified that they are tailless, erect, and have human attributes, even to building huts and *throwing* things. This is an altogether different matter to de Loys' asinine "ape." We are, in fact, once again confronted with the strange fact that great numbers of people of all manner of tribes, nationalities, and even races, insist that ABSMs are wild men, as opposed to manlike animals. This is the one theme that runs consistently through all ABSMery.

9. Africa—the "Darkest"

Africa was crossed for the first time just a century ago. Today, it virtually controls the UN, but great hunks of it have still not been mapped.

In some respects the continent of Africa has much the same structure as that of South America, but in reverse, or rather, mirrored, the Congo Basin being equivalent to the Amazon Basin; the great upland chain of its east side being comparable to the Great Andean chain on South America's west side; and there being a number of isolated mountain blocks dotted about the rest of it. However, if the sea flooded in to the 200-meter contour, Africa though looking a bit smaller would retain its present shape. Much more of it is composed of uplands above the 500-meter contour; and upon these are raised many mighty mountain ranges. Yet, the Congo is like the Amazon in one essential respect. It also was once, and until fairly recently, flooded, but it appears to have formed either an enormous lake or a completely landlocked sea. Finally, it broke out to the Atlantic by cutting a deep and narrow gorge through the Crystal Mountains that join the lower Gabun with the Angolan uplands.

The Congo Basin today is the home of several very unique and ancient forms of animals—a strange Water-Civet (*Osbornictis*) found only once, in 1916, which seems since then to have vanished; the forest giraffe or Okapi (*Okapia*); and the famous Congo Peacock (*Afropavo*)—but, like the Amazon, it seems nevertheless to have been repopulated comparatively recently from the slopes of the surrounding mountain blocks, and all the funniest plants and animals in it tend still to be found around its edges rather than in its middle.

Then, there is another very important thing about Africa. This fact is that, actually, very little of it is forested, and especially by true lowland Equatorial Rain Forest (or T-E-F; see Chapter 18). Apart from the Congo, there are really only two such areas; the west Guinea coast, and the Nigeria-Camerun-Gabun coasts. That over on the east coast is not typical T-E-F, and a lot of it is not even closed-canopy forest. Moreover, even on the west coast there are open areas within the main lowland forest blocks, while the center of Nigeria has now been almost entirely cleared, and there are large orchard-bush and savannah areas even in the Congo itself. Nevertheless, there is a great deal of montane forest in Africa. This clothes the slopes of the mountains to comparatively great heights up the sides of all those blocks in the equatorial belt, which face those lowlands clothed in rain-forest. This is so of the south face of the Guinea block; both sides of the Gabun-Camerun block; the south face of the Ubangi-Shari; and a swath running south from the Bar-el-Ghazal to Kasai. It is in these forests that the more ancient and retiring creatures make their abode. They, also, are the least known parts of the continent. (See Map VII.)

ABSMs have been reported from three areas in Africa—the southern face of the Guinea Massif; the east side of the Congo Basin; and the eastern escarpment of Tanganyika. This makes very clear sense both from a geographical and zoological point of view because each has an adjacent mountainous area as a retreat in case of general land subsidence or of flooding by the general sea level rising. The Gabun-Camerun west face might be expected to be included in ABSM distribution but it is not, so far as I know. Before coming to the details of ABSMery hereabouts, I must point out a new factor in our story that now appears for the first time and which will be with us through the Orient, and until we go to Eurasia.

It is that we now have interjected into the scheme of things the *Apes*. There is a point here that puzzles everybody and which must be cleared up if possible. This is a hang-over from the initial pronouncement of Darwin's theory of our origin

MAP VII. AFRICA

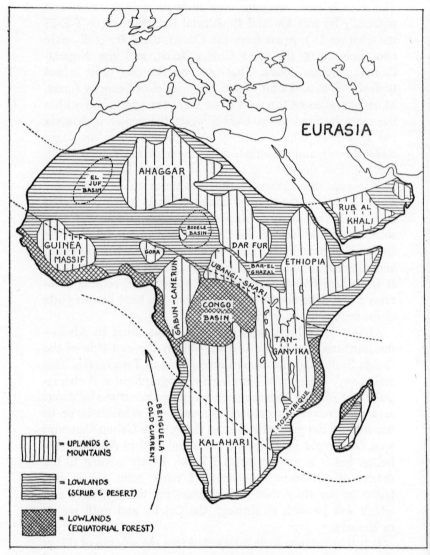

See facing page for legend

and that of Men generally. While it is pretty universally recognized that Darwin never did say that we were *descended* from apes, but that both we and the apes had a common ancestor, there is still an almost universally held belief that, nonetheless, "monkeys" came first, then "apes," and finally "men." This is partially correct in that the monkeys do seem to be of "older" stock and they are certainly more primitive or lower down the scale; but when it comes to apes and men, ever more evidence is piling up to indicate that we both started off at the same time; but out of step. Following this line of reasoning, the two lots of us were at one time very alike but, as time went on, and both our branches proliferated into various kinds, we drifted ever farther apart. Just because some "men" (or, better, Hominids) retained a hairy coat and small brains, does not actually mean that they remained apelike. Conversely, just because the living apes still have a complete furry coat and small brains does not mean that they have *evolved* any less far from their origins than we have. They just have changed in other directions. And,

---

MAP VII. AFRICA

Africa is the second largest and most compact of the continents. Apart from Madagascar, which is hardly a part of it, and Fernando Po in the west, it also is singularly lacking in peripheral islands. Its northern rim is really a part of the continent of Eurasia and, lying as it does today north of the great Sahara Desert, has not only a different climate but, to a large extent, flora and fauna differing from that of the rest of the continent. Africa is divided into three parts—a northern desert belt; a central forested belt; and a southern desert belt. However, further confusion is produced by there being extensive and almost over-all highlands all the way down the eastern half of the continent and other extensive upland blocks in the northern deserts and in the western half of the forested equatorial belt. Africa also has three large basins, but all are inland and without any real outlets. Two—El Juf and the Bodele—now lie in the deserts. The third is that of the Congo, the second biggest riverine system in the world. The tall, closed-canopy, equatorial forests are actually rather limited in extent, and are clearly divided into four parts, in the west and center. The lowland coastal forests of the east are not T-E-F and are more arid. ABSMs are reported from three forest areas, with a rumor from the far southwest.

along our lines, there were just such (probably) hairy chaps with very small brains—*vide: Zinjanthropus* from East Africa, and the other Australopithecines from South Africa.

Nonetheless, as of now—and if we preclude the possibility of ABSMs—the Hominids have changed a great deal, but lost all but one of their branches; while the apes have sort of got stuck, but still exist in about twenty distinct forms.* There is no reason, however, why there should not still be other kinds of *apes* still living today that we have not yet found. The Pigmy Chimp (*Pan paniscus*) was a long time being accepted; there is supposed to be a pigmy species of Gorilla of which we think we have skins and skulls, but which has never been seen by scientific collectors in the wild; and there is the extremely odd Lesser Siamang of South Pagi Island, of the Mentawi Group, off the west coast of Sumatra (*Symphalangus klossi*), that was not recognized till 1903. Then, in 1955, a professional American animal-collector brought back parts of a preserved specimen of a small kind of ape from the Gabun that is now lodged in the museum in Zurich, Switzerland. He also had photographs of the animal when alive, and it was certainly the oddest-looking creature; like a tiny orang-utan, with a high-domed forehead and quite unchimplike face, clothed in black hair, but having *no thumbs!* The collector insisted that the local natives know these animals well; that they are not chimps; and unlike chimps, they are completely arboreal, travel in parties of about 40, and never come to the ground. This specimen has been tentatively put down as an extremely abnormal baby chimp but it has a complete set of adult teeth!

There may indeed still be not one but several Apes (i.e. Pongids) to be discovered, and more than one in Africa. This tends to muddle the issue and more especially because many Africans regard the gorillas—though never the chimp, it seems —as a form of degenerate race of *men*. Thus, when they speak of any equivalent of an ABSM, they often do not make any

---

* The popular belief is that there are but four living apes—the Gorilla, Chimpanzee, Mias (Orang), and Gibbon. This is not so. See *The Monkey Kingdom* by the author.

distinction between men and apes in describing it. This, however, is particularly the case only with man-sized or larger alleged creatures. One and all are insistent that the pigmy types are little men, *not* animals. Therefore, we must be on guard from now on against interpreting all reports as those of potential ABSMs, and we must bear in mind that other apes—perhaps even bipedal, terrestrial forms—could exist on this continent and, though not so likely, in the Orient as well. Some certainly seem to exist in Africa.

Reports of what appear to be pigmy types of ABSMs turn up on the west and east sides of the continent—that is on the southern edge of the Guinea Massif and on the eastern side of the main upland area about Tanganyika and Mozambique. Those from the former consist of a single series of statements made to Bernard Heuvelmans by a scientific correspondent who investigated some reports in what used to be the Ivory Coast. As these are original and unique, I have sought permission to reproduce them here in their entirety. Apart from them, I have only vague folk tales from the Nigeria area of the former existence of pigmies in that country.

In the Ivory Coast, almost at the opposite end of Africa, there is a legend of reddish hairy dwarfs identical in every respect with that in Tanganyika, as I learn from private information kindly supplied by Professor A. Ledoux of the Faculty of Science of Toulouse University. In 1947 he was the head of the Zoological Department of the Institute of Education and Research at Adiopodoumé, which was then being formed 12 miles from Abidjan.

One evening a young African who worked in his laboratory came and saw him after dinner and went rather furtively about asking him the simple question whether there were pygmies in Africa. The professor told him that they were found in Central and Equatorial Africa and lent him a book on the subject. All the same he was puzzled at this conspiratorial manner and asked him why he wanted to know. Because, the African replied, one of his colleagues in another scientific department of the Institute had seen one not far away on the previous day.

The professor raised his eyebrows.

"A pygmy, here?"

"Yes, about 500 yards from here."

The professor nearly fell out of his chair in surprise. The Institute was surrounded by forest, but though it was thick in places it was very well known and natives were constantly passing through it. The story seemed most suspect.

"Why didn't your friend come and tell me at once?" he asked suspiciously.

The young man explained that since the whites disbelieved the rumours about pygmies in the neighbourhood his friend had been loath to be laughed at or thought to be mad. But he knew the man well enough to know that he had not made it up, so he decided to make inquiries for him himself.

Professor Ledoux was more and more intrigued and insisted that the eyewitness should come and see him. He promised that he would not make fun of him and would not tell anyone his story.

The next day [the Professor tells me] I had a visit from the boy responsible for the observation. He was well-educated and had a certificate for primary studies. I asked him about the circumstances in which he saw the "pygmy."

It happened near the Meteorological set when they were taking their daily readings at 8 o'clock in the morning. Among the roots of a silk-cotton tree (*Bombax*) there suddenly appeared a little man with long reddish fur and long hair on his head—"same like white man"—but also reddish. [The long hair on the head, like a European's, was reported by all the Africans Professor Ledoux questioned. This feature could obviously not apply to true pygmies.]

At once the little red man and the large black one took to their heels in opposite directions. For, according to the legends, the little forestmen brought bad luck. You only saw them once in a lifetime and you had to be alone.

I went to the place with my two informants. It lay in the shadow of thick forest, but was not too overgrown since the silk-cotton tree grew near a path. It was very likely that if there had been anything there it would have been easy to see.

I asked to be informed at once if a similar meeting occurred again, but this never happened.

Professor Ledoux admits that he was then very incredulous. It seemed unthinkable that within 12 miles of a big town like Abidjan, and 500 yards from huts inhabited by 6 Europeans and some 300 Africans, there could be unknown creatures in forest which though thick was far from virgin. Moreover the African who claimed to have seen the mysterious pygmy did not come from the Lower Ivory Coast but from the Middle Coast, which is wooded savannah country. Perhaps the great forest, which

[188]

is most impressive, had over-excited his imagination. And in his solitary walks in the forest the professor had often put up bushbuck. If the African had seen the russet back of one of these antelopes among the bushes it was not impossible that his imagination should have led him to think that it was one of the "little hairy men" of the legends. At all events when the professor showed him a book containing pictures of Central African pygmies he insisted that his creature was not like them.

Despite his scepticism Professor Ledoux decided to make discreet inquiries about the native legends and what they were based on. He questioned several Africans who trusted him, and he pretended that he fully believed their stories, thus putting them at their ease and getting much more detailed information. In this way he came to visit most of the neighbouring villages.

As a matter of fact I did not obtain any important information, for while there were plenty of men who "had seen" (?) them, they were reticent on the subject, always concluding that they were probably mistaken for all the encounters had taken place at nightfall. This is likely enough.

There was one relatively exact fact. In March 1946 a team of workmen under one Djaco—who later became my lab-boy and my informant and who died of poisoning in 1949—together with a European of whom I can find no trace, were supposed to have seen one of these little red men, at about 8 in the morning, in a tall tree in a very wooded little valley about half a mile from the future site of the station. The European asked what it was and the Negroes explained what a rare thing it was to see such a creature and the evil effects of doing so.

I was at once deluged with stories of dwarfs with their feet back to front, people who lived half in the lagoon and half on land (I think that manatees must be responsible for this legend). These tales were of no interest to me, but I mention them so that the record should be complete.

He then questioned the Europeans who had travelled in the Ivory Coast: One of them told me the following:

During one of his expeditions in the course of 1947 the great elephant-hunter Dunckel killed a peculiar primate unknown to him; it was small with reddish-brown hair and was shot in the great forest between Guiglo and Toulépeu, that is, between the Sassandra and Cavally rivers. Its remains disappeared while it was being carried home, no doubt having been disposed of by superstitious porters. Dunckel even offered to take my informant to the place and he in turn invited me to go with them.

In 1951 the professor's new boy, in reply to his usual barrage of questions, told him that when he was young, probably around 1941, he had himself seen a hunter at Seguéla bring back a little man with red hair in a cage. The local official had put clothes on it for decency's sake and sent it to Abidjan by way of Bouaké. The boy did not know what happened to the little prisoner afterwards.

This tale seems to me to have been embroidered somewhat. If the creature was really human it would not have been put in a cage, and if it was an ape the official would not have worried whether it was decently dressed. Either it was a creature half way between man and an ape, or more likely, it was an unknown primate which had been dressed up as a joke, as tame monkeys so often are.

Professor Ledoux remarks that these tales of an unknown reddish-brown primate in the Ivory Coast are overlaid with the very firmly held belief that there are pygmies in the forest between the Sassandra and Cavally rivers.

According to an African technician of mine from Toulépleu called Méhaud Taou, an intelligent boy keenly interested in these questions, there was recently a system of barter between the negroes and these forest creatures; various manufactured goods were left in the forest in exchange for various fruits. This was supposed to have gone on until 1935. The little men who practised this barter were hardly known even to the negroes themselves. The Guérés called them *Séhité*.

It is possible that these *Séhités* may be true pygmies like those in Central Africa.

The professor's inquiries among the Europeans brought out a significant fact. Those who had never spent any length of time between the Sassandra and the Cavally denied out of hand that there could be any little men in the forest, whether they were true pygmies or unknown primates. On the other hand those who had lived in this area were seriously prepared to consider that pygmies might have lived there in the past and also that there might be a real basis for the legend of the red dwarfs. His own impression was that the legends and rumours in the Ivory Coast were based on the fairly recent presence of pygmies and the present existence of reddish-haired primates whose exact nature was still problematical.

The reports from the east side of the continent are more numerous and varied, and come from more separate sources. Central to these is a statement, that has been repeatedly republished, by one Capt. William Hichens in *Discovery* for December, 1937, included in an article entitled "African Mystery Beast." This goes as follows:

Some years ago I was sent on an official lion-hunt to this area [Ussure and Simbiti forests on the western side of the Wembare plains] and, while waiting in a forest glade for a man-eater, I saw two small, brown, furry creatures come from dense forest on one side of the glade and disappear into the thickets on the other. They were like little men, about 4 feet high, walking upright, but clad in russet hair. The native hunter with me gazed in mingled fear and amazement. They were, he said, *agogwe*, the little furry men whom one does not see once in a lifetime. I made desperate efforts to find them, but without avail in that wellnigh impenetrable forest. They may have been monkeys, but, if so, they were no ordinary monkeys, nor baboons, nor colobus, nor Sykes, nor any other kind found in Tanganyika. What were they?

Subsequent to the publication of this observation, a gentleman by the name of Mr. Cuthbert Burgoyne wrote to the publication *Discovery*, seconding Captain Hichens' story with the following:

In 1927 I was with my wife coasting Portuguese East Africa in a Japanese cargo boat. We were sufficiently near to land to see objects clearly with a glass of 12 magnifications. There was a sloping beach with light bush above upon which several dozen baboons were hunting for and picking up shell fish or crabs, to judge by their movements. Two pure white baboons were amongst them. These are very rare but I had heard of them previously. As we watched, two little brown men walked together out of the bush and down amongst the baboons. They were certainly not any known monkey and yet they must have been akin or they would have disturbed the baboons. They were too far away to see in detail, but these small human-like animals were probably between 4 and 5 feet tall, quite upright and graceful in figure. At the time I was thrilled as they were quite evidently no beast of which I had heard or read. Later a friend and big game hunter told me he was in Portuguese East Africa with his wife and three hunters, and saw a mother, father, and child, of apparently a similar animal species, walk across the further side of a bush clearing. The natives loudly forbade him to shoot.

Once again Bernard Heuvelmans has brought to light two further reports, albeit brief. The first appeared in *The Journal of the East Africa and Uganda Natural History Society* in 1924, from the pen of one Mr. S. V. Cook. This states that:

Fifteen miles east of Embu Station there rises from the Emberre plains the lofty hills of Dwa Ngombe, nearly 6,000 feet high. They are inhabited, the Embu natives say, by buffalo and a race of little red men who are very jealous of their mountain rights. Old Salim, the interpreter at Embu, tells me with great dramatic effect how he and some natives once climbed to near the top when suddenly an icy cold wind blew and they were pelted with showers of small stones by some unseen adversaries. Happening to look up in a pause in their hasty retreat, he assures me that he saw scores of little red men hurling pebbles and waving defiance from the craggy heights. To this day even the most intrepid honey hunters will not venture into the hills.

The final scrap of confirmation comes secondhand from Roger Courtenay who tells in his *The Greenhorn in Africa* a story related to him by his guide named—as is almost invariable, all down the east side of Africa—*Ali*. Using Courtenay's own words, this goes:

"But have you heard of the little people who live in the Mau—small men, who are less men than monkeys? Less than *shenzi* (i.e. loathsome foreigners), these little men, and almost monkeys in their lives and ways." And he went on to tell how his own father, who was driving his sheep to pasture on the slopes of Mount Longenot, fell into the hands of these gnomes when he went into a cave, following the trail of blood left by one of his cattle that had been stolen. He was stunned from behind, and when he came round he found he was surrounded by strange little creatures. "The Mau men were lower even," he told his son, "than those little people of the forests [the pygmies] for, though they had no tails that I could see, they were as the monkeys that swing in the forest trees. Their skins were white, with the whiteness of the belly of a lizard, and their faces and bodies were covered with long, black hair." To his great surprise the shepherd noticed that his spear was still lying at his side. "The Mau men who are so nearly monkeys did not know what was the spear. It is possible they did not know I could have fought with it and killed many of them."

The first reaction to reading these reports is, perhaps naturally, to suggest that all the reporters, both local and foreign, had stumbled upon a group of true Pigmies, the race of little men who are so well-known in the Uele District of the Congo, and at some other points about central Africa. It is true that the skins of these people are not by any means lacking in a fine, yellowish, downy hair, and that they also make a practice of painting themselves white or red for certain ceremonial purposes. Also, they are tiny and primitive enough to fit the bill. Further, there is no doubt that they were once very much more widely distributed almost all over Ethiopian Africa. Then also, we must remember that there was once—and there are some still living in the Kalahari area—also another completely different race of men that were spread all over the continent, and perhaps even into what is now Europe in the middle Stone Age. These are the yellowish-red-skinned Bushmen. Some of them are, and were, very small but they have no body hair.

The Negroid peoples are apparently the most modern or newest development among human beings, and have very specialized characteristics. It seems that they did not even appear on the scene until just about the beginning of historical times, and their point of origin appears to have been about the headwaters of the Nile. Thence they spread outward in all directions possible but in two main streams, one to the west across the three Sudans; the other to the east and then south around the great lakes and down the eastern uplands. As a matter of fact, the Negro peoples only reached South Africa just about the same time as the white man did from Europe by sea in the 16th century. These eastern tribes, by interbreeding with early Caucasoid types, produced first the Bantu peoples, and later the Hamitic. The former wheeled west and crossed the Congo, reaching the Cameroons. The tribes that went directly west through the Central and into the Western Sudan, encountered a different state of affairs. North of the Congo Basin, and all the way to the bulge of Africa to the west, there were no Caucasoids to intermingle with but there were apparently lots of peoples of the Bushman variety, living in the forests at a very low stage of culture. These, the Negroes did not absorb to any substantial degree. Instead, they either exterminated or completely enslaved them. This is a most important fact that is not customarily known about forest West Africa.

I was once greatly surprised when, upon inviting a whole large village of the Akunakuna tribe on the Cross River in Nigeria to gather for an evening of music and other festivities, to see the community drawn up in four very clearly separated groups; two in the foreground with the Paramount

Chief and assorted Chiefs and Sub-Chiefs in front, and two other groups on either side, far behind. It was still daylight when they assembled and to our greatest surprise we suddenly saw that we were looking at two quite different peoples composed of: tall, very dark-brown skinned Negro men, and sturdy tall women of the same cast of feature and skin color, *and* of very short, almost pigmy men, with pale reddish-brown skins, flat faces, broad noses, hugely everted lips and little bandy legs on the one side, and a mass of tiny but very fat women of the same type and color on the other, all of whom had tremendous bottoms. Demanding to know from the local Headman "What be those?" and why everybody was not mixing it up in the truly democratic way, I was solemnly informed that, as I was not a government official, I would be pleased to know that those others were "slave-man." Slavery being absolutely taboo in that [then British] Protectorate, I sought further information and learned that there are whole enslaved peoples living within the body of many tribes in the general area, who are hewers of wood and drawers of water and with whom it is, and always has been, absolutely ver-boten to interbreed. These people were as good "Bushmen" types as I have ever seen.

We must therefore bear in mind that really extremely primitive peoples do still exist all over Ethiopian Africa, and that these have manifestly been either enslaved, or actually hunted by the tall, proud Negro peoples for centuries but still survive. Those not enslaved must be pretty wary and adept at concealment. Nonetheless, both the Sudanese in the west and the Bantus in the east seem to insist that such as the *Sehités* and *Agogwes*, though *men* alright, are even more lowly and ancient than these Bushman-like primitives. I do not think that we have to go so far as to dredge up the Australopithecines to explain them [though that, of course, is by no means impossible] because there must have been innumerable races and subraces of men, submen, and apelike-looking men (or Apemen, if you will) in the intermediate 500,000 years. Relics of goodness knows how many races could still be lingering on in the montane forests of Africa. Let us not forget that it was not till 1910 that the second largest land animal in the world was found in Africa (Cotton's Ceratothere, or *Ceratotherium cottoni*)—a kind of Rhino—and the fabulous Okapi (*Okapia johnsoni*) turned up. The Congo Peacock had to wait till 1936! To say that there is no place where creatures, even of the size of pigmies, could still lurk unknown on this continent is outright stupidity, as evidence the arrival upon the zoological horizon this past years of the large *Ufiti*. This ought to be an object lesson to all skeptics.

This story broke in February, 1960, with a news report that sounded as wacky as any we have so far encountered. It read, in one version (*The Sunday Mail*, Zomba, Nyasaland, February 14, 1960):

Nyasaland game rangers, investigating reports of a "black, shaggy monster" seen in the forest region of Nkata Bay, Lake Nyasa, have discovered more than 30 mysterious tree-top structures in the area. In an official report, the Chief Ranger, Mr. Oliver Cary, says they are believed to be lairs [sic] built by these strange creatures. Known locally as "Ufiti"—the ghost of the supernatural—various reports have described the animal as black with long hair, a colorless posterior, no tail, broad-chested, and about 5 ft. tall. One was photographed recently by a Public Works Department employee, Mr. D. McLagen, in the vicinity of Limpasa Bridge which crosses a stream near Nkata Bay on Lake Nyasa in the Northern Province of the territory.

This is a very good example of the sort of report with which we have to deal as a normal course of events in ABSMery. It is nonetheless shot full of blather; so let us just look at it critically before we come to the windup of this story.

First, as usual, anything not previously and definitely known of the animal kind is invariably and immediately called a "monster." This is totally irresponsible and especially in a case like this when, as it turned out, the object involved was something that every foreigner should have seen many times in any zoo. Next, the designation of certain objects as "mysterious tree-top structures" is deliberately misleading. Why didn't the reporter state what sort of structures they were, and why they were mysterious? Birdhouses made of plywood, or Amerindian-type wigwams in Nyasaland *would* be mysterious, but these crude nestlike platforms of twisted branches and twigs were not—they were typical. Then, if the creatures were known locally by a perfectly good name, there was nothing really mysterious about the platforms. To call their makers' "ghosts" is going a bit far. Since when have ghosts been reported making treetop nests? But, when we contemplate the phrase *"the ghost of the supernatural"* one—at least one whose native tongue is the English language—stands aghast. Aren't ghosts supernatural anyway? Or am I mistaken? And what the heck is *the* ghost of the supernatural? What is more, the writer immediately goes on to call the thing an "animal" with a "colorless posterior"! What is "colorless"? Was the damned thing's fundament black, white, or yellow, like peoples' or was it just a great "nothing." "Words," as a famous British parliamentarian once said, "should convey meaning." Why this obvious axiom should not apply also to official reports and news-stories I cannot for the life of me see. Finally, we are informed that one was photographed; but no reproduction is attached to the story. Was the picture "classified"; was it so bad you couldn't see what it was; or was it so bloody obvious that nobody dared show it for fear of being called an ass. Or, alternatively, was it so clear but "out of context" that nobody wanted to admit it?

The whole story, as it was subsequently unfolded, is a classic; and it may well serve as an example of the functioning of the modern world in face of anything unexpected and frankly unwanted. It had the usual red herrings; some ridiculous, others most extremely interesting. The first were bandied about infinitely; the latter have been totally ignored. Then everybody, at first, said that it was a "native myth"; next, they got it as an animal; then they affirmed that it had run away from a circus [what circus in a patch of forest not previously penetrated in "darkest" Africa?]; and then the real "wipe" began. The "experts," having been confounded by the production of photos and the insistence of "authority" that the thing existed—they having said that it couldn't—could not "explain it away." This time, however, photographs seem to show clearly that the creature seen was a representative of a race of chimpanzees indigenous to this large patch of closed-canopy forest isolated from the nearest of their race, or any such forest by no less than 700 miles. [This conclusion has, nonetheless, been nicely covered by a firm order that, in no circumstances, is one of these creatures to be caught or killed for proper examination.]

This is all very well but it has a number of singularly unpleasant aspects. First, the natives thereabouts seem to have known the thing quite well and to have had a name for it; yet, the nearest whites and even the game people treated the thing as a "story." True, this creature, like ABSMs in many places, was only brought to light when the first road was pushed into this forest—shades of Jerry Crew—but then everybody indulged the most ridiculous folderol about "ghosts of the supernatural."* Then, the alleged photos have not been published. They arrived in Salisbury, capital of the Federation, on February 6 and were said next day to have "puzzled anthropologists and zoologists." We then get "A spokesman for the Victoria Memorial Museum in Salisbury" saying that the pictures were not sufficiently clear for positive identification. He pointed out, however, that "the picture [singular, this time] and description tallied with a Bushman painting found in the Ruwa region that had been thought to be a 'bear.'" [This is a near classic in that no bears now live in nor have any fossils of any one of them ever been found anywhere in Africa south of Morocco and Algeria.] Be that as it may, we then read on—and I quote from *The Rhodesia Herald,* of February 7, 1960:

An eminent Rhodesian zoologist, Mr. R. H. N. Smithers, of the National Museum, was able, even from the poor pictures available, to point out several unusual features. He said:

"From the statements I have heard from Nyasaland, and from the pictures, the animal would at first appear to be a chimpanzee. There are,

* I am wondering if by this expression the writer meant that a ghost of something invisible and probably nonexistent might, by inference, be presumed to be something visible and substantial. I cannot quite conceive of a ghost of a ghost.

[197]

however, two facts that do not support this contention. The animal has a distinctly muzzle-like face, while the chimpanzee has a flat 'pushed in' type of face. Secondly, the animal is, as far as I can recollect, more than a thousand miles from where it should be if it was a chimpanzee. The beast is obviously not a baboon, even though it has a baboon-like face, as baboons have tails and are not black in colour. In addition, a baboon does not have this animal's posture and bearing. Then there is the enormous size of the animal, which does not agree with either the chimpanzee or the baboon." Mr. Smithers said it was most unlikely that the animal was of a new species, and added that, if better photographs could be obtained as well as plaster casts of the feet, it would probably be possible to identify it.

Just what the worthy zoologist meant by "a new species" I cannot determine.

This is all very splendid but then history began to change. I have a set of press releases on the subject, issued with the compliments of the Nyasaland Information Department, P.O. Box 22, Zomba, Nyasaland, and numbered 28/60, 38/60, 51/60, 69/60, 73/60, 81/60, 93/60, and 106/60. These constitute ten legal-sized mimeo sheets of most fascinating reading. I wish only that I could reproduce them for you in full as they constitute a most exemplary public relations procedure and a most typical example of what a press officer has to contend with when dealing with "experts." Here is the whole story, told officially, starting with a report from two pragmatic Public Works Department officials on December 16, 1959, of "an unknown animal seen on a road" to a final pronouncement by two "game experts" from the Rhodes-Livingstone Museum on the following March 17, 1960. This last I herewith reproduce.

PRESS RELEASE NO. 106/60.
(Issued Wire & Telephone services)

UFITI STILL ELUSIVE

Zomba, Thursday.

Nyasaland's rain-forest monster, Ufiti, has been identified as a new sub-species of chimpanzee by two game experts from the Rhodes-Livingstone Museum.

Mr. B. L. Mitchell and Mr. C. Holliday, who are keeping the creature under almost daily observation, have not yet been able to obtain any photographs.* Ufiti remains as elusive as ever, vanishing as soon as she is approached, and thick bush and poor light add to the difficulties of getting clear pictures.

Ufiti, who is believed to be in season, has returned to her favourite observation point at the Limpasa Bridge after an absence of about a fortnight. The Chief Conservator of Forests, Mr. R. G. M. Willan, who is touring the area, was among several people who saw the creature when it reappeared near the road on Tuesday.

The two game experts, who are collecting photographs* and other forms of visible evidence, hope to arrange a bigger expedition to explore the whole rain-forest area.

It is unlikely, however, that any scientific expedition will be allowed to capture Ufiti for closer examination until it can be established that more of the creatures exist in the rain-forest.

Issued by the Press Section,
Nyasaland Information Department,
P.O. Box 22,
Zomba, Nyasaland   March 17, 1960

This would at first sight all appear to be more than satisfactory. For once, it would seem, the mystery has been explained, the "monster" identified, and zoological knowledge enhanced. But unfortunately and quite apart from the fact that nothing further has been done about anything, a number of most pertinent questions have either been left hanging or neatly buried. Let me dredge up some of these from the official releases first. For instance, in Release No. 93/60, we read the curious statement that "Although reports indicate that *Ufiti* is likely to prove a giant subspecies of Chimpanzee, her *pug* [sic] marks are said to be more human than animal. *She* is unusually large for a chimpanzee and her mouth is much smaller." Then, in Release No. 81/60 we find "*He* appears to be almost 6 feet tall with short legs and powerful arms, and most observers estimate his weight in the region of 150 lbs." In the same issue it goes on to say "Plaster casts of *its* hind footprints reveal *three* [italics mine] toes and a

* Please note! (*Author*)

[199]

large thumb." I had better cut in here to point out a few items.

Either footprints were obtained or they were not; if they were, they were either more human, or more animal; but no human has only three toes and a large "thumb" on its foot, while gorillas and chimps have four toes and a widely separated and enormous big toe. Then, no chimp ever stood 6 feet tall, or even 5 feet; chimps of those dimensions being unable to stand on their puny hind legs alone. What is more, if this is a chimp, and of that size, it would weigh more in the neighborhood of 300 pounds, by the very construction of the beast. These are official conundrums. Others come from nonofficial sources. The first is in the form of a letter to the *Rhodesian Herald,* of February 24, 1960, from a Mrs. Ida P. Wood. This reads:

Sir,—Further to your article on the unknown animal photographed by "Lofty" McLaren, perhaps the following would be of use to you and to the authorities, who seemed doubtful of the identity of the beast. During an explanation to my houseboy on the picture of a tiger on a certain breakfast cereal packet, I told him that this animal did not live in Africa, and the animal he calls tiger was not in fact a tiger at all and that the one on the box was very strong. The word strong seemed to strike a bell because, cutting the story as short as possible, he asked me then, did I know the "Strong Man." After much hand-waving indicating height and breadth, and after being told that it was like a baboon only much, much bigger, I came to the conclusion he meant a gorilla.

I said, yes, I did know it, but had not seen one out here. He, it seems, has seen them in Nyasaland. He went on to describe them, said there were two kinds, a grey and a black one, the black being slightly smaller than the grey, about his own height to be exact, 5 ft. 8 in.–5 ft. 9 in. The boy comes from Nyasaland and says he saw the first black one in the forests near a village by the name of Nazombea in 1952. The other he saw in P. E. A. in 1953 by the village of Kurriwe.

Both these names are the Chinyanja pronunciation and the animals in Chinyanja are called Fireti. I questioned him closely about the possibility of the black one being a different kind but "no," he said, "they are the same, only one black, one grey, and only ever one at a time."

[200]

It makes a bed as a gorilla* does, large and untidy, usually sleeps in it only once, I suppose to eliminate the possibility of discovery.

I should be interested to know what the anthropologists interested in the previous article think of this information if you would be good enough to pass it on. Have the photographs been printed yet?

(signed)   Mrs. Ida P. Wood
Sinoia

This clearly indicates that these creatures *are* known in the area, and I have no doubt that a little ingenious and patient inquiry among the "benighted local natives" would disclose the fact that they have always been very widely known. I should explain that this business of "there is a black one and a white one" is almost universal in Africa and usually denotes marked sexual dimorphism, which is displayed by so many animals. [Incidentally in many parts of Africa only three colors are recognized—black, white, and red. The last is all the earth colors from deep orange, through all the browns to deep red. Everything, including blues and greens are either white or black according to whether they are in strong light or in shade. All shades of color are "so-so" red, black, or white.] This African's insistence that, although there are two kinds, they are the same beast, would indicate that the differences are either sexual or due to age.

The other concurrent oddity was from quite another part of Africa, 1400 miles distant, and in an area from which we had not previously had any reports. This is actually a very astonishing report and one that should be taken most seriously in view of the almost constant surprises that are coming out of Southwest Africa and Angola. This part of Africa is rapidly assuming the guise of truly "The Darkest," for big game never even known to exist there is turning up, and among it are many record specimens, while it is the home of the otherwise nonexistent, Giant Sable Antelope, and so forth. This apparent ABSM was originally reported in the *Evening Standard* of Salisbury for November 18, 1959 but did not become

* How did Mrs. Wood (or her houseboy) know of this?—*Author.*

[201]

fully recognized until after the *Ufiti* or *Fireti* affair broke. It reads as follows:

Windhoek, Wed.—The authorities in South-West Africa and farmers in the vicinity of Outjo are wondering whether a large shambling ape or monkey which has been seen on farms near Outjo is not a gorilla. People who have seen the animal state emphatically that it is not a large baboon. According to their descriptions, the animal closely resembles a gorilla. Its footprints are also like those of a gorilla. A farmer, Mr. Thuys Maritz, who's Ovambo herdboy reported that the animal had stolen his blankets and food, tried to track the animal down but lost the trail over rocky ground. The spoor clearly showed that the animal walks on two legs. Occasionally, prints resembling knuckle impressions were found next to the spoor. The footprints are about 5½ in. wide and resemble marks made by a human hand. The five fingers or toes are clearly defined. The authorities have appealed to farmers not to shoot the animal but to try to capture it alive. The nearest place where gorillas are known to live is in the Belgian Congo, nearly 1000 miles from South-West Africa.

Disregarding this report, and reverting to the *Ufiti* for a moment, it should be pointed out that there is something very wrong with the whole thing. I cannot bring myself to believe that game wardens, forest officers, and such other solid citizens could all have been absolutely ignorant of chimpanzees as specific animals. Had none of them ever been to a zoo where one was housed, or seen so much as a picture of one in a book? Even a fleeting sight of such an animal ought to have been enough for them to recognize it—if it *was* a chimpanzee. Zoologists and anthropologists ought, almost to the same degree, to have been able to spot such an animal from any photograph that displayed anything even approaching an outline. That any could be in doubt about the identity of a picture which was clear enough to ascertain that the animal depicted was *not* a baboon, is frankly amazing. There is nothing impossible in a sub-species of chimpanzee turning up in this forest and having been there all along though in several respects it might be considered unlikely, but there is absolutely no doubt about the footprint of an ape. It is utterly different from that of any Hominid. There can be no doubts here. The matter of the *Ufiti* is a most damning indictment of "the experts" for, from the published record on file, they would obviously then be shown not to know the first thing about their claimed specialties. I now have copies of the photographs mentioned together with some other most clear close-ups taken later. All, and even the foggiest, clearly shows a robust, and typical chimpanzee in very fine coat, either peering intently down from a tree in typical Pongid fashion, or standing stolidly on all fours in the preferred Great Ape stance. [No walking about on *her* back legs, mark you.] The photographs of two footprints, part of a track in soft earth, are at first rather startling as they look almost human but have only four toes. However, it is the photo not the prints that is startling for, viewed from other angles the "missing toe," namely the great one that is very widely—and properly for a chimp—separated is quite plain. This is a tale of woe but most important to our search, because it goes far to show just what appalling mistakes can be made, misconceptions built up, and fantasies conceived in a matter such as this.

When we come to the last great area for alleged "unknowns" in Africa we do not, thank goodness, have to deal with experts. However, we have to rely on travelers, big-game hunters, and other nonexperts who are sometimes almost as bad. [Oh, for the good old days of bulldozer-operators, and timber-cruisers!] However, there is one very bright gleam ahead and this—and almost for the first and only time—is a real, honest-to-goodness, fully trained, truly expert, and also successful professional "animal collector"; none other than Charles Cordier, the Swiss, who has persistently brought back to museums and zoos what they really want; properly housed and fed, or properly preserved. Here at last is a man whom we can not only rely on for common sense reporting, but who really knows his animals and his zoology, as well as a great part of the world. You may place more reliance on what he says than upon almost all of the rest of the involuntary and even the voluntary ABSM hunters combined.

This information comes to me once again from Bernard

Heuvelmans who had just [at the time of writing] received it from Cordier who (January, 1961) was somewhere in the Congo. Charles Cordier wrote Bernard: "We met three tracks of hind feet—no knuckle marks—in soft mud near water. The tracks were most unusual," he says. Also, they were not those of a gorilla; and, Cordier goes on: "I ought to know, I have a silverback. These 12″ tracks were no gorillas." [The gorilla imprint should be compared with the photograph of one made in plaster in Appendix B.] This find was made, as far as we know of now, somewhere in the Bakavu area. The track-maker is obviously some new form of large Pongid.

This brings up a whole string of stories from less reliable sources. These begin with something that has been named the Tano Giant, and was first described by one, Louis Bowler, half a century ago. It has some funny features, and some illogicalities. It states:

Far away in the primeval forests of the Upper Tano, in the Gold Coast Colony, a strange tale is told by the natives of a wild man of the woods, which would appear from the description given to be a white ape of extraordinary stature and human instinct. The natives who live in the village near to the haunts of this freak of nature are terrified out of their wits. They barricade their doors at night, and place broiled plantains and cassava on the jungle paths leading into the village to propitiate him and appease his hunger. They declare he comes to the village at night, and only runs when fire is thrown at him. The women especially are almost scared to death, and go in a body to their plantain farms. It appears that two women while gathering plantains were confronted by this creature. One he seized and flung over his shoulder carrying her off; the other ran screaming with fright back to the village. No trace of the other woman has been found. Several children have been taken by this creature, their mutilated bodies being found with the whole of their bowels devoured.

The hunter and women who have seen this animal describe him as "past all man" in size; his arms they describe as thick as a man's body; his skin "all the same as a white man," with black hairs growing thereon. The hands have four fingers but no thumb, the head is flat, and, as they describe it, "left small for big monkey head," meaning that it was very near or like a large monkey's head. They say the mouth "was all the same as monkey with big teeth sticking out, and he carries a skin of a bush cow," which the natives say "he carries for cloth when small cold, catch him," meaning he wraps himself up in it when feeling cold. A hunter tried to shoot him, but he smashed the gun and broke both the hunter's arms. Many other incidents are related of this terror of the Upper Plains.

The most outstanding aspect of this report is, to me, that once again it is of something definitely Hominid and that came out of a montane forest onto orchard-bush, as in the Southwestern case from Windhoek. This is indeed unusual. The other outstanding fact is the mention of the absence of a thumb. I understand that it is believed that the thumb of *Plesianthropus* was exceptionally small for the size of its hand, and was placed very high up on that hand. Is it possible that it might have been carried pressed against the side of the palm and so not be apparent? The fact that this creature was alleged to have a light skin covered with black hairs is also novel. The whole account is actually more than just aggravating in several respects because it stands absolutely alone as far as I have been able to find out. Naturally, one presumes that it is but a traveler's tale picked from native imagination to give it a tone of authenticity; yet, among such tales—and I have hundreds—it is one of the few that seems in some way to have validity.

Perhaps this is because I got to know the West Africans rather well myself once; and, while I fully appreciate their great storytelling abilities, I did find them essentially most down-to-earth people when it comes to the question of their native fauna. West Africans told us some of the wildest-sounding things about their local animals but, in almost every case, they made good on their words by producing the darned things. They are not the sort to think up "thumblessness," a white skin, or a head "left small for big monkey head." If they said that—and these purport to be firsthand accounts, not traditional tales—they meant it, and precisely. The disemboweling of the children also seems to smack of the real thing. I know just what the teller meant to imply: namely, "Don't try and tell me this was a leopard because I know, even if you don't, that that is not the way they start to eat you."

The only other African ABSM that has been mentioned, and this several times, and by several different travelers, is the *Muhalu*. This is a muddled issue as may be seen from the following extract from the book *Hunting We Will Go* by Mrs. Attilio Gatti. This reads as follows:

Then there are rumors about strange anthropoids. One is a large ape which is said to live in the Rainy Forest, the pygmy tribes call it the *Muhalu*. Commander Attilio Gatti, the well-known African explorer, has repeatedly declared that he, for one, believes in the existence of the *Muhalu* and willingly accepts the descriptions of the pygmies who say that it is exceptionally large, walks erect habitually, and is covered with very dark, possibly black, fur, except for the face, where the hairs are white.

Another again, and the worst of all, is a big animal with a coat of long hair, black on the back, white on the other parts of the body. And it is enough to be seen by this monster, for one to die in the most atrocious agony.

We found awaiting us a man from Soli's to say that the pygmies had been on the trail of a Bongo mother and young one, and that if the Bwana would come they were sure they would capture the little one.

So Tille decided to have one more fling. He also decided to take a group of our own boys with him to act as porters. Before they could start, however, an event occurred which reduced all Kalume's men to panic.

Ever since we had been in the Ituri we had heard repeated tales and rumors of a great animal called by the Bondande, "muhalu." Of all things that could arouse terror, this muhalu was the King Bee. Tille had been extremely interested in the matter and believed that the creature really did exist and was a hitherto unknown fifth anthropoid or subhuman.

At this time, however, he had done no more than talk about it now and then. Now, on this morning, one of our men rushed into the clearing, his face gray with fright, babbling about the dread muhalu. His stories were conflicting. First he said it had knocked him down, and this seemed odd because the natives firmly believed that a muhalu had only to look at a man and that man would instantly die. Then the boy said he had seen the muhalu first and ran away. No matter what had actually happened, the news that a muhalu was in the vicinity nearly paralyzed our men.

Tille insisted on going to investigate at the point where the boy claimed to have seen the beast. I don't know how he succeeded in dragging that boy, half-dead with fright, or in flicking the pride of Lamese and two of the other men until they agreed to accompany him.

He did find enormous footprints, and several stiff black hairs in the hollow of a tree where the evidence showed the brute had been sitting. Neither hairs nor print corresponded to any other known ape.

But the panic of our natives had grown so fast that Tille could not stem it. Even Kalume begged us, with all his heart, to leave Tzambehe and come down to his village. All of our natives, though they had no wish to abandon us, were preparing to leave.

In this area, namely the southern face of the Ubangi-Shari Massif, it would seem that we have to deal with two quite separate entities—one an unknown Pongid, and another a Hominid, or ABSM. Despite the rather obvious exaggerations of the descriptions given by locals—and notably by the Pigmies, with whom it is extremely difficult, if really at all possible, to communicate—neither appear really to be too outrageous. Perhaps one is the terrestrial ape that leaves the odd prints now recorded by Cordier, while the other is something akin to the Tano Giant. There are a set of tracks recorded from Bakumu which the locals say were made by what they call the *Apamandi,* which they there describe as a very heavily built small man, clothed in black hair, but having a light skin. These prints are approximately eight inches long, very short and broad, and have the strange distinction of having the second toe longer than either the first or third, and being somewhat separated from the first or big toe. The significance of this toe proportion will become apparent when we come to investigate the *Meh-Teh,* or Snowman of the Himalayas.

The accounts of these two [or is it but a single] creatures are very vague, fragmentary, and rare. Yet, if you visit the northwestern edge of the Ituri Forest you will find that it (or they) are taken quite for granted as being rare, but by no means excessively rare, units of the local fauna; living in the upper montane forests to the north, and from time to time coming down on to the lowlands. I have talked to many people who have been into this, previously unadministered, area but only those who were specifically interested in its fauna, or who spent time investigating the ideas and knowledge of the locals, had ever heard of it. Those who did so, however, all seem to be of the opinion that there is a race of gorilla in the area, or that there is at least some large terrestrial ape there. When I asked if, in their opinion, it could be a primitive Hominid rather than an advanced ape, the opinions have been violently divided. Most returned my query with a perfectly blank stare; but some said "Yes" and invariably went on to talk about the possibility of some larger form of Australopithecine having survived thereabouts—and they usually pick on *Plesianthropus*, probably because that form has been so well publicized, along with reconstructions of it.

Africa is undoubtedly the land of Pigmies and of some Great Apes, but it does not seem to sport any giant Hominids. At least the Africans don't imply this, even if they do refer to the Tano character and the *Muhalu* (or one of them) as being very big. Our real *Oh-Mah* types would be the perfect target for African bogeyman stories, but they just don't appear here, and we shall not meet them again until we reach just the place where they ought to be.

10. The East—the "Mysterious"

The "East" has always puzzled everybody in the "West." We talk about the Orient, but what really is it? Much more important; what's in it for us?

We are now going to make a major hop across an ocean, from East Africa to what is commonly called the Orient, and specifically to southeast Asia. This may look like, and in point of fact is, a long hop spatially, and it may seem doubly exaggerated because we are going also to skip over all that lies between the two points specified, such as Arabia, India, and Ceylon, though they manifestly form sort of steppingstones along this route. This is nevertheless justified on more than one count.

First, there is no current ABSMery to be discussed in those intermediate areas, though there is quite a lot of myth, legends, and folklore, especially in Ceylon. Second, geologists tell us that there was once a great land-connection between the two extremes (Africa and southeast Asia), which they have named Gondwanaland, and it is obvious that lots of primitive animals still living today are represented by different but either comparable or obviously related kinds on the two sides of the Indian Ocean. Whether individual examples of these emigrated from one side to the other, or vice versa, is no concern of ours, but it is certain that there was from very early times such a connection between the two sides of this ocean. A good example is the Lorisoid Lemurs of Africa, and of the Orient *; another is the flightless birds called Ratites, including the Ostriches (*Struthio*), on the one hand,

* The Pottos (*Periodicticus*) and Bushbabies (*Galago*) of Africa and the Lorises (*Loris*) of the Orient.

MAP VIII. MALAYA AND SUMATRA

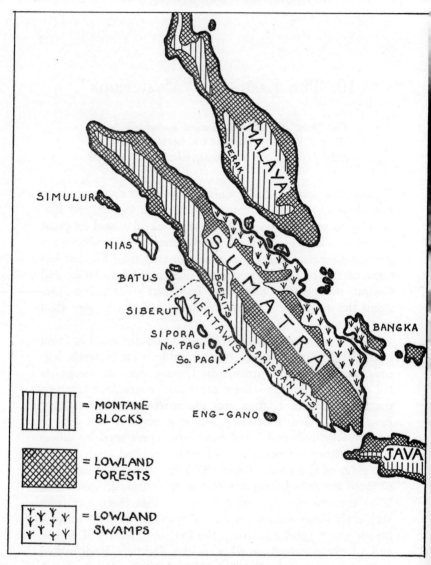

See facing page for legend

the Emu (*Dromiceius*) and the Cassowaries (*Casuarius*) on the other. Then again, the Great Apes are found on both sides, as are different forms of the very specialized Leaf-Monkeys or Coloboids—the Guerezas in Africa; the Langurs in the Orient. These each represent different ages at which this land connection existed.

Primitive men and the Hominids generally, seem also to straddle this ocean. Whether the land-connection still remained above sea level when the most primitive of the latter were evolved—such as the Australopithecines of South Africa, and the Pithecanthropines of Indonesia—is not yet known, but it is almost certain that it did not do so when the first races of True Man were spread all over both sides (or, alternatively, passed from one to the other). These most primitive peoples are today the Pigmies of which there are representatives in forest Africa, on the Indian Ocean, in the Massif on the Malay Peninsula, and in the Philippines. [It should be noted that the pigmy people of the west end of New Guinea are now thought to be merely "pigmy" breeds of the otherwise tall Papuans of that island.] These little people have much in common on both sides of the Indian Ocean, and they are now thought to constitute a real sub-species of the human race.

These Pigmies are indeed primitive, but even they say that they were not the first people in the countries they now inhabit, and the Semang of Malaya state that there remain some living representatives of these still earlier people in

---

MAP VIII. MALAYA AND SUMATRA

This small area is one of the most peculiar in the world. In it there are living a large number of animals not found anywhere else, while the only relatives of these are found far away. Most odd, and still least known of all, are the Barissan Mountains of south Sumatra, in and around which ABSMs, particularly in the form of the pigmy *Sedapa*, have been for centuries alleged to exist. These and other unknown primates are reported from the east Sumatran lowland forest and the swamp belt, and from the inner montane block of the Malay Peninsula. The Mentawi Islands have unique apes and monkeys.

[211]

their country. Malays call these "Devil Sakai," * (Hantu Sakai) and say that they live in and move about through the trees; an astonishing statement since the Senoi also readily take to the trees, and are highly agile therein. There is evidence that these proto-Pigmies [which simply means, Those-who-were-before-the-Pigmies] once were spread very widely in East Africa, southern Arabia, India, Ceylon especially, and throughout Malaya and Indonesia. We will find allusions to them cropping up all the way through our story for some time from now on and we must watch out for them because in this area (i.e. eastern Orientalia; namely, the whole of that subcontinent apart from India and Ceylon) there is really no clear line of demarcation between fossil sub-hominids that *are* known, really primitive Men, and what we are calling in this book ABSMs.

This is a point that I would like to stress forthwith. On account of that awful expression "the abominable snowman" and all the fuss that has been made over "it" in the Himalayas, not only the popular concept of such creatures, but our whole thinking from a purely scientific point of view also is colored by a picture of some mythical exaggeration pounding about on a snowfield, ripping apart yaks or hapless Sherpas. Actually, if one comes to examine the matter more closely, and in its entirety, as we are trying to do in this book, it should be apparent that what we are dealing with is really the whole history, past and present, of the Hominids, and the origins of Man per se. Frankly, our term "ABSM" really means *hominid, other than known kinds of modern man;* no more and no less; and it is my firm belief that in due course, the whole business will be lifted clean out of the "mystery class" and simply become a part of physical anthropology. Even if no example of any of the (as it now seems) dozen or so ABSMs is ever caught, I further think it will be found that all which has been reported upon them throughout the world may legitimately be taken into consideration in trying to reconstruct the past history of man, and fill in some of the vast gap in that

* The term *Sakai* means degenerate and is not the real name of a people though applied to the Vedda-like *Senoi* of Malaya.

history that at present lies between little *Oreopithecus* of the Miocene coal strata of Italy and, say, the Bushmen or the Pigmies. Moreover, it is in this Oriental Region that we are going to come closest to the chain of stages that linked, and that still links, those two extremes.

Our first port of call in this new region is perhaps one of the oddest, oldest, and from a zoologist's point of view, the most exciting in the world. This is the southwest portion of the great island of Sumatra and a string of islands off its west coast called the Mentawis. The whole of Sumatra is odd in several respects and not entirely due to its enormous size, dense forests, comparatively small human population, and virtual neglect throughout history. It, with the foot of the Malay Peninsula, Java, Borneo, and some associated smaller islands [and possibly Palawan, which is usually grouped with the Philippines] forms a zoogeographical sub-area with most special aspects (see Map X). Not only does this sub-area contain elephants, rhinoceroses, tigers, and other mainland Asiatic animals, it has some even odder and more ancient animals—the Malayan Tapir, the Orang-utan (or *Mia*), the Siamangs, the Tarsiers, and the little, most primitive of all living Primates, the Pen- or Feather-tails (*Ptilocercus*). Actually, the list even of mammals is extraordinary, and there are here unique birds, reptiles, amphibians, fishes, and invertebrates of all kinds. It is a sort of leftover land to which all manner of creatures have at times of climatic change, crustal shift, or oceanic flooding, retreated. But, within this limited area, there is an even more peculiar sub-sub-area. This is the Mentawi Island chain and the immediately opposite Barisan Mountains of southern Sumatra.

Here there are absolutely unique and really very strange animals. To exemplify, I need mention only what is called the Mentawi Islands Langur, and the South Pagi Island Pigmy Siamang. The first is not really a langur monkey at all but a short-tailed Snub-nosed Monkey (named *Simias concolor*) that constitutes a genus all by itself and which is completely unlike anything known anywhere else. The Pigmy Siamang (*Brachytanites klossi*) is a diminutive ape, classed with the Gibbons and standing somewhere between them and the much bigger and more "advanced" Siamang (*Symphalangus*) of mainland Sumatra and Malaya. It seems in fact that this bottom bit of Sumatra is a retreat within a retreat, and the animals which retreated thereto are really relics. You will notice from the map that the Barisan Mountains, though continuous with the Boekits and the rest of those of west Sumatra right up into Achin, are coastal. Also, they culminate in the northwest in Mount Marapi, north of Padang, beyond which there is a distinct break. The flora and fauna of the Barisans has more in common with the Mentawais than with the mountains of northern Sumatra. [Eng-gango Island is even more odd.] This sort of fossil attic is the headquarters of a group of Oriental ABSMs and notably one that is called locally the *Sedapa* or, in kitchen-Malay, the *Orang Pendek* (Little Man) or *Orang Letjo* (the Gibbering Man).

Here, we come to a pretty problem. There is spread all over what is called by zoologists the Malaysian Subregion— i.e. that described above as encompassing the foot of the Malay Peninsula, Sumatra, Borneo, and Java—a wealth of folklore concerning not just this *Sedapa*, but also a man-sized ABSM, and, in Sumatra, a giant type called very simply and logically the *Orang Gadang*, or Great Man. This folklore is very specific. In Java, it is buried, and deeply so, in pre-Hindu mythology; that island having been so highly civilized and so thickly populated for so many millennia that, although there still remain in it some really wild areas and even relic animals such as a special rhinoceros, any primitive hominid that may have lingered there since the time of *Pithecanthropus* and *Meganthropus* (see Chapter 16) was long since exterminated. Borneo, on the other hand, has remained very wild and forms a special case. It too has its zoological oddities (like the Proboscis Monkey) but not apparently even any folk memories of ABSMs—though a very strange story of one such having arrived there not too long ago on a boat as a captive of pirates was published! Sumatra and Malaya proper, on the other hand, are rife with not just hints but most definite reports of at least three kinds of primitive hominids or ABSMs.

[214]

The Philippines constitute another zoological sub-area; and the Celebes and their associated islands, still another. Both have unique animals, and the latter, though lying on the Australian side of Wallace's Line, the great divide between that continent and Asia, has a mixture of marsupial mammals and other typically Austral fauna and forms with obvious Asiatic affiliations. Among these are the small black baboon, known as the Black Ape (*Cynopithecus*), and two species of a Macaque Monkey (*Maurus*). Of ABSMs there are none reported from either of these sub-areas, but there are genuine Negrito Pigmies in the Philippines, and there are constant references to "men with tails" from there and especially from the Island of Palawan. The whole question of tailed hominids is a sorry subject and has been going on throughout the ages. Many peoples have attributed tails to their neighbors or more distant foreigners with the sole implication that they were a lowly lot of rascals. Others mistook crude accounts and pictures of monkeys for lowly forms of humans in other lands. Finally, people are sometimes born with fairly decent tails. [There was a very nice fellow at school with me who had a 3-inch job clothed in reddish-brown, fine hair about an inch long.] This is said to be an "atavism." This is hardly the right word for it, as it would then be a throwback to the time before either apes or men got started. [I show a photograph of a Malayan-Filipino gentleman so equipped—see Fig. 54.]

Let us, then, return to Sumatra and investigate the matter of the *Sedapa*. The existence of wild men in this island has been rumored since ancient times. It was mentioned by Marco Polo [though he also had tails on the brutes, and naked ones at that]. Its existence was first definitely reported by an Englishman named William Marsden who was resident at Benkoelen on the west coast of Sumatra in 1818, but it was not till this century that definite reports were made by Westerners. As everywhere else, both the veracity of the reporters and the possibility of the existence of any such creatures was heatedly denied by just about everybody who did not reside in Sumatra, and particularly by those who had not even been there. This attitude to the matter was taken to

great extremes by the Dutch curator of the museum at Buitenzorg in Java, Dr. K. W. Dammerman. Most, but not all
scientists followed his example until World War II. Then,
when Indonesia gained her independence, there was at first
a very noticeable change in opinion, especially as displayed
in the Indonesian press. However, the general attitude has
reverted to type more recently, so that the present professor
of anthropology at the university at Djakarta wrote to my
friend Prof. Corrado Gini of the Institut International de
Sociologie in Italy, stating flatly that the "*Orang Pendek* is
only a variety of the Orang Kubu, a primitive people, quite
human in character, of whom the Indonesian Government
takes special care."

While I am glad to hear of the Indonesian Government's
special concern for the Kubu, something that must be somewhat difficult to exercise in the political circumstances, I
would point out that while Sumatra is Indonesian territory,
the Indonesian Government is actually Javanese and really
knows extremely little about Sumatra—rather less, in fact, than
the Hollanders once did. Also, I am not interested in the
Kubu people who have been well known for centuries but
rather in the Orang *Gugu*. The Kubu are not hairy; the *Gugu*
are said to be, whether they exist or not. As Marsden first
clearly pointed out, the Kubu are hairless humans at a primitive stage of culture but great hunters, and live in the Barisans. The *Gugu* are not human, were even then very much
rarer, and lived in the depths of the montane forest, and had
no language. The Malayan peoples of Sumatra called them
by various native names such as *Atu*, *Sedabo*, or *Sedapa*.
They often appended their word *pendek* or *pendak* to these
to indicate that they were refering to a small one, of two—
the other being *gadang*, which simply means large.

On the validity of the *Sedapa* I cannot offer anything but
the accounts as published. That such a creature could exist
is not only quite possible but, I think, almost probable—and
especially if the local native and indigenous peoples say that
it does—and the Barisan Mountains area is just the place

where ancient forms of Hominids might most likely have been able to survive. As we shall see, there is no dearth of candidates for the *Sedapa* along the Hominid branch of the family tree; and then, we have the near presence of the Pithecanthropines of Java. Also, the existence of the Malayan Tapir (*Tapirus indicus*), whose sole remaining relatives live in tropical America, shows just how safe a retreat this corner of the world really is. When it comes to "available space" for any such creatures to live more or less unseen, words almost fail me. I spent many happy months wandering about Sumatra in my youth accompanied by an Achinese (with the very sensible name of Achi, as it happened) and all I can say is that its forests put most others in the world to shame, and they seem just to go on and on forever. The known population is comparatively minute, and the amount of the country that is opened up is quite minor. Apart from the rivers, the great swamplands are not penetrated at all; the lowland forests are tall and dense, and the montane growth is intolerable.

The history of the *Sedapa*, as far as the Western world is concerned, is due mostly to the researches of Drs. W. C. Osman Hill of the Zoological Society of London, and, once again, Bernard Heuvelmans of Paris. There were certain Hollanders who somewhat earlier devoted themselves to the pursuit of this matter in Sumatra. Notable among these was a Dr. Edward Jacobson, who first brought the subject up in *De Tropische Natuur* [once published in Weltevreden, Java] in an issue of 1917. However, Dr. Jacobson's investigations went back to 1910 and it was under his aegis that some facts collected by Mr. L. C. Westenek, once Governor of Sumatra, came to light. The earliest of these is the report of an overseer of an estate, who was staking out a newly acquired and large tract of virgin land in the Barisans near a place called Loobuk Salasik. This man left a carefully worded written statement. This was that, at a distance of only 15 yards, he saw "a large creature, low on its feet, which ran like a man, and was about to cross my path; it was very hairy and it was not an orang-utan; but its face was not like an ordinary man's. It silently and gravely gave the men a disagreeable stare and then ran calmly away. The workers ran faster in the other direction." The overseer remained where he stood, quite dumfounded.

The significance of this statement centers around the definite statement that the creature was *not* an orang-utan, that it stood on its hind legs and ran on the ground, and that it was "low on its feet." This latter seems to indicate that it had short legs, which is really another way of saying that it had overly long arms in proportion to its torso and legs; and all this, in turn, emphasizes that it was not an orang-utan; an animal that, except when young, cannot even walk on its hind legs alone. Dr. Jacobson became greatly interested in this matter when camping on the slopes of Mount Kaba in the Boekits in early July, 1916. Two hunters came to him there one day and said that they had seen a *Sedapa* breaking open a fallen tree at a distance of only some 20 yards from them. It was apparently looking for beetle larvae—a delicacy relished by many peoples the world over, but when it realized that it was being observed, it ran off on its hind legs. Otherwise, this description agreed in every other respect with the traditional one of the *Sedapa*. It was clothed all over in short, black hair.

I should point out here, and rather strongly, that the larger Siamang, a really big and sturdy ape, intermediate in many respects between the Gibbons and the Great Apes, though highly adapted for life in the treetops, quite often comes to the ground upon which it runs along on its hind legs, swinging its arms instead of holding them aloft as the gibbons do when *running* as opposed to just walking. Also, I have myself come across Siamangs going meticulously over fallen rotten logs collecting the insects that often crowd into their cracks. I owned a Wow-wow Gibbon (*Hylobates moloch*) during the whole year that I was in Indonesia. It had been raised in a human family and it traveled all over the Indies with me. I happened to be collecting insects on that trip, and the majority that I obtained were actually found, caught, and then handed carefully to me by this small anthropoid companion. It used to run ahead on its hind legs in the forest, holding its long chain off the ground with one hand, and upon locating a rotten log climb aboard and start probing into all the cracks with its long forefinger [he was left-handed] and producing all manner of rare specimens that I simply never could find by myself. It was uncanny, as was the manner in which he used to offer me the first and all subsequent ones of the same kind until I indicated that I had enough specimens: then he ate the rest. Gibbons may be Pongids but they certainly are "almost human" in many respects. The related Siamang is almost more so; and, in fact, the Malays often treat them as such.

Later, Dr. Jacobson was shown some tracks of the alleged *Sedapa* on Mt. Kerintji. These were definitely not those of a gibbon, siamang, or any other ape, all of which have a widely opposed and very large great toe; it was exactly human but tiny, very broad and short. Quite a number of alleged *Sedapa* footprints have been recorded. These vary rather bewilderingly. In 1958 some plaster casts of some prints were obtained about halfway between the Siak and Kampar Rivers by Harry Gilmore. These, however, are almost undoubtedly those of the small, Malayan Sun-Bear (*Helarctos*). This animal stands erect and even walks along, though it never runs, on its hind legs more frequently than any of the other bears. It is about 4 to 5 feet tall, is covered in short black hair, and has surprisingly broad shoulders. It may even swing its arms when walking. Also, it has a pale face which, when seen head-on in the poor light of the high forest floor, may give it a startlingly human look—I know, I was nearly scared out of my wits by these animals, standing silently watching me, on more than one occasion. The hind footprints left by this animal are nonetheless fairly distinctive and are not like the drawings, tracings, and casts taken of alleged *Sedapa;* like all bears, their toes increase, albeit in this case only slightly, in length from both sides to the middle toe; they are packed together, not splayed; and claw marks are almost invariably present. The Siak River, moreover, is somewhat out of the range of the *Sedapa* proper, though there is plenty of tradition about it in those parts.

In 1917, according to Westenek, a Mr. Oostingh, while in the Boekits and near the same mountain where Dr. Jacobson had been when the hunters said they saw a *Sedapa,* became "bushed." He wandered around in circles for several hours, as one invariably does if one gets lost in high forest. Suddenly, as his account goes, he came upon what he thought was a local man sitting on a log with his back toward him. Overjoyed to see any human being, as one also invariably is when so exhausted, he went forward but then got a profound shock. I let him tell about it in his own words, as taken from Westenek's account in *De Tropische Natuur,* and translated by Richard Garnett. This reads:

I saw that he had short hair, cut short, I thought; and I suddenly realised that his neck was oddly leathery and extremely filthy. "That chap's got a very dirty and wrinkled neck!" I said to myself.

His body was as large as a medium-sized native's and he had thick square shoulders, not sloping at all. The colour was not brown, but looked like black earth, a sort of dusty black, more grey than black.

He clearly noticed my presence. He did not so much as turn his head, but stood up on his feet; he seemed to be quite as tall as I (about 5 feet 9 inches).

Then I saw that it was not a man, and I started back, for I was not armed. The creature calmly took several paces, without the least haste, and then, with his ludicrously long arm, grasped a sapling, which threatened to break under its weight, and quietly sprang into a tree, swinging in great leaps alternately to right and to left.

My chief impression was and still is: "What an enormously large beast!" It was not an orang-utan; I had seen one of these large apes a short time before at Artis [the Amsterdam Zoo].

It was more like a monstrously large siamang, but a siamang has long hair, and there was no doubt that it had short hair. I did not see its face, for, indeed, it never once looked at me.

Here again, the most obvious suggestion is, just as Mr. Oostingh himself says, that the creature was an enormous Siamang, perhaps a lone old one somewhat short on hair.

That it was more likely an ape than a Hominid is also perhaps further impressed upon us by the remark that it had
"ludicrously long arm[s]." I do not know what to make of
this report but I certainly wish that the creature had left some
footprints.

Meantime, there was a Mr. Van Heerwarden timber-cruising
from the other side (the northeast) of the Barisans in Palembang province, but down in the swamp forests by the coast
near the Banjoe-Asin River. In 1918 he spotted two series of
tracks on the banks of a small creek in the Musi River district;
one larger than the other, as if of a mother and child, as he
remarks. These were perfectly human but exceedingly small.
Later he discovered that a Mr. Breikers had also found such
tracks in the same area. He then started making serious inquiries among—and this is of considerable significance in view
of the Indonesian Government's statement given above—the
Kubus; and he found three who had all, but unknown to
the others, seen *Gugus* (i.e. *Sedapas*, or *Orang Pendeks*) in
that region. Their descriptions agreed perfectly in that they
were about 5 feet tall, walked erect, were clothed in black
hair that formed a mane, and had prominent teeth. Van Heerwarden later heard that a hunter had found a dead one and
tried to carry it back to his village but its body was much
decomposed and the hunter himself died shortly afterward.
Another, he learned, was said to have been spotted in a river
and surrounded by locals in canoes but it dived adroitly and
escaped.

By this time Mr. van Heerwarden was convinced that there
really was some small hairy Hominid in these forests and he
devoted much time to inquiries among the local hunters as
to where they were most frequently seen. In time he was
directed to a particular spot and decided to do exactly the
right thing—namely, go there, sit down, shut up, and wait.
And, he appears to have been well rewarded for, unless he
is not only a complete but most adept liar, he got an extremely
good look at one of the elusive creatures. He tells us that he
was wild-pig hunting in an area of forest surrounded by rivers
named Pulu-Rimau, in October, 1923, and, having failed to

come up with the sounder (herd) decided to do this quiet sitting, and so went into hiding. For an hour or so nothing happened and then something in a tree caught his attention. He says:

Then I happened by chance to look round to the left and spotted a slight movement in a small tree that stood alone. By now it was time for me to be going home, for it was not advisable to journey through such country after sundown. But all the same I was tempted out of curiosity to go and see what had caused the movement I had noticed. What sort of animal could be in that tree? My first quick look revealed nothing. But after walking round the tree again, I discovered a dark and hairy creature on a branch, the front of its body pressed tightly against the tree. It looked as if it were trying to make itself inconspicuous and felt that it was about to be discovered.

It must be a *sedapa*. Hunters will understand the excitement that possessed me. At first I merely watched and examined the beast which still clung motionless to the tree. While I kept my gun ready to fire, I tried to attract the *sedapa's* attention, by calling to it, but it would not budge. What was I to do? I could not get help to capture the beast. And as time was running short I was obliged to tackle it myself. I tried kicking the trunk of the tree, without the least result. I laid my gun on the ground and tried to get nearer the animal. I had hardly climbed 3 or 4 feet into the tree when the body above me began to move. The creature lifted itself a little from the branch and leant over the side so that I could then see its hair, its forehead and a pair of eyes which stared at me. Its movements had at first been slow and cautious, but as soon as the *sedapa* saw me the whole situation changed. It became nervous and trembled all over its body. In order to see it better I slid down on to the ground again.

The *sedapa* was also hairy on the front of its body; the colour there was a little lighter than on the back. The very dark hair on its head fell to just below the shoulder-blades or even almost to the waist. It was fairly thick and very shaggy. The lower part of its face seemed to end in more of a point than a man's; this brown face was almost hairless, whilst its forehead seemed to be high rather than low. Its eyebrows were the same colour as its hair and were very bushy. The eyes were frankly moving; they were of the darkest colour, very lively, and like human eyes. The nose was broad with fairly large nostrils, but in no way clumsy; it reminded me a little of a Kaffir's. Its lips were quite ordinary, but the width of its mouth was strikingly wide when open. Its canines showed

[222]

clearly from time to time as its mouth twitched nervously. They seemed fairly large to me, at all events they were more developed than a man's. The incisors were regular. The colour of the teeth was yellowish white. Its chin was somewhat receding. For a moment, during a quick movement, I was able to see its right ear which was exactly like a little human ear. Its hands were slightly hairy on the back. Had it been standing, its arms would have reached to a little above its knees; they were therefore long, but its legs seemed to me rather short. I did not see its feet, but I did see some toes which were shaped in a very normal manner. This specimen was of the female sex and about 5 feet high.

There was nothing repulsive or ugly about its face, nor was it at all ape-like, although the quick nervous movements of its eyes and mouth were very like those of a monkey in distress. I began to walk in a calm and friendly way to the sedapa, as if I were soothing a frightened dog or horse; but it did not make much difference. When I raised my gun to the little female I heard a plaintive "hu-hu," which was at once answered by similar echoes in the forest nearby.

I laid down my gun and climbed into the tree again. I had almost reached the foot of the bough when the sedapa ran very fast out along the branch, which bent heavily, hung on to the end and then dropped a good 10 feet to the ground. I slid hastily back to the ground, but before I could reach my gun again, the beast was almost 30 yards away. It went on running and gave a sort of whistle. Many people may think me childish if I say that when I saw its flying hair in the sights I did not pull the trigger. I suddenly felt that I was going to commit murder. I lifted my gun to my shoulder again, but once more my courage failed me. As far as I could see, its feet were broad and short, but that the sedapa runs with its heels foremost is quite untrue.

This has always seemed to me to be a most straightforward report so it is interesting to note the reception it received when poor Mr. Van Heerwarden finally told of it. Even the equable Heuvelmans cannot restrain himself from quoting certain of these expressions by people who were neither there nor, in some cases had then ever been anywhere near Sumatra, and most notably those of the same Dr. K. W. Dammerman of Buitenzorg. This is so delightful that I herewith re-reproduce it for your edification and guidance as a glorious example of the sort of rubbish spouted by experts and for which you have to be constantly on the lookout. This savant, after saying that no white man except Mr. Van Heerwarden had ever so much as said that he had seen a *Sedapa,* goes on to say: "But this writer is almost too exact in his description of the animal, so it does not seem impossible that the incident was either based on his imagination [i.e. that he was a liar—*Author*], or, that he has written it strongly impressed by the stories about the *Orang Pendek.* But, even while admitting the general truth of the story [i.e. not daring to say that he *was* a liar—*Author*], would it not be more likely that the animal in question was an Orang utan?" No it would *not.* I am wondering if Dr. Dammerman knew any zoology; I can hardly credit it.

This is by far the most complete account of the *Sedapa* but it was by no means the last. The matter has been going on ever since, and plenty of people, both native and foreign, have said they have seen the creatures. There were also other events. In 1927 one was said to have been caught in a tiger trap, and once again the irrepressible Dr. Dammerman gets into the act: this time as serological (blood) and trichological (hair) expert but without any better results. In fact, he becomes quite blathering, for, of some blood and hair found in this trap, he stated that "it was impossible to obtain any results with regard to the hair [this is indeed plausible, as identification of hairs is not easy—*Author*], but the blood pointed *faintly to human origin* [italics, mine]. However, we may not accept for a fact that the blood found came from the escaped animal: it is quite possible that it came from some native who had injured himself while handling the trap." I may just point out here that if you have a large enough specimen for any analysis there is no question as to whether it is human or not, so that it cannot "point faintly" to anything. Secondly, the "natives" of that area are Malays, of the mongoloid branch of humanity, who have no body hair but most distinctive head-hair. Thirdly, who said that an "animal" had been caught in the trap? At this point words do fail me.

Our principal trouble with the *Sedapa* is that, not only has there been a great deal of double-talk of this nature on the one hand, but that, on the other, there have been not a few obvious and deliberate hoaxes. The worst occurred in 1932 when local newsmen in Sumatra attributed the shooting of a mother *Sedapa* and the taking of its infant to the much respected local dignitary, the Rajah of Rokan. The world press went a bit mad about this, but only a little local inquiry elucidated the fact that the Rajah had had nothing to do with the incident—though he had for some time been interested in the matter, and had offered certain inducements to anybody who could produce definite evidence of the existence of these beings—but that two hunters had produced a *"baby Sedapa."* Dammerman said it was a mutilated young Socrili (*Semnopithecus*), although he gave the name of the *Javanese* species. More reliable sources indicate it to have been a Lutong (*Trachypithecus* sp.). This was said to be dead; about 17 inches long; with a skin the color of an *Orang blunda* (or White Man); and, naked, but for a thick topknot. Said "baby" was obtained by purchase and sent to the same Dr. Dammerman who was able actually to demonstrate, for once, its complete lack of authenticity. It turned out to be a young monkey of the genus known as *Presbytis* (or the Leaf-Monkeys) that had been shaved; had its long tail cut off; and its skull crushed and face remodeled with bits of wood inserted under the skin of the nose to make it look more human.

This making of "incubi" is an age-old practice in Sumatra, having been mentioned by Marco Polo, and being one of the principal sources of those horrible little homunculi that were exhibited at museums and displays of curios in Europe from the 17th to the 19th centuries. These were monkeys' dried bodies to which were sewn all manner of unpleasant heads and appendicularia, and which were sold to sailors. It is possible therefore, that the very strange affair of the "Sindai" of 1957 may have had a similar origin.

This began with an international wire-service story that some form of subhuman had been "captured" in southern Sumatra. This was said to be a young female (about 17 years old) "Sindai," which, it was implied, was a rare "something" well-known to the natives and which was considered very important by them. It was hinted, or rather queried, that this might be the first real example of a "missing-link" yet caught. This report came out of Palembang just when a local revolution was in full swing in that area. News from those parts, thereafter, tended to be somewhat unreliable and garbled.

I have definite statements about this "Sindai" teen-ager, stating that it was clothed in short, fine, pure white hair, and had no tail, walked on its (her) hind legs, and in every other way behaved like a tiny human being, but apparently had no speech and ate only raw foods. It was then stated that it had been shipped to Java for "examination by leading scientists." And that, frankly, is the last that was ever heard of it. It was also said to have been taken to Palembang, en route. The only thing I can add to this bizarre news-story is that there is a form of Coloboid Monkey named the *Simpai*, or Banded Leaf-Monkey (*Presbytis* to zoologists).* As far as I am concerned, therefore and in the meantime, I preserve not a little restraint in trying to assess the matter. I feel that there are sufficient reports that look genuine enough to warrant a lively interest in the affair; but, there is the presence of the little, sometimes bipedal, Malayan Sun-Bear, and of the Siamangs. Both certainly muddle the issue. Yet, the thing has been going on too long, and I only wish that I had had the opportunity to talk at length to the local people—as I have had the privilege of doing in so many other countries—even in a debased form of kitchen-Malay, so that I could have assessed for myself the depth of their sincerity; the position that they assign to it in the general scheme of "things"; and could have learned some more details about their notion of it from a biologist's point of view. [Biologists can ask the damnedest questions!]

Traveling on to the mainland of Malaya we encounter quite a different and, in many ways, exactly contrary state of affairs. Here, the actual reports are extremely limited; the local native knowledge is very extensive; and the creatures concerned

---

* It should be carefully noted that modern nomenclature has now adopted the name *Trachypithecus* for the Lutongs, *Semnopithecus* for the true Langurs, *Kasi* for the Purple-faced Monkeys, and *Pyagathrix* for the Doucs. One of the *Presbytis* does sometimes display an almost pure white form. Philologists tell me that a conversion of *simpai* to *sindai* is almost natural!

could not possibly be mistaken for any of the local fauna. This is what has so stimulated even the natural skeptics—and has been the cause of the British Army being called out on two occasions to try and do something about it. Here, however, we are going to run head-on into the problem of men versus sub-men that we mentioned above as becoming troublesome in this area.

There is a most remarkable book entitled *The Pagan Races of the Malay Peninsula* by Messrs. W. W. Skeat and C. D. Blagden, published in London in 1906, that is a real eye-opener. This is a solemn, ponderous, and somewhat pompous, scientific account of the peoples inhabiting this somewhat limited area, done in the painstaking and slightly Germanic style prevalent at the turn of the century. There is nothing excitable about it. It is simply a sort of official statement of the facts, as then known, about the peoples of the area. It makes most astonishing reading.

In this book, not only are the Malayan peoples now settled in the country fully discussed, but the Sakai (i.e. the *Senoi*), those strange retiring mountain peoples are fully examined, and then the Semang, a really primitive Negrito group. The way of life of the last, as herein described, is really hardly human. It should be read in the original. Then, the *Santu Sakai* (or Devil Sakai) are brought up, and are stated [though admittedly second hand from the Sakai; the Semang being almost uncommunicable-with] to be hairy, and definitely not human. The authors then go into the "myths, legends, and folklore" of the various people, including the little Senoi; and they dredge up from these tailed men; men with razor-bones on the outer back sides of their forearms; *and* a larger type that stinks. These are said to be "men" all right, but to be wilder than any of the rest of the line-up. There is a curious tradition about this last type that needs airing.

It is reported that they live (and only) in the upper montane mist forests of the higher mountain ranges, both in the boot of the Malay Peninsula and in the next bit north—*vide:* Tenasserim (see Map X): and that they customarily stay up there. However, it is likewise reported that they do sometimes come down on to the lowlands and that, at that time, they are highly carnivorous, rapacious, and what is commonly, but perhaps inaccurately, called "cannibalistic"; meaning that they catch, kill, and eat humans. Also, and note this, it is absolutely affirmed that these descents occur only after unusually prolonged periods of cloudy weather or a succession of very rainy and overcast seasons; and that, then, said creatures attack *only thin people*. This may at first sound absolutely absurd but I would urge a note of caution.

In Norway, perfectly good "werewolves" are on medical record. They are teen-agers—and usually males—mentally deficient; with a grotesque growth of head and body hair often growing right up to the tops of their cheekbones and down to meet their eyebrows; prognathous jaws; and sometimes even short bowed legs and enlarged irregular teeth. They are nothing more than kids who grew up in the almost perpetually sunless and rainy climate of the upper mountain valleys of the western side of Norway and, before the discovery of the existence of vitamins, had gone into a physical decline due to a lack of what are called the "sunshine vitamins" (E, and its concomitant, D). These poor wretches, cast out of the community, or having run away due to their abnormalities, sometimes managed to maintain life by hand-hunting and gathering, and one and all seem to have an insatiable desire for raw meat. At the same time, they show a very pronounced intolerance to fats of any kind. What they wanted and apparently needed was lean meat and entrails.

We may now reconsider the status, condition, and the sometime plight of a race of Hominids; driven way back up into the upper montane forests in an equatorial region. Deprived of many of the foods to which they had formerly been accustomed and to which they had been evolved, they did the best they could; but, when the climate continued in such a manner that some of the few essentials that they needed did not flower or seed, their whole metabolism went haywire. To counterbalance this, their bodies demanded that they do something; so, overcoming their natural racial fear, they descended upon their old homelands looking for what they needed—i.e., what we call "red meat." And, to take this to its end, let us say that, fats nauseating them, they picked the lean—and what easier than thin people?

This is one of the most abstruse niches in all ABSMery but it has intrigued me for years. Anybody can make up any kind of story but why anything which sounds to us so utterly bizarre? There ought to be a reason. There may be others, and many of them, but, in the meantime, this one *could* make sense.

Yet, these ultra-primitive humans or sub-humans, or other even more lowly forms of Hominids, do not seem to be the only conundrums in this small but extremely esoteric area. Maybe they are the "Stinking Ones": maybe they are something else. Nevertheless, the former turned up in a very definite manner in 1953, and so concretely so, and so many times in rapid succession, that not only the benighted natives, but the European overseers, the local militia, the museum authorities, and even the "Government" itself became apprised of the matter and lent a hand. This is really a rather unusual turnout in ABSMery. It now transpires that just the same sort of thing had been going on throughout peninsular Malaya a few miles back from the few main roads since way back. These incidents had been either not reported, reported but not listened to, disbelieved, ridiculed, or actually suppressed, and, perhaps, latterly because of Communist guerrilla activities. However, this one got out—and, as the colloquialism goes, "but good." Looking over what published accounts of this incident there are, a really extraordinary number of quite baffling things come to light. I would say that this too is a classic example of what happens when a good case of ABSMery —or any other matter that is not at present accepted—occurs. But, first let me give the facts, as reported, chronologically.

It appears that on Christmas Day, 1953, a young Chinese girl by the name of Wong Yee Moi was engaged tapping rubber trees on an estate run by a Scot named Mr. G. M. Browne, in the Reserve that is called variously the Trolak, Trollak, or Trolek, in south Perak State, northern Malaya. According to her account, she felt a hand placed lightly on her shoulder and, turning around, was confronted by a most revolting female. This poor character wore, according to Moi, only an abbreviated loincloth of bark, was covered with hair, had a white (i.e. Caucasoid-type) skin, long black head-hair and a mustache; and she stank as if "of an animal." Half hysterical, Moi fled for the compound, but not before spotting two somewhat similar types which she said were males [no loincloths?] standing in the shade of some trees by a nearby river. These, she said, had mustaches hanging down to their waists. Up till this point, the account is fairly rational, even including Moi's addendum to the effect that the female grinned and showed long nasty fangs in what she (Moi) seems to have considered, despite her panic, to have been a friendly gesture. After this report, everybody became slightly insane.

Analyzing all the published reports that I can lay my hands on, it seems that this Mr. Browne immediately called up Security Forces' local headquarters—there being a continuing Communist emergency in the whole area—and, in response, a posse of the Malayan Security Guard was dispatched immediately under the leadership of one Corporal Talib, who seems to have been an extremely intelligent and also sensible man. He immediately deployed his forces and made search of the estate, in due course coming to the river mentioned by Moi and spotting three just such hairy types on its banks. However, upon bringing his platoon's arms to the ready, said creatures dived into the river, swam under water, emerged on the far bank, and forthwith vanished into the jungle. Subsequent to this, the only concrete facts in the case are that a Hindu Indian worker, named Appaisamy, on the same estate, the next day, also while squatting to shave the bark to bring on a flow of rubber latex, was suddenly encircled by a pair of hairy arms. He became completely panic-stricken; broke loose; headed for the compound, but fell down in a dead faint on the way. As he revived, the same trio were nearby and laughing at his discomfiture. He admitted this. That same day, a patrol of Corporal Talib's Guard again spotted the trio on the same riverbank.

That is all we have, apart from a few further anatomical details of the creatures given in retrospect by the various witnesses. Then, however, the experts, and other nonpresent commentators got into the act. And they provided the international wire services with some pretty interesting material. All kinds of previously unheard-of official departments came to light such as that of "The Aborigines" at Kuala Lumpur; the "Federation's Department of Museums and Aboriginal Research" and even "Radio Malaya" in the person of its Assistant Director, one Mr. Tony Beamish. These people made various suggestions. They ridiculed an idea put forward some years before, when an almost exactly similar incident had occurred, that the creatures seen were AWOL Japanese soldiers, tired of the war, and who had managed to survive life in the jungle; though they did dredge up the old one about having "white skins because they had lived in the dim light of the jungles so long." [This is, of course, rubbish; though it is true that a white man will get a lot whiter in such an environment.] But, some people came up with some really startling ideas.

Most prevalent were hints that these things could be, or might have been "primitive humans trying to get away from British aerial bombing, or flooding of their jungle abodes"; or again, "that they might be descendants of a race of hairy aborigines who, according to old legends, once roamed the forests of northern Malaya." What I would like to ask is, what had the Department of Aboriginal Affairs been up to prior to this astonishing suggestion, and why had they not turned up some evidence [other than that of Messrs. Skeat and Blagden] of the necessity for protecting them? Also, as that excellent radio person—Tony Beamish—is alleged to have said, this could be "one of the most valuable anthropological discoveries for years." (Actually, it would have been *the* greatest of *all* times.) It is really rather remarkable that nothing was finally done about it. Experts of the same "Department of Museums, etc." did state that they were trying to organize an expedition and they made a statement. Statements are always good; and they are often good for a laugh. This one was a near classic. It stated:

1. The creatures apparently had seen rifles because they

[231]

fled when a security force corporal raised his rifle. Some of the "things" jumped in the river and swam away. Another ran into the jungle.

2. Their light skin probably indicates they have lived for years in the dark, overgrown Malayan jungles where sunlight rarely penetrates.

3. They recognized a crop of tapioca on one estate as food, pulling up roots and munching.

4. They spoke a language that was clearly not Chinese or Malayan, but more of a series of guttural grunts.

And this, mind you, from persons who were not only scientists and experts but officials. We stand amazed; but we make certain notes and reservations.

The number of ABSMs that jumped into the river has now changed from "all" to some; they are now alleged to have pulled tapioca roots and eaten them; * they had a language. I cannot find any of these facts in the original reports of the Christmas, 1953 case but they do indeed appear in earlier cases, and in other parts of Malaya. In fact, it appears quite obvious from these latter that there had been quite a lot more information on this unpleasant subject in the files of the Department of Museums, etc., long before this time.

The most outstanding aspect of this case is perhaps the alleged "stink" of the creatures, as recorded by all witnesses who were near enough to them, and included in similar statements that emerged later about others, reported to have raided crops in different parts of Malaya. This single fact is exactly in accord with the age-old statements of the locals about such creatures. It is also in accord with some of the statements of the Amerinds about their large ABSMs in Canada and the northwestern United States. It accords, too, with remarks passed about them, almost casually, by Kurds, Sinkiangese, Mongolians, and others. Apart from this, the fangs, hairiness of body, but ultra-long-hairedness of face and head, the sug-

* "Tapioca" is made from the juices squeezed from the root of the Cassava, a woody-stemmed herbaceous plant. The roots are deadly poisonous unless macerated and the juice pressed out of them. A few animals, however, manage to eat them raw.

gestion of primitive clothing, and the general "come-hither-ness" of these creatures speaks a great deal.

It is interesting to note, anent this matter of a powerful stench exuded by ABSMs, that when the last of the Mau-Mau leaders—Dedan Kimathi—was finally tracked down and captured along with some of his men, in Kenya, not only the white men present but the local natives—the same people as Kimathi —agreed that to smell the band was so sickening as almost to prevent handling of the captives. This is the more odd because any real "bush man" (as opposed to *Bushman*) never washes, though of course he may bathe, when in the forest simply because by so doing, and especially with soap, he removes all the natural oils from his skin and these oils are among the most powerful insect repellents and anti-fungus spore deterrents known. [And this goes for white men who really know the forest and have to work therein for periods.] It is the sweat itself that causes the smell, and this by going putrid in clothing, so that a real bushwhacker changes his clothes at least three times during the twenty-four hours. Kimathi's gang wore untanned animal skins. So did the mustachioed manlike ABSMs that invaded the Malayan rubber estates.

Another fascinating fact appeared from the prolonged Kimathi hunt. This was that Kimathi himself developed a sensitivity, not only of his five major and some twenty (now recognized) other senses, but some other unknown attribute so incredibly acute that he became almost unapproachable. It is said that he would awake from sleep on the (unauthorized) cracking of so much as a single twig at great distances and immediately vanish. Sometimes even his own men just found him gone. If men—and many of Kimathi's, and even he himself, had not previously been true bush men—can develop such acute senses in so short a time, how much more may not ABSMs that have for hundreds of millennia been as much of the wild as nondomesticated animals. This is one of my strongest arguments against trying to hunt them: I personally think the idea worthless on this account. It is also one of the reasons why I think that the employment of dogs is the worst idea of all. Dogs are purely "artificial" animals, as well as being domesticated, and they have an odor which is instantly spotted by any truly wild creature. Then again, there is still another point.

It has been observed that animals, such as antelopes, which are born to and used to being hunted, do not even bother to move aside when for instance a cheetah rushes a group. Only one animal takes off and the cheetah goes straight for it. [It is often old or sick.] Also lions may be seen lying almost back to back with their natural food-animals in the daytime. But animals that are *not* used to being food for other animals are excessively wary. So are the predators themselves. Just try hunting a marten or any other weasel for that matter. ABSMs are neither born to, used to, or prepared to be hunted, any more than men are; *and,* they have both some intelligence *and* the senses of the wild predator to boot. In order to "collect" one therefore, methods quite other than hunting must be employed. Personally, I suggest an appeal to their inquisitiveness —it almost never fails.

By the accounts, these are no hairy, gibbering monsters, or even pigmies, but man-sized and, at least partly, man-thinking entities who seemed above all to want to "make friends." Could it really be that Communist-hunting, bombing, and general modern military maneuvering since the Japanese invasion, had caused some otherwise amiable primitives to move, and come looking for handouts?

11. The Great Mix-Up

Sometimes everything—like geography for instance—really does get involved. Strangely, this is just when people show up at their best.

We now turn northward and start climbing, and we are going to need maps as we never needed them before. The area of the world which we are now approaching is perhaps the most puzzling and, to us, seemingly the most mixed-up in the world. The political situation is bad enough (see Map IX) but the topography is frankly awful, so that even a physical map is utterly confusing. This confusion, moreover, is worse confounded by our use of "feet" for measuring altitude. In this area they just aren't big enough, and maps showing the usual contour changes of color at 600, 1500, 3000, 12,000, and 18,000 foot levels end up as one glorious mishmash in which the main and basic features of the land are obscured. If, however, we do our measuring (and coloring) in meters, matters become much clearer. I have therefore constructed the map showing this province (Map X) on the 500 and 5000 meter contours, with a special shading for one particular area (Tibet) for reasons that will be explained later (see also Map XI). This device brings out at a glance more or less all that we want to know, and makes it possible to attempt a more detailed explanation of the more difficult parts. We are now approaching the summit of our interests and we will have to take our cue from the mountaineers and initiate what they succinctly refer to as an "attack" upon the problem. In order to do this we have first to try and sweep away a whole handful of misconceptions.

The first and most basic of these is to attempt to get rid of

MAP IX. NORTHERN ORIENTALIA (POLITICAL)

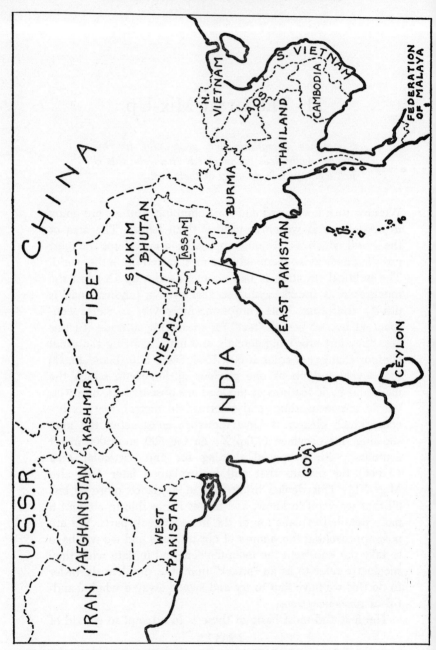

See facing page for legend

the notion that the Orient [or *Orientalia* as it is better called; see Chapter 18], is just a part of Asia. To the contrary, just as what we call Europe—the boundaries of which have been in dispute throughout history, and still are—is *not* a separate continent but merely one large peninsula of a much larger land-mass called Eurasia, so, conversely is Orientalia not a part of that land-mass but a quite separate continent. Its climate, past history, geology, and, above all, its vegetation are quite different from that of Eurasia. Also, it is almost absolutely separated from Eurasia by a continuous physical feature that is every bit as divisive as an ocean. This is a tremendous mountain barrier that runs from Baluchistan in the west to the plains of Kian-Su in north China in the east. However, here comes the second and most important point of all.

The mightiest mountain range in the world, the Himalayas, lies well *south* of this line and is in Orientalia and *not* in Eurasia. Further, the massif which mounts to the highest peak on land in the world, Mt. Everest, straddles the division and is north of and *not in the Himalayas*, as is almost universally supposed. It stands on the equally vast and high southern rim of Tibet which forms the southern boundary of Eurasia.

MAP IX. NORTHERN ORIENTALIA (POLITICAL)

Orientalia is today, from a political point of view, an appalling hodge-podge. This continent lies to the south of the southeastern end of Eurasia, all of which is Chinese. (Most countries have for centuries recognized Tibet as being a part of that hegemony.) To the west, it is bordered by Iran and Afghanistan, and a thin eastward extension of the latter separates it from the U.S.S.R. The greater part of it is covered by India and the two pieces of Pakistan. The eastern half is about equally divided between southern China proper and eight other sovereign states—Burma, Thailand, Laos, Viet Min, Vietnam, Cambodia, Malaya, and Indonesia. In addition, there are sundry territories (such as Sarawak) of other status, and some small colonial possessions. Right in the middle are the independent kingdom of Nepal, the territory of Sikkim, and the indeterminate Autonomy of Bhutan. The island of Ceylon is an independent country; and there are also sundry tiny enclaves, such as the Portuguese colony of Goa.

Using the 5000-meter contour, we see that between this barrier and the Himalayas there is really a great gutter at comparatively much lower level. It is in this gutter that the small state of Nepal lies. Thus, when we come to the Himalaya province, we must bear in mind that we will still be in Orientalia, and that we shall continue to be as we approach Mt. Everest from the south side until we top this Tibetan barrier. This is frightfully important because the flora and fauna of the Himalayas—and there is a great deal of vegetative growth forming massed forests that run up almost to the snow line all over them—is quite unlike that of Tibet but has relationships with that of the Indo-Chinese Massif. And this brings us to our third problem.

It will be seen from Maps X and XI, that this Indo-Chinese block terminates abruptly to the north against a monumental barrier of towering mountain ranges that are confluent with Tibet. These actually form a small "peninsula" of Eurasia that sticks down into this part of Orientalia as shown. On an ordinary physical map it will be seen that the Indo-Chinese block is formed of endless sub-parallel mountain ranges and strings of ranges, with very narrow deep valleys between them, running roughly from north to south. These form fingers going south into the Annams, down central Thailand, and down the Tenasserim peninsula. There also depends from them the parallel sub-massif composed of the Naga Hills and the Arakan. In the northwest, this block is very clearly and widely marked off from the east end of the Himalayas by a horn of true lowland equatorial forests. This, contrary to expectation, instead of separating the two upland masses actually cements them firmly together, from the plant and animal point of view, for the same forests cover both facing slopes to form a perfect bridge for both the migrants, emigrants, and immigrants from one side to the other. To the northeast, affairs are quite different.

Here, there is first, to the north, a small enclave of comparatively modest uplands running down from the Tibet plateau to the Red Basin of Szechwan (see Map X). These have a Chinese flora and fauna. Then, south of these, the Indo-

Chinese Massif abuts on to the more modest upland mass of Southern China, from the south of which a long thin chain of mountains—in Si-Kiang—runs east. This funny little promontory is rather important because a lot of myth, legend, and folklore pertaining to ABSMs extends that way. The south China uplands are of course densely populated and have been for a very long time so that their vegetative cover is now quite different from that of the Indo-Chinese Massif. If the two were ever alike is questionable, for the south China uplands really form quite a separate biotope, or florofaunal area. The Indo-Chinese province is therefore really rather isolated and distinct. It is also quite unique in many other ways. Armed with these facts we may now enter this Indo-Chinese country from the south, and immediately run into difficulties. This is the country in which the second largest form of the bovine or ox-cow tribe turned up in 1938—the Kouprey (*Bos sauveli*)—to the great consternation of established zoological thinking. It looks like a large edition of the now extinct Aurochs (*Bos primigenius*) of Europe, with widespreading horns, but the bulls have large tassels, the strands of which go upward, just short of the tips of their horns. This was an astonishing discovery in a land inhabited, and thickly, since most ancient times. Of more interest to us, however, are the Primates of the area.

These include a lot of strange types. First, it is the headquarters of the little apes called gibbons, one species of which, the Hoolock (*Hylobates hooloch*) reaches north and into the Himalayas. Then, there are also there the Doucs (*Pyagathrix nemaeus* and *nigripes*) which is one of the most brightly and variegatedly colored of all mammals; and, the Snub-nosed Monkeys. There are two distinct genera of these, one found in Tonkin (*Presbytiscus avunculus*); the other being the large and very extraordinary Rhinopithecus. Of the latter there are three species: Biet's Monkey (*R. bieti*) from Yunnan which forms a part of the Indo-Chinese Massif; Brelich's Monkey (*R. brelichi*) a really enormous form with a large white cape over its shoulders which lives in the Van Gin Shan mountains in west central China; and the Golden Monkey (*R. roxellanae*) of the upper end of the Indo-Chinese block and extending up that small enclave facing the Red Basin mentioned before. This animal is a glowing metallic gold all over but has a sky-blue face. These monkeys really are gigantic and look even bigger since they are clothed in long thick fur that forms a cape.

This is not to say that there are not other monkeys in this province; to the contrary, there are dozens of Leaf-Monkeys and Langurs, while there are also lesser Primates. It is, in fact a sort of hotbed of Primates, in and around which most of the living apes reside, a large proportion of known fossil Hominids have been found; and quite a "coterie" of different ABSMs are rumored. Here, what is more, we have a state of affairs comparable to that which we encountered in Africa, but compounded, for, in addition to having apes (or Pongids) to contend with as well as fossil Hominids and alleged ABSMs, we have also lots of large terrestrial or semi-terrestrial monkeys as well—i.e. the Macaques (which include the Rhesus) and these Snub-nosed jobs. Nor is that all, for the local folklore is full of allusions to "men with tails," on the one hand; and to giant, bipedal monkeys, on the other. This is all very muddling to the layman but seems also to have thrown the specialists—and even those few in the field of ABSMery—into confusion. Then lately, the Chinese have still further muddied the picture by coming up with an exceedingly ABSM-like race of *people* in Yunnan; while anthropologists and ethnologists generally have unearthed all kinds of primitive and most unexpected nations, tribes, and groups in this province.

One of the most extraordinary of these is a group of tribes in the central mountainous region of this territory, who have very pale brownish skins, Caucasoid features, and wavy hair. They keep strictly to themselves and have one curious custom that may be of great significance to those investigating Malayan folklore where there is said once to have lived a race of tailed men who had a cutting edge of bone along the outer (hind) edges of their forearms. These tribesmen possess practically nothing that is traded from outside but they always carry a large sharp knife of a certain shape; and they always carry this pointing backward up the arm and with the blade turned outward. With this they make their way through thick undergrowth at great speed by a curious down-slashing movement of the arm, so avoiding endless entanglements with vines by swiping at them.

Straight ABSMery in this province is not extensive until we get to the extreme northern end of it. In fact it amounts really to some legends and rumors, except in Yunnan and in northern Burma. Of the first, a Russian writing in *Tekhnika Molodyzhi* (Vols. 4 and 5, 1959) a science magazine for the Youth Movement states:

In 1954, the Province of Yunnan in China was visited by a representative of the USSR Society for Cultural Relations with Foreign Countries by the name of Chekanov. Speaking to Ma-Yao, the assistant chief of the National Minorities Department of the Kunming City Committee of the Communist Party of China, he learned that at the beginning of that same year, 1954, some people had been found in the mountains of Western Yunnan who in Ma-Yao's opinion were only at the pre-historic stage of their development. They led an animal existence, wore no clothes and had no articulate speech. It seems that Ma-Yao had also mentioned that their bodies had been covered with hair, and that one of them had been captured and brought to Kunming.

Chao Kuo-hoi, head teacher of the Yunnan National Minorities Institute also told Chekanov that the mountains where the people of the Khani nationality lived in the Hung Ho District were also inhabited by some strange people who belonged to no nationality whatever, that they wore no clothes and hid from ordinary people. One of them was captured and brought to Kunming. When he was dressed in human clothes he seemed quite satisfied and smiled.

According to what he had heard, Chekanov recalled that this captive wild man was finally sent to Pekin to be studied by the scientists. All this evidence, however, stems from people who had only heard about the wild inhabitants of the mountains from others.

Of the central area there is not much to be said and actual reports are neither numerous nor extensive. What there are concern a very large form of ABSM called locally the *Kung-Lu* or "Mouth-Man." This was first, as far as I can discover, mentioned by Hassoldt Davis, the well-known American traveler and author, in his book entitled *Land of the Eye*, which is the account of the Denis-Roosevelt Asiatic Expedition to Burma, China, India, and the lost kingdom of Nepal. In this the author says:

Jack (John Kenny) was the only one of us who could be called a hunter. He had shot bear and moose in Maine, and here it was his heart's desire to try his skill with tiger or Binturong or the Bear Cat (*Artictis*) or the great rhinoceros which is now found only in this wild corner of Burma near the Siamese Border. And more exciting even than these was the report of a creature, the *Kung-Lu* (or Mouth-Man), which had terrified the people for centuries. The *Kung-Lu*, according to Thunderface,* was a monster that resembled a gorilla, a miniature King Kong, about 20 feet tall. It lived on the highest mountains, where its trail of broken trees was often seen, and descended into the villages only when it wanted meat, human meat. We were told also that no one in Kensi † had been eaten by the *Kung-Lu* for more years than the eldest could remember.

It is perhaps permissible to speculate on the fact—could it be coincidence—that Chief Thunderface described a rather typical *Sasquatch-Oh-Mah* creature? This was my first reaction; and it was a pretty strong one; but, then, the same thing crops up much more extensively but with less exaggeration farther north where there are not, as far as we know, any Amerinds.

There, there is either a similar creature or a closely related one named the *Tok*, which I am told also means "mouth." My account of this originally came in the form of a personal communication from a gentleman who had heard me discussing ABSMs on the air. He gave me the name and address of a young American, then in the service of his country, who had been born in the Shan States and brought up there, his

* This "Thunderface" turned out to be a North American "Indian" by the name of Chief Michael Joseph Thunderface, a graduate of the California Mission College, of 1921! He had gone to the Orient as part owner of a small circus that had disbanded, and he had settled down in this Burmese village and in time been elected chief.

† The village of Kensi is now called Kawmyo and is near the Thai border. It is noted for its Naga (the King Cobra snake) worship attended only by priestesses.

parents having been missionaries. In turn, I got in touch with this young man, whose name I was asked not to publish, and he told me of two personal encounters—in fact actual physical contacts—with *Toks*, while he gave me several other reports, and passed me on to others who also in turn wrote me their stories. All were Americans with much experience of the country. In the end, it seemed to me that this ABSM may be the same as the Kung-Lu reported from so much farther south, about which there is, once again, that most curious detail of all in ABSM reports; namely, that it, also, attacks *only thin people* and ignores fat ones.

My young American correspondent states that he actually had a *Tok* in his arms twice and when it broke loose it left handfuls of long, coarse, shiny black hairs in his hands. The occasions were when it broke into his family home which was deep in the hill jungles and some distance from the nearest small, permanent settlement. On both occasions it chose a bright moonlight night and both times it crashed about apparently looking for food. Both times the young man tackled it thinking that it was a native thief or marauder and, being a powerfully built man and an athlete, and since his parents refused to possess any firearms, he did so with his bare hands. On each occasion it did not attempt to attack him in return, but only to flee, and being immensely strong and well over 6 feet tall it easily broke away, once running straight through a screen door. As it crossed to the forest in the moonlight, my informant had a very good look at it. He tells me that it had very wide shoulders, small head, was covered with jet-black hair, but had straight legs like a man and very pale soles to its feet. From this correspondent, and some of those others he put me on to, emerged various local names for this creature all of which must be translated as "mouth man" or "the man with the incredibly big mouth."

Hassoldt Davis' *Kung-Lu* is from the southern end of the Indo-Chinese mountain area, the *Tok* from the northern, where it would seem to merge with the *Dzu-Teh* of Eastern Tibet (the area that was once called Sikang) on the one hand, and the *Gin-Sung* or Bear-Men of central China on the other. These areas are all adjacent to the places where the teeth and bones of *Gigantopithecus* have been unearthed, and if they are all the same creature, it would bear out Bernard Heuvelmans' theory that they are indeed *Gigantopithecus*. But we will come to the *Dzu-Teh* and *Gin-Sung* later. We must now turn aside for a moment to try and clear up something that is really very puzzling. In doing this, I am going first to have to jump backward a little and then leap forward right into the middle of the Himalayas and also into the middle of the chronological sequence of events there. This I have to do as we will never make any sense out of the situation in this area unless we get this sort of "appendage" out of the way.

It begins way down in the plateau of Kontum, in what used to be northern Indo-China. There, the locals say they have a kind of enormous *monkey* that walks on its hind legs and which is actually vicious and is quite willing to attack people. They call it the *Kra-Dhan*. In the neighboring territory of the Jölong it is called the *Bêć-Bôć* (Bekk-Bok). The mountain people of the south also insist that it is a *monkey*, and not a man or an ape. This is odd, for there are virtually tailless monkeys thereabouts, the Stump-tailed Macaques (*Lyssodes*). At the same time, the locals are equally insistent that these creatures are not ghosts, departed spirits, demigods, or anything nonmaterial; all of which, though they often speak of them, they most clearly distinguish from real physical beings.

There is a report that one of these creatures either committed a murder, or was responsible for a murder near Kontum in 1943. Unfortunately the matter was tried by the local native court, of which no records were sent to the central French authority, while the French Resident of that area at the time is no longer alive, and the native Commune has been dispersed since the retirement of the French. This is not by any means the only report of these *Kra-Dhan* to be made to foreigners, and we have heard of similar entities in areas far to the west of Kontum. There would be nothing unexpected in reports of an unknown *ape* in this area, and I personally would not be a bit surprised if someone told me of an alleged ABSM thereabouts; and for all the same old reasons—ample, unexplored montane forests; small and isolated human communities; and appropriate geographical position. But, the insistence on the "monkey" theme is novel.

Now, as we have said, there have been countless stories throughout the ages about tailed men. However, I know of only one case of a possible ABSM ever having been stated to have one. This is one of the most peculiar of all reports, and is unique in many respects. It happened right smack in the middle of what has now become virtually traditional ABSM territory—namely, on the main route to Katmandu, Nepal from the north. It is alleged to have taken place in June, 1953. Those involved were two Americans, Dr. George Moore (M.D.) and Dr. George K. Brooks, an entomologist. The former was Chief of the Public Health Division of the U.S. Operations Mission, under the Foreign Operations Administration, and was public health adviser to the Nepalese Government. Dr. Brooks was on his staff. Dr. Moore had been in the country 2 years. They were descending the Gosainkund Pass (of some 17,000 feet) on their way back to Katmandu, the capital, from a trip to the north, and had entered the upper montane forests, there mostly coniferous, leaving their pack-carrying porters far behind. There was a thick mist. But it is better that Dr. Moore tell the incident that then occurred in his own words. It goes:

The forest was deathly still. Fog banks, raw and cold, drifted through the tall pines and left their boughs dripping and slimy.

Rounding a sharp turn in the trail, Brooks stopped abruptly. He leaned against a large rock to extract a leech which was on the point of disappearing over the edge of his boot. I stood there watching Brooks and fumbling for my pipe when an almost imperceptible movement in a clump of tall rhododendron caught my eye. Something had moved, I was sure. There it was again! This time, a few leaves rustled, more than mere chance could move. Brooks, sensing something was wrong, quickly forgot about his leech. Almost simultaneously we both slipped our revolvers out of their holsters. On our right the slope was dangerously steep. Behind us the slope climbed upward. There was a large boulder by the side of the trail and we eased over to it, glad for the protection from the rear it afforded. We waited, tense and expectant. The stillness was awesome. The fog and mist seemed to form weird shapes writhing and twisting through the dense foliage.

Suddenly, from in front of us a raucous scream pierced the air. Another followed from the right. The ghostly quality of the mist and the unreality of the situation had a nightmarish tinge.

"God!" Brooks whispered, "what was that?"

My spine was tingling in high gear now. I gripped my .38 S&W more firmly. About 20 feet away, somewhat in front of our rock, was the clump of rhododendron where the first scream had come from. We fastened our gaze on the leaves, trying to peer through them. Another scream broke the stillness. This time it seemed as though it was behind us.

"Brooks," I managed to whisper, "let's get on this rock in a hurry!"

Brooks did not need a second invitation. In an instant, we had scrambled on top of the massive boulder. From our new perch, we carefully searched in all directions for the next move. Our movements must have been closely watched, for a loud chattering immediately assailed us from the bushes in front. The angry chatter filled the raw air as new cries joined in the chorus from all sides. We were definitely surrounded.

Brooks muttered, "Oh my God, how many of them are there? And *what* are they?"

We got some idea of what was there when a hideous face thrust apart the wildly thrashing leaves and gaped at us. It was a face that I shall not long forget. Grayish skin, beetling black eyebrows, a mouth that seemed to extend from ear to ear and long, yellowish teeth were shattering enough. But those eyes . . . beady, *yellow* eyes that stared at us with obvious demoniacal cunning and anger. That face! Weird ideas were beginning to force their way into mind. Perhaps . . . but no . . . damn it . . . it has to be! This is the Abominable Snowman!

A chill sent gooseflesh along my back. The thought of these creatures had often been in my mind when we had trekked over the snows and high place. No European or American had ever proved the existence of the snowmen, although the natives certainly believed in them. Our boys had entertained us many an evening around the campfire with horror tales of the snow beasts, or "yeti," as they called them. They told how solitary travelers had been found torn to bits in the vast reaches of the mountains; how huge footprints had been found leading away from the murders. A few Sherpas had even met the monsters face to face and lived to tell the tale. We considered these accounts unlikely "hill stories," although I admit now they had left us somewhat uneasy.

[246]

No, I insisted to myself, there is no such creature as an Abominable Snowman. This face has to be an ape . . . or a man . . . or a demon . . . or the SNOWMAN!

A hand pushed through the leaves. Then, a quick movement and a shoulder. There, before us, appeared the semblance of a body. Sweat was visible on Brooks' face now as we crouched lower, hugging the rock for what it was worth. My hands looked white in the semi-darkness.

As the creature emerged through the dark leaves, we strained to make out his form. I felt blind panic start through me. Then I stopped. "Balls of fire," I thought, "I've got to get a grip on myself."

The creature was about 5 feet tall, half-crouching on two thin hairy legs, leering at us in undisguised fury. Claws—or hands—seemed dark, perhaps black, while his bedraggled, hairy body was gray and thin. It shuffled along with a stoop the way a neolithic cave man might have walked. Well-built and sinewy, it could prove to be the most formidable opponent. Teeth bared, it snarled like an animal. Two long fangs protruded from its upper lip . . . Suddenly, a sharp flicking movement behind it caught our eyes.

"George! A tail! Look there," Brooks cried.

A thousand thoughts raced through my mind at once.

"Well, Brooks," I replied, "this thing could be the Abominable Snowman but it also could be an ape . . . a langur ape, perhaps."

Truthfully, I was more concerned with survival than identification. The band of animals was certainly aggressive, giving every indication that they meant to destroy us. But I couldn't help thinking about the creatures themselves. They didn't look like the common langur monkeys I'd seen in India. At the same time they had apelike characteristics. Scientific possibilities crowded their way into my mind even as I checked my revolver for the attack. Higher altitudes, less minerals in the water could produce less hair. Lack of heavy timber in the high regions, which would make climbing ability relatively valueless, could produce an erect species. Mutations—the methods by which new species are created—have occurred, and are constantly observable in laboratories. Variations within a single species over a period of time can produce animals greatly different from the parent strain. I had no time to share these thoughts with Brooks. The best I could mumble was an unsteady, "Get ready."

Other figures were approaching now from several directions. We could make out 6 or 7 of them through the mist. One appeared to be carrying a baby around its neck. They seemed to mean business as they growled at each other. The one that had pushed through the foliage first was the leader. There was little question as to his authority as he led the attack.

[247]

"Brooks," I said hurriedly, "let's try firing over their heads to see if we can scare them. Don't hit them, for heaven's sake, or we may have them in a frenzy. A wounded animal—if they are animals—won't stop. And if they are demons, the Sherpas will never forgive us if we kill them. The Sherpas, superstitious as they are, would rather be killed than offend their gods, especially here."

"Okay, George, you say when," he replied softly.

We sighted carefully through the fog and waited until the repulsive faces were about 10 feet away. We squeezed the triggers almost together. The blast swirled the fog in front of us. Splinters of wood and torn leaves fell through the foliage. The creatures stopped abruptly.

The original account, which appeared in the magazine *Sports Afield,* May, 1957, concludes with quite a long passage relating the purely human reactions on the part of the author, his companion, and their Sherpa porters. It is indicated that the latter seem to have assumed that they had met some *Yetis*—the general Nepalese term now used by the Press— and they were greatly relieved that their employers had not been harmed. However, they did not resort to any exaggerated expressions and, it seems to me at least, they were singularly lacking in observations of any kind. In fact, I have an impression that they were somewhat mystified, and perhaps even unbelieving, but too polite to so comment. The account and the locale do not jibe with anything said by any natives of ABSMs on either count.

This is one of the most factual reports we have of anything [be it of ABSMs or not] to come out of Nepal as we shall most abundantly see in the next chapter. Moreover, it was made by a highly trained medical man, a person of all classes of educated men least likely to panic in face of bodily abnormality, and who must also have had some training in comparative anatomy if nothing else. Also, it occurred at less than 11,000 feet so that there cannot be any accusation of mental fatigue producing illusions that *can* be brought on by very high altitude and rarefied air if one is not acclimatized to them. In addition, the teller had a witness of equally high mental caliber and training. Moreover, if they had wanted to turn the creatures they saw into the traditional "abominable snowman,"

[248]

[of the giant or *Dzu-Teh,* the bestial or *Meh-Teh,* or even of the little forest *Teh-lma,* variety] they could quite well have done so, simply by neglecting to mention the tails. Tails just don't fit onto ABSMs.

There are also some extremely pertinent remarks in this account that have not, as far as I know, been commented upon nor even perhaps noticed. The first, is the very definite statement that their eyes were bright *yellow.* Not much is known or recorded about the color of wild animals' eyes, and quite a number of the stuffed specimens in our great museums have completely the wrong colored irises. One of my duties when I was a collector was to record the colors of the eyes of the animals. No ape has a yellow eye: they all have dark brown eyes; though I have seen an abnormal chimp with pale gray eyes. Many monkeys, on the other hand, do have bright yellow eyes—in fact, this color is rather common among them and it seems to go with lighter coat color. Some of the Langurs have yellow eyes, as do also at least two of their African relatives among the Mangabeys (*Cercocebus*).

Pertinent to this story also, is that I was once "attacked" by a large band of Red-topped Mangabeys (*C. torquatus*), in a mist, on the ground, in an upper montane forest, in West Africa. I say "attacked" advisedly because they ran at me threateningly—and particularly the big males, one of which I was forced to kill and which proved to be the all-time record in length for that species [its skin and skull are in the British Museum]. As I could not run away [which I admit is my natural instinct and invariable practice in face of any such danger], due to the density of the lower-level forest growth under which I had to wriggle along on my stomach, it was manifest that this action was not concerted or carried through. The one I shot did come most alarmingly close and was screaming and grimacing at me, and showing its very long yellow fang-like canine teeth. When it stood up on its hind legs, it seemed almost to be looking at me eye for eye, and I thought it was actually going to jump me. When I shot it, the others just renewed their howling, and they kept this up for about 10 minutes while rushing at me in simulated onslaught. Eventually I just went away, backward on my stomach, and left them.

Another point that Dr. Moore makes is the thinness of the animals' legs. This is a monkey feature, as is also the slimness of their bodies when they stand up. But most significant of all is that he says that "One appeared to be carrying a baby around its neck." This is an odd one. Young baboons and macaques at first hang under their mothers' bodies—they being quadrupedal—but they later ride astride their mothers' hind backs, holding on to her back fur. Almost all other monkeys carry their young in their crooked forearms or in one arm, but some of the Lutongs (*Trachypithecus*)—very near relatives of the Langurs or Semnopithecines—wrap them around their necks like feather boas or mink scarfs, *and especially* when they descend to the ground.

The whole attitude of the creatures in this story seems, indeed, to savor much more of a kind of monkey than of an ape or sub-hominid. As of now, I class them as such, but with reservations. Yet, monkey or not, I feel that the report is the truth and that we have therefore to be keenly on the lookout for the interjection of "evidence" presented for the existence of some ABSMs in this area being the result of the existence of giant monkeys. It is clearly manifest that these creatures, and such as the *Kra-Dhan*, actually have nothing to do with ABSMs. They, like the local bears, are just another side issue, and a complication. And this brings up the next of our problems in this mixed-up area. This is the *known* fauna. The trouble here is that none of the people who have been to the Himalayas seem ever to have known anything of what *is* known of the mammalian fauna of the region, while most of those who really do know that fauna are few and far between; either in museums or zoos in Europe or America, and almost none have ever been near the Himalayas. There is thus a most appalling muddle as to just what mammals do live there and which don't.

The worst confusion is over the bears. There are representatives of three genera of bears actually known to live in the Himalayas—the Himalayan or Moon Bear (*Selenarctos*), black with a white V-collar; the Sloth-Bear (*Melursus*), a strange aberrant type with a long nose, which eats mostly insects and honey; and the Brown or Dish-faced Bears (*Ursus*). Of the last, three species, sub-species, or races distinguished by color, have been recorded. These have been called the Red Bear, the Blue Bear, and the Isabelline Bear. There is an appalling muddle over the scientific names of these, apart from the Red Bear, which everybody agrees is not red and is simply a local variety of *Ursus arctos*, the Brown Bear of the rest of Eurasia. There is a name, *Ursus isabellinus*, which was once bestowed upon an almost white specimen of Brown Bear from the Karakoram, but which was jauntily and popularly called the "Snow Bear." Later, bluish-gray pelted specimens appeared from other localities and were either so called, or named *Ursus pruinosus* or *Ursus arctos pruinosus*. Some were creamy, others almost white, but most were gray. Nobody today is prepared or can say just how many races of Brown Bears there are in the Himalaya range of mountains, nor what their exact ranges are; whether they are full species, sub-species, or merely races; nor even whether they breed true. In other words, this "Isabelline Bear" is a lovely bogey to be waved at people who are not only not specialists in zoology, but particularly not specialists in mammals—and Oriental mammals at that! In my opinion the thing is a myth, just like our North American so-called "Grizzly Bear" which is and can be any Dish-faced Bear [as opposed to one of our Black Bears] that happens to have a grizzled pelage. One almost white specimen of a bear was killed in Tibet, and immediately called an Isabelline Bear but turned out to be an albino Himalayan Black (*Selenarctos*) not a Brown Bear (*Ursus*) at all.

But this is not all. While most bears can stand up on their hind legs for brief periods and can wobble along for a short distance on two legs, they happen to have a certain most peculiar feature. This has already been most ably demonstrated by Dr. Bernard Heuvelmans, and is that all bears are pigeon-toed and thus leave tracks that look as if they were bipedal, but walked with their feet put on backward. The toes point a little inward, the heels outward; in men, it is the other way, except for the Amerinds and some others who often walk

MAP X. EAST AND SOUTH ORIENTALIA

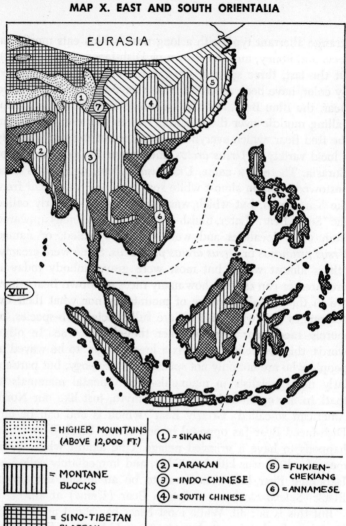

EURASIA

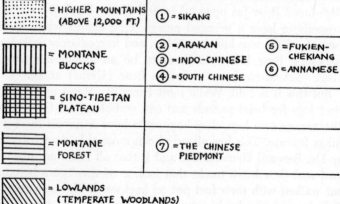

![stippled] = HIGHER MOUNTAINS (ABOVE 12,000 FT.)	① = SIKANG
![vertical lines] = MONTANE BLOCKS	② = ARAKAN ③ = INDO-CHINESE ④ = SOUTH CHINESE ⑤ = FUKIEN-CHEKIANG ⑥ = ANNAMESE
![cross-hatch] = SINO-TIBETAN PLATEAU	
![horizontal lines] = MONTANE FOREST	⑦ = THE CHINESE PIEDMONT
![diagonal lines] = LOWLANDS (TEMPERATE WOODLANDS)	
![cross hatch] = LOWLANDS (TROPICAL FORESTS)	

BOX VIII = MAP VIII

For legend see facing page

absolutely straight ahead. Apart from this, nobody, and least of all a "native," could ever mistake a bear's track, or print, for that of a man, and even more especially that of an "abominable snowman." In bears, the middle toe is the longest, the *outer* one the largest; they leave claw marks in any material into which they sink lower than the hairline on their feet. Finally, they cannot go on their hind legs on anything but level, unencumbered ground, and even then, only for short distances. Bear tracks have been mistaken for ABSM tracks *but* ABSM tracks have never been mistaken for bear tracks. Bears as an explanation of ABSM tracks, have also been brought up in North, Central, and South America, and in Malaya and Sumatra where species of bears do exist. However, they have not, of course, been able to be used in Ethiopian Africa where this group of animals has never been found or reported. (See Appendix B for tracks.) When it comes to animals that could possibly be the origin of Himalayan ABSM reports, the bears are not alone. However, all other kinds of animals so far suggested as being the true origin of ABSMs are absolutely ridiculous. Several have been suggested, such as Langur Monkeys of the species *Semnopithecus entellus*, which happens to be a purely Indian form, the Giant Panda,

---

MAP X. EAST AND SOUTH ORIENTALIA

The eastern half of Orientalia is also enormously complex from the topographical and phytogeographical points of view. Its central core is the huge Indochinese Peninsula—a vast mass of mountain ranges running from north to south—that lies between the Indian and the Chinese lowlands. This abuts southward onto a vast lowland which constitutes Thailand. From this depends the Malay Peninsula. Around it lie a diadem of islands, starting with the Andamans and Nicobars in the Bay of Bengal on the west; encompassing the greater Indonesian islands of Sumatra, Java, and Borneo on the south; and continuing on via Palawan to the Philippines and Formosa (Taiwan) on the east. Between and among these are literally hundreds of thousands of other smaller islands; plus another string along the coast, terminating in Hainan. The southeastern end of the continent is "Wallace's Line"—running between the Philippines, Borneo, and Java on the one hand, and the Celebes and the Australoid islands on the other.

wolves, the snow-leopard, and even large birds! But when it comes to candidates for scalps and hairs, the list is very much greater. [I append a list of the larger mammals found in that area and in Tibet as Appendix D.]

In leaving the Indo-Chinese province we omitted to stress one point; this was, simply, that its northern part is a meeting place of three outstanding ABSM areas. Each of these appear to have different indigenous kinds. These are of the usual four main forms—namely, a giant, a *Meh-Teh*, a human type, and a pigmy. There is evidence of a very manlike, man-sized one in the south, as we have seen (*vide:* in Malaya); to the north in the eastern end of Tibet and the eastern Eurasian area generally there is a very large one, the *Dzu-Teh, Tok, Kung-Lu,* or bearlike *Gin-Sung,* Mountain Man of the Chinese; in the west [that is to say in the Himalayas themselves] there are two kinds; first there is the little 4-foot tall *Teh-lma* of the lower montane forests; and secondly, the heavy-set *Meh-Teh* (the original Abominable Snowman) with a conical head, and very large and widely separated first *and* second toes, which often treks over snow-covered passes from one valley to another. The giant, with almost human-type feet, is *not* found in the Himalayas nor along the Tibetan barrier but is confined to the mountains between Tibet, China, and Burma.

12. Anyone for Everest?

By all accounts one would think there'd hardly be standing room in the Himalayas. As usual, this is quite wrong.

Having now reached the summit, I wish to ask your indulgence. Personally, I am so sick and tired of "Abominable Snowmen" per se, and of foot-tracks in the snows of "Tibet" [sic], and, most of all, of poor, old, long-suffering Mt. Everest, that I simply cannot bring myself to go over the whole dreary business again in detail. Yet, for all that has been published on the subject, which includes really quite a number of books as well as a veritable cascade of news-stories, one thing is most notably lacking to date. This is any real semblance of order upon which the whole picture may be assessed. On this occasion therefore, I ask to be excused for compressing my purely reportorial duties to the limit—in fact, into a chronological list, as you will find a few pages farther on—and thus reserving my energy and what mileage is left for some background information and, I hope, some legitimate comment. Before we tackle the issue, however, a few points should be stressed.

The first is a reiteration, and one that cannot be too often repeated or too strongly stressed. This is that the Himalayas are *not* a part of Tibet, or even in the same continent. Further, the racial, national, political, cultural, and all other aspects of humanity pertaining in this area are extremely complex, most muddling, and very little understood. For once, national boundaries hereabouts serve some really useful purposes (see Map IX): also, some of them even have some actual physical validity and coincide with natural boundaries.

Perhaps the single most interesting fact to emerge from this

MAP XI. ORIENTALIA

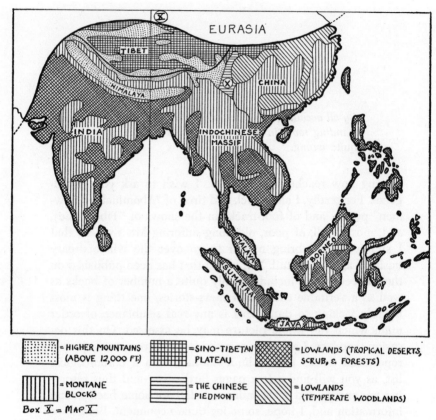

Box **X** = MAP **X**

Geographically, Tibet is a part of Orientalia but, due to its extreme altitude, it is connected with Eurasia. Orientalia is divided into six parts—India, the Himalayas, Indochina, southern China, Malayo-Indonesia, and the Philippines. (For the last three subdivisions, see Map X.) In this continent we have three major levels—lowlands, mountains, and super-mountains. Most of the first are clothed in equatorial forests but there is a large desert area in western Pakistan, and most of southern China lies in the temperate forest belt and has a distinctly Mediterranean flavor. The mountains fall into seven major and many minor blocks: there are two complexes in India, the Arakan, the Indochinese Massif, the Annams, the southern Chinese Massif, and the Fukien complex. Lesser blocks are on the peninsulas and islands. The Himalayas constitute a special region. The

[256]

is that the block of extreme mountains that is peaked by Mt. Everest really lies in Tibet, and is thus more than half in Eurasia. From all the hoopla as well as the very real and legitimate interest that has been engendered by the "attack" on and "conquest" of Everest, a general impression has been gained that either this mountain is more or less synonymous with the Himalayas, or that it is at least the only important one therein. Apart from not even being *in* the Himalayas, it is only one of a very great number of monsters on both sides of the Great [Brahmaputra] Gutter, and dominates several others by only a rather modest height. Further, a mountain was seen, and fixed for altitude, by more than one American military plane flying The Hump during the war, that was stated to be very much taller than Everest. So vast is the triangle of uplands between the Pamirs in the west, the Nam-Shans in the northeast, and the mountains of inner Yunnan in the southeast that, despite a healthy expedition led and financed by Mr. Reynolds [of ball-point pen fame] this monstrous thing has never been found again. This mountain has, however, recently been downgraded considerably.

Turning now for a moment to the human element in this chapter of our story, it should be noted that the inhabitants of Tibet are quite distinct from most of the peoples who inhabit the Himalayas, though the famous Sherpas, Ghurkhas, and Lepchas, of Nepal, were originally Tibetans, and are of that group of peoples. However, the true Tibetans inhabit quite a different land, having more intimate connections to the east with the Sikang region [now incorporated into the Chinese Province of Szechwan]. They therefore take quite a different view of things, and this is most noticeable when they come to talk about and describe ABSMs. Another point that is worth bearing in mind is that many of the inhabitants of both areas are most highly educated people,

---

range of that name is immensely long but narrow, and it is erected upon a huge upland. To the north, it is separated from Tibet by the great gutter of the upper Brahmaputra.

especially the monks of the Lamaist Buddhist faith, and the religious mystics and ascetics of the Hindu. An impression has been disseminated that, though the most excellent chaps for mountaineering, fighting, and other forms of endurance, the Nepalese are a poor bunch of uneducated hillsmen, and that all that Tibetans do, apart from spinning prayer wheels, is tend yaks. Some scholars in the monasteries of these countries speak, read, and write a dozen languages, both dead and living ones at that, and they possess vast treasure troves of documents and whole libraries of record. Books published by them five centuries ago on such subjects as history, medicine, and zoology, are as precise and objective as any of our own, as we shall see later when we visit the northern side of their country. Do not, therefore, sell the locals short on either common sense or outright knowledge. They can also be sharper than we are.

Finally, still another note of warning. Sportsmen, in the form of mountaineers, big-game hunters, and so forth, are not the only outsiders who have penetrated and wandered about the Himalayas and the southern rim of the Tibetan Plateau. Sometimes they almost appear to be, because of the Everest business, and the enormous volume of their published works. In these, however, you don't learn much about the country as a whole, whereas you do get a tremendous amount about the mountains (per se; and usually above the snow line) and about mountaineering. As I remarked in the first chapter, if you really want to get at what facts *are* known about the area, the best place to go is to the reports of the various British Government Surveys—political, topographic, and biological—which continued for years with the utmost precision and most painstaking persistence and care. Since the conclusion of those surveys, it is notable that the only people who seem to speak boldly and rationally on quite a number of matters pertaining to these countries have been anthropologists, ethnologists, and botanists, who have really traveled the country at lower levels, and taken the trouble to talk to the local people, learn their customs, and understand their languages and outlook. Of these I shall speak further after my reporting job is accomplished. Some of their expressions have been quite delicious.

Now I shall tackle the facts so that we may try to gain some kind of mutual understanding as to what everybody has been talking about. In my brief introductory history of ABSMery I had of necessity to mention not a few of the items that appear in the following chronology because until now the whole history of this subject has been mostly in and of the "Himalayan" area, and the general popular conception is that it is exclusive to it. By now, however, it should be plain that this is so far from actuality that the Himalayas have really been reduced, if not to a secondary status, at least to only one of three in the major class—the other two being the northwestern North American region, and the central eastern Eurasian. Nonetheless, these facts are important and must be re-emphasized, for I am constantly having to remind myself that hardly anything has been published on all the other areas, while intelligent people still say to me almost daily: "Do you really think there is *an* abominable snowman?" with the same old implication that there is just one individual hairy giant, who has been pounding about the upper Himalayan snowfields for centuries. This impression has, of course, been deliberately fostered in the mind of the general public by press and science alike, since nothing is better than a good debunking and a great number of people don't want anything of this nature found.

It comes as quite a shock, therefore, when one presents a proper list of those who have said not only that they have found foot-tracks of Himalayan ABSMs, or bits of their fur, or their excrement, but who have stated, and in most categoric and detailed terms, that they have seen them, have hunted or been hunted by them, or who know of people killed by them. Actually, the numbers of persons in all these classes runs into the tens of thousands, and has been going on for millennia. In fact, European and American travelers are the only group who *don't* seem to see ABSMs regularly when going through this area, and even some of *them* do. Perhaps the following inventory may make this a bit clearer. I take the famous year 1920 as my real starting point.

S = Seen by Foreigners
NS = Seen by Local Native Persons

1887. Major Lawrence A. Waddell, LL.D., C.B., C.I.E., F.L.S., F.A.I., comes across large tracks in Sikkim.

S. 1906. Mr. H. J. Elwes, well-known botanist and explorer, sees an ABSM run over a ridge.

NS. 1915. J. R. O. Gent, British Forestry Officer in the Darjeeling Division reports tracks in the Phalut area, India, and ABSMs seen by local inhabitants.

S. 1920. Stanley Snaith states in his book *At Grips with Everest* that one Hugh Knight, a British explorer, came face to face with an ABSM carrying a crude bow in this year.*

S. 1921. Lt. Col. C. K. Howard-Bury, on approaching Everest, watched a group of ABSMs on a snowfield at 20,000 feet, through binoculars. Later found their tracks on the spot.

1922. Tracks found on the Bireh Ganga Glacier by Englishman who signed his report "Foreign Sportsman."

S. 1923. Members of an Everest Expedition saw "great hairy, naked, man running across a snowfield below," at 17,000 feet.

1924. ——

S. 1925. A. N. Tombazi, Member of the Royal Geographical Society and leader of a photographic expedition to Sikkim, saw an ABSM grubbing for roots with a stick near the Zemu Gap at foot of Mt. Kabu. Later found humanoid footprints at spot.

* The search for this Hugh Knight continues. It (or He) is proving every bit as elusive as any ABSMs. The latest comes from Prof. W. C. Osman Hill, since this was written, and states: "I find a book listed in the Royal Geographical Society [of London] library catalogue by a Captain Knight (no initials given) entitled:—Diary of a Pedestrian in Cashmere and Thibet and dated 1863. It may well be the one we are after and if so antedates Waddell."

1926–30. ——

1931. Wing-Commander E. Bentley Beauman, RAF, found tracks at headwaters of the Ganges.

1932–35. ——

1936. (1) Ronald Kaulbach, botanist and geographer, found tracks at 16,000 feet, on pass between the Chu and Salween Rivers near Bumthang Gompa, Nepal.

    (2) Eric E. Shipton, famous mountaineer, found tracks on way back to Katmandu from Everest.

1937. (1) A British traveler who signed himself "Balu" found tracks on the Biafua Glacier in the Karakorams.

    (2) F. S. Smythe, reported tracks from the Bhyundar Valley, in Garwhal, India. [These were said to have been made by a bear, but there was some reasonable doubt.]

    (3) Sir John Hunt found a set of tracks of something apparently wearing boots on the Zemu Gap, also steps cut in the ice, though nobody had passed that gap at that time.

1938. (1) Cairn on top of sacred mountain, taboo to locals, above Rongbuk Monastery, and placed there by climbers, found to have been destroyed and stones moved.

    (2) H. W. Tilman, famous mountaineer, finds apparently booted tracks crossing the Zemu Gap, near Menlung, on Darjeeling side.

    (3) First American Karakoram, report calls, falling rocks, etc.

NS. 1939. [Reported by Prince Peter of Greece.] Locals got ABSM drunk by leaving liquor at wellhead; captured and bound; but creature revived and burst bonds.

1940–41. ——

S. 1942. One Slavomir Rawicz and four companions on flight from Siberian prison camp to India, reported meeting two ABSMs.

1943. ——

S. 1944. A Mr. W. W. Wood, in company with a Major Kirkland and Capt. John B. Maggs, at Liddarwat, near Srinagar, Kashmir, saw a creature bounding down a hillside with zigzag motion. (See below for possible explanation.)

1945-46. ——

NS. 1947. A Yak breeder named Dakhu, a resident of Pangboche, saw one at 50 yards distance. It walked away.

S. 1948. A very strange story of an encounter and fight with a pair of ABSMs near the Zemu Gap, by two Norwegian uranium prospectors, named Aage Thorberg and Jan Frostis. [Suspected fabrication.]

NS. 1949. (1) A villager of Pangboche named Mingma, heard yells, saw ABSM, took refuge in stone hut and observed.

NS. (2) In November an ABSM came out of the forest and played about near the monastery of Thyangboche until driven away by the monks beating gongs and blowing trumpets.

NS. 1950. (1) Sherpa Sen Tensing in company with others saw ABSM at 25 paces near Thyangboche.

NS. (2) One Lakpa Tensing saw a small one sitting on a rock.

NS. (3) Tibetan Lama Tsangi reports having seen one.

1951. Eric Shipton comes across tracks on the Menlung Tsu Glacier, in the Gauri Sanka Range on the way to Everest. Photographs.

1952. (1) Sherpa Pasang Nyima in company with others went to look for an ABSM seen near Namche Bazar, and observed it at 200 yards.

(2) Sir Edmund Hillary with George Lowe find hair on high pass.

(3) Swiss Expedition. Dr. Edouard Wyss-Dunant, with Tensing Norgay, find tracks.

NS. (4) Villager Anseering and wife of Thamnu, see one by forest.

NS. 1953. (1) A Tibetan Lama named Tsultung Zangbu, traveling in Assam, meets one carrying two large rocks. It passed by.

(2) Edmund Hillary finds tracks in Barun Khola range.

1954. (1) The *Daily Mail* Expedition. Sets of tracks found in four widely separated locations. [See Ralph Izzard's account.]

(2) Two Britishers of Hillary's outfit find tracks in the Choyang Valley.

(3) Swiss Expedition; Dr. Norman G. Dyhrenfurth photographs tracks in company with others.

1955. (1) French Expedition on Makalu, find tracks, photographed by the Abbe Bordet, geologist.

(2) Argentinian Mountaineering Expedition, led by famed climber Huerta, reported that one of their porters was killed by an ABSM. No further details available.

(3) RAF Mountaineering Club Expedition, found tracks.

S. 1956. John Keel, author of "Jadoo" claims to have followed ABSM for 2 days and finally seen it in a swamp.

NS. 1957. (1) First Slick Expedition. Three sets of tracks, excrement, and hairs found at three widely separated locations.

(2) Two Sherpas told Tom Slick they had seen ABSM early that year.

S.   (3) Peter and Bryan Byrne, of the Slick Expedition saw ABSM in the Arun Valley.

NS. 1958. (1) Second Slick Expedition. Two Sherpas with Gerald Russell at low altitude meet *Teh-lma* (Pigmy-type ABSM) by river; numerous tracks seen by Russell.

S.   (2) One Godwin Spani meets an ABSM.
    1959. (1) Third Slick Expedition. Numerous tracks
              found, and ABSMs followed.
          (2) Japanese Expedition under Prof. T. Ogawa,
              finds tracks.
          (3) Fukuoka Daigaku Japanese Expedition finds
              tracks.
    1960. Seven separate parties [but not the Hillary expedi-
          tion, which saw nothing but tried to debunk scalps]
          of foreigners and numerous locals reported finding
          tracks, and caves inhabited by ABSMs.

In addition to this somewhat impressive list I have detailed
records of many other sightings by both foreigners and natives, but for which no definite date is given or for which I
have been unable to obtain a definite date. Then, I have also
some delightful expressions by the ethnologists. These scientists seem not to be in the least interested in the grumblings
and mutterings of their confreres in other sciences—notably
zoology—and seem to have gone merrily on their way and with
their work, adopting a slightly amused attitude, at the discomfiture of others. As a fine example of this calm common
sense, one cannot do better than quote Prof. C. von Fürer-
Haimendorf of the School of Oriental and African Studies,
who wrote: "By coining the picturesque name 'The Abominable Snowman' Westerners have surrounded the *yeti* with
an air of mystery; but to the Sherpas there is nothing very
mysterious about *yeti;* and they speak of them in much the
same way as Indian aboriginals speak of tigers. Most Sherpas
have seen *yeti* at some time or other, and wall-paintings in
monasteries and temples depict two types of them—one resembling a bear and one resembling a large monkey. It is
generally known that there are two such types, and in hard
winters they come into the valleys and prey on the Sherpas'
potato stores, or even on cattle. The idea that it is unlucky
to see a *yeti* may be due to an association between the hard-
ships caused by an abnormally heavy snowfall and the ap-

pearance of *yeti* near human habitations on such occasions. No particular virtue is ascribed to the headdress of *yeti*-hide in Pangboche; it is freely handled and treated neither with reverence nor with any superstitious fear." *

This is one of, if not the most, refreshing statements that I have come across in over a quarter of a century of investigation of the matter of ABSMs. It also stands out as a statement by any scientist on any subject, and on its own merits, quite apart from ABSMery. Would that a zoologist might just once have so pronounced; but then, none who have made pronouncements have ever been to the Himalayas or considered the matter from the local point of view. Almost equally pragmatic is a passage written by Prof. René von Nebesky-Wojkowitz, after a 3-year sojourn in Tibet and Sikkim devoted to ethnographic studies. This reads:

It is a remarkable fact that the statements of Tibetans, Sherpas, and Lepchas concerning the Snowman's appearance largely coincide. According to their description a warrant for the arrest of this most "wanted" of all the inhabitants of the Himalayas would read as follows: 7 feet to 7 feet 6 inches tall when erect on his hind legs. Powerful body covered with dark brown hair. Long arms. Oval head running to a point at the top with ape-like face. Face and head are only sparsely covered with hair. He fears the light of a fire, and in spite of his great strength is regarded by the less superstitious inhabitants of the Himalayas as a harmless creature that would attack a man only if wounded.

From what native hunters say, the term "snowman" is a misnomer, since firstly it is not human and secondly it does not live in the zone of snow. Its habitat is rather the impenetrable thickets of the highest tracts of Himalayan forests. During the day it sleeps in its lair, which it does not leave until nightfall. Then its approach may be recognized by the cracking of branches and its peculiar whistling call. In the forest the *migo* moves on all fours or by swinging from tree to tree. But in the open country it generally walks upright with an unsteady, rolling gait. Why does the creature undertake what must certainly be extremely wearisome expedi-

* Tom Slick, seconded by Peter Byrne, now tells me that the inhabitants of Pangboche never claimed that this was the scalp of an ABSM, but that it was made in imitation of one held in a monastery elsewhere, and made from a goat skin. Anent this, see Appendix E.

tions into the inhospitable regions of snow? The natives have what sounds a very credible explanation: they say the Snowman likes a saline moss which it finds on the rocks of the moraine fields. While searching for this moss it leaves its characteristic tracks on the snowfields. When it has satisfied its hunger for salt it returns to the forest.

This is not only founded on good common sense and some proper investigation, it is also truly scientific in that it is "imaginative" in its mention of the search by the creatures for "a saline moss." Actually, there are certain lichens, not mosses, in this area, not saline, but veritable vitamin factories, notably of Vitamin E. It is strange that this report had to wait for an ethnologist's mention, since a similar matter has been known to botanists and zoologists for almost half a century, having been the key to Professor Collett's famous and definitive work on the causes of lemming swarmings and emigration. This, that researcher had shown, was that the cause of the sudden great increases in virility and resulting swarms of these small rodents is due to the continuous excess of these vitamins in their diet, which consists of these lichens for which they dig under winter snow.

Nor are lemmings alone in making a mad dash to get at this vitamin-rich food—the principal reason why birds take the trouble to fly annually for thousands of miles to the edge of the melting polar snows to breed is that the vegetation coming out from under that snow in the spring, and the insects that feed on it, are so rich in vitamins that young birds can be raised healthily on a very limited area. The ABSMs of this very cloudy area periodically need such vitamin and so go up to grub under the rotting snow for it, led by their age-old knowledge, or what is sometimes called instinct—just as some humans have a mad craving to eat certain earths and know exactly which ones and where to dig for them.*

But what, you may still want to know, exactly did all these people say they found or saw. I could quote you their actual statements but am not going to do so for two reasons. First, they are almost all already in print and most of them in read-

* Geophagy is widespread and cropped up in New York City a few years ago.

ily accessible books as listed in the bibliography. As a guide to them, you should read Bernard Heuvelmans' *On the Track of Unknown Animals,* which fully covers the issue. Second, I refrain from so doing because, frankly, even I find them somewhat boring, for they are all so absolutely alike.*

The great majority of the reports are of a roughly man-sized—though of a very large and sturdy man compared to the wiry little Sherpas—ABSM, with a conical head, bull-neck, prognathous jaw, and very wide mouth but no lips, clothed in reddish-brown, thick, short, hairy fur often grizzled in larger specimens and almost black in the smaller, which goes naked but uses sticks on occasion. Its excrement indicates that it is omnivorous, but feeds mostly on small mammals, insects, young birds that it can catch, snails, and various softer vegetable substances. It lives in the upper montane forests but comes out from time to time to grub under old snow, and in very severe weather it may descend into inhabited valleys and maraud. It has short, very broad feet, with a second toe larger than its big toe while both of these are much wider than the other three and are separated from them. It is shy and retiring unless provoked or imagines itself cornered, when it will put up a terrific display just like a great ape, but seldom carries through its threats.

This is not just the pattern but the identity of the vast body of the reports. However, it is not by any means the only one. There appear to be at least four if not five quite distinct creatures involved in this general area, only two of which are certainly indigenous to the Himalayan ranges themselves and to the "Great Brahmaputra Gutter" north of it. These two are the man-sized ABSM described above which is clearly distinguished by the local inhabitants as the *Meh-Teh,* and the little, pigmy type, only from 4 to 5 feet tall, that inhabits the lower and warmer valleys, eats frogs and insects and is generally omnivorous, and which the natives call the *Teh-lma.* This is clothed in very thick red fur with a slight mane, and

* The appearance and significance of the foot-tracks and prints is fully discussed in Appendix B; that of the creatures' possible relationships in Chapter 16.

leaves tiny, 5-inch-long footprints. The third ABSM appears only *to be spoken of* in the area, being an inhabitant of eastern Tibet, Sikang, and the northern Indo-Chinese Massif. This is the *Tok, Kung-Lu, Gin-Sung* creature called by the Sherpas the *Dzu-Teh,* or "The Hulking Thing" (see Appendix A). This by all accounts is immensely taller and bulkier than the *Meh-Teh,* with a black to dark gray, shaggy and long coat, a flat head, beetling brow with a sort of upcurled bang on it, long powerful arms and huge hands, and very human-type feet that leave imprints like those of a giant man but with *two* subdigital pads under the first toe just like the *Sasquatch* and *Oh-Mah.* This is the creature that Bernard Heuvelmans long ago (1951) suggested might be a descendant of, related to, or even actually a *Gigantopithecus,* which at that time was thought to be a pongid rather than a hominid. [That *Gigantopithecus* could be a very primitive sub-hominid and still have *hominoid* feet, will become apparent when we come to discuss fossil Anthropoids as a whole.]

The little *Teh–lmas* present a fine problem all their own. They are the least known and the most neglected by everybody. In fact, it was not really until 1957 that even the most ardent ABSM hunters acknowledged their existence, and only one man has done anything about them—W. M. Russell, commonly known to his countless friends all over the world as Gerald. Yet, this is probably the commonest of all ABSMs with an enormous distribution and is certainly "the *Yeti* most likely to succeed," if only somebody would do something about him.

Philologists, such as Sri Swami Pranavananda (see Appendix A) and others, in attempting to debunk the whole of ABSMery through their specialized methodology, have created a positive shambles of the Nepalese languages and dialects thereof, and quite apart from calling them all "Tibetan" [sic]. They have tried to show that *teh* has two stems and meanings: one being *treh, t(r)e* or *dred* which they state means a Brown Bear; the other, *te, dey,* or *da,* meaning a ghost. It transpires that they are wrong on both counts and in both cases. The crypto-esoteric details of all this will be found in the previously mentioned appendix; suffice it to be said here that *teh* turns out to mean "manlike *creature*." The ending *lma* is actually a Buddhistic inversion of *m//la*, which might be written for us phonetically as *m'ghoola*. This, in turn is a southern form of a phrase that sounds something like *me-ulléēr*, meaning originally an "incarnate vehicle." When used as a qualifying word attached to the name of an animal or other living creature, it implies "a being" or "thing." Thus, the little *Teh-lma*, is actually called—and rather simply, as it turns out in the end—"The Manlike Being." Nothing could be more pragmatic and appropriate.

There is a wealth of information on the form and behavior of this creature to be gleaned from all the native peoples from the western border of Sikang in the east to the feet of the Pamirs in the west, throughout the Himalayas. Practically nothing of this has been recorded simply because nobody realized that there was more than one "abominable snowman" and, even when they did aspire to this obvious intelligence, they simply could not stomach more than two types. As "the other" place was pre-empted by the mighty *Gin-Sung* or *Dzu-Teh*, the poor little lowland *Teh-lma*, got lost again. It was Gerald Russell who first spotted it as a quite separate species or type and, due to his long experience in collecting animals, prompted him to concentrate all his efforts on it—and down in the forests. I give the results to you in the words of Peter Byrne, Deputy Leader of the 1957 Slick-Johnson Expedition to search for ABSMs in Nepal. [This is herewith reproduced in full by the kind permission of Peter Byrne and the North American Newspaper Alliance]:

The first sighting was made by a Sherpa villager who said he was hunting edible frogs by the river at night with a torch hung on a bamboo pole. Moving upstream about 300 yards from Gerald's blind the man came upon a wet footprint on a rock. As he swung his torch low to examine it he saw a snowman squatting on a boulder across the stream, 20 yards away. The Sherpa was terrified, for tales of the *Yeti* in these mountain villages are full of accounts of the creature's strength and habit of killing and mutilating men. He shouted in fright. The beast slowly stood on two feet and lumbered unhurriedly upstream into the darkness.

The following night Gerald's Sherpa guide Da Tempa, a veteran Himalayan tracker from Darjeeling, went out with the villager at midnight, the note relates. While Gerald remarked it was "sporting" of the villager to venture out again, he noticed the fellow was trembling with fear and kept behind Da Tempa as they left the camp. After more than an hour of scouting up and down the Choyang River banks, Da Tempa and his companion were making their way back to Russell's camp when Da Tempa saw movement ahead on the trail. He thought it was probably leaves of a bush rustling, but shone his flashlight at the spot.

There, not more than 10 yards away, stood a small ape-like creature, the Snowman! The Snowman advanced deliberately toward the light, and Da Tempa turned and ran. Next morning Gerald said he found four very clear footprints in the gravel trail, which he has photographed. From questioning Da Tempa and the villager these facts emerged about our elusive quarry:

He is about 4 feet 6 inches high, with hunched shoulders and a very pointed head which slopes back sharply from his forehead. He is covered with thick reddish gray hair. His footprints are about 4 inches long. The villager was shown our pictures of bear, orang utan, chimpanzee, gorilla and prehistoric man. He unhesitatingly pointed to the gorilla picture as being most like the creature he saw, but he emphasized the head was more pointed.

As we trekked up the Choyang Valley to meet Gerald, Bryan and I are speculating what this description of the Snowman may mean. Is the beast sighted by Da Tempa the smaller variety of Snowman known as the *Meti*? Or is it a young of the giant *Yeti* which has been described as more than 8 feet tall? The footprints are certainly much smaller than the 10-inch tracks left by the animal that twice visited our camp by night in the Barun Valley. The tracks our expedition photographed last year measured 13 inches.

Peter writes again on June 5th (1958) from Gungthang, Nepal:

Frogs are the clue to the Abominable Snowman, and now we are using them as bait for our elusive quarry. Twice our party has seen the Snowman when he came into the dark gorge of the Choyang River at midnight to catch the foot-long yellow frogs for food. Now we have set out live frogs, tied down by fine nylon fishing line, as a lure. We have built a bamboo "machan," or hunter's blind, in a tree commanding a stretch of river baited with frogs and have a second blind of rocks along the bank farther down. From these points of vantage my brother Bryan and I are watching nightly.

We decided on this tactic after a reconnaissance showed where the Snowman had overturned huge river boulders in his search for food. Some were so large it took two of us to move the stones. And we found two footprints in river sand leading to a flat rock on which were the remains of a half-eaten frog. Toe prints were clearly visible in the sand, but the 4-inch prints were smaller than the ones we photographed in the Barun Valley snows some weeks ago. We have been dogged by foul weather, moonless skies and relentless rain.

Heavy rain, light rain, torrential rain and dreary drizzle. This has been the "Chinese water torture" endured by our expedition for more than a month now. The rain begins at 9 a.m., continues all day and night until the dawn sun breaks through the forest with golden streams of light at 5 a.m. It has hampered our plans for tracking the creature.

At midnight, with the rain pouring down in pitch blackness and waterfall drowning out even the sounds of breaking twigs and falling stones we hunters learn to follow in the dark, our nightly vigil has been a nightmare.

The *Dzu-Teh* is *not* a Himalayan inhabitant. However, there does appear to be still another creature in this province and on the southern rim of Tibet. Now, there seems to be some evidence pointing to this really being a giant *monkey*. [I am for now ignoring the tailed creatures reported by Drs. Moore and Brooks, which would constitute the fifth local unknown, and which I frankly believe to be some huge species of Coloboid Monkey and thus related to the Mangabeys and Guerezas of Africa, and the Langurs, Leaf-Monkeys, Lutongs, Proboscis and Snub-nosed Monkeys of Orientalia.] The Abbé Pierre Bordet has dredged up a tiny gem that is of great significance to this monkey problem. Namely, that the mountain massif that contains Mt. Everest is called by Indians, *Mahalangur Himal*, or "The Mountains of the Great Monkeys" —and *not* of great apes, please note. Then, there is also the fact that the Tibetans, as opposed to the Himalayanese peoples, talk freely of a monstrous monkey in their territory that has nothing to do with either the *Dzu-Teh, Meh-Teh,* or *Teh-lma* (which, incidentally, they call in various parts of their

[271]

country by numerous other names). It is, they say, nothing more than a monkey and has all the habits and character-istics of a monkey, even to a sort of totalitarian bravado and insufferable provocativeness combined with blind cowardice that in extreme cases of defeat may lead to its turning into a completely insensate homicidal maniac. There is but one group of monkeys that so very well fits this billing.

To me it is very strange indeed that neither this whole idea nor the possibility of this particular group of monkeys being involved seems ever to have even been so much as mentioned. The group concerned is the Cynocephaloids or *Cynocephalidae*, the Dog-faced Monkeys, which includes the Gelada and Hamadryad, the Drill and Mandrill, the Baboons, the Black Ape of the Celebes, and the Macaques and Rhe-suses. Not only are the largest monkeys members of this group; they are mostly terrestrial; most of them walk on the whole soles of their feet and hands; they have extremely manlike hands; they are certainly of high sagacity and, despite small brains, have a highly developed "social" (or at least com-munal) system. They are also strongly xenophobic, and, finally, they are in many cases extraordinarily ingenious, facile, and adept at manipulation with their hands. The ancient Egyp-tians trained some of them (Hamadryads) to weed gardens, stack cordwood, sweep temples, and serve at banquets: a S. African railroader supposedly taught one to throw switches in a signal box and water the engines, and this animal is al-leged to have saved a train wreck by pulling the right switches when its master had had a heart attack. That was a baboon. Even more intelligent and amenable to co-ordinated activities, however, are the Giant Rhesus and the strange Stump-tailed Macaques (*Lyssodes*), to which the Japanese "Ape" belongs. The former are customarily trained to collect coconuts on plantations, and the Malayan Forestry Service trained them to collect botanical specimens from the tops of tall trees. As to the mastery of human affairs on the part of the latter I can personally attest from many years' companionship with several individuals. Some of the things they learned to do altogether surpassed anything I have ever seen an ape do,

[272]

and they work at it with much greater persistence and reliability than do apes. They are, at the same time, incorrigible "slobs," unpredictable to strangers, and terribly dangerous. But, as if this were not enough, there is a positively enormous species that lives in the mountain recesses of that little enclave of Indo-Chinese territory that runs up the eastern face of the Tibet-Sikang Plateau and mountains. This is known as *Lyssodes (Macaca) thibetanus.*

These huge monkeys inhabit the fastnesses that are also inhabited by the Giant Panda—and which concealed this animal for so long—and these have never been explored. The species of Dawn-Trees (the *Metasequoias*) discovered not so long ago came from there, as also did the very odd Thorold's Deer (*Cervus albirostrus*), as well as other rare creatures like the Royal Chinese Sable (*Mustela liu,* a sort of enormous mink) and a small spotted cat just like an Ocelot. These great monkeys have no visible tail, that object being a tiny, flattened, naked twist concealed in the long, rich reddish-brown to orange overcoat that clothes these animals. Sometimes they descend in hordes upon the cultivated valleys of the hill peasantry and completely devastate everything, even attacking and tearing down houses made of mud and wattle, and not, it appears, being in the least frightened of men, even if they use firearms. And there is another interesting point about their behavior. When there is snow on the ground, they sometimes walk on their hind legs, which are very sturdy, albeit with an arm-swinging and staggering gait but which, I was told by an observer, seems to be due more to the deep snow than to any imbalance. Apparently, like apes, they do not like to get their hands cold by putting them on the snow.

These monkeys have rather short faces that are naked and pink, going bright red in heat and bluish when cold. Their other naked parts are dirty gray. The head is very curiously shaped, having practically no forehead but beetling brows, is flattened from side to side and comes to a point above but then has great domes of long hair running from the corner of the eyes back to the neck to join a profuse mane. Normally, these animals walk on all fours with a kind of strutting pace, the four limbs being of about equal length. One I saw in Hong Kong had a head and body [the head is carried straight ahead but the face does not point downward] length of three foot six, measured directly and *not* along the curvature of the body. My Chinese traveling companion, who had collected in outer Szechwan, told me that this was but a moderately small male and that if a really big leader-male stood up on his feet, as they sometimes do, he would look me eye for eye—I am exactly 6 feet. These monkeys go in snow.

My comment here is that, in view of the existence of these huge, tailless monkeys in the province concerned just east of Tibet, and in view also of certain remarks made by the great 19th-century explorer, General Pereira, who was and is still just about the only Westerner really to have crossed this territory and, again, to passing references made by the Abbé Père David [discoverer of the otherwise extinct primitive deer, named after him, in the Manchu royal parks, and in a way of the *Bei-Shung* or Giant Panda], there could well be a giant species of mountain Macaque in eastern Tibet that may occasionally enter the Himalayan Oriental Province and then become extremely "difficult" if met by a lone yak-herder. [I have a record of a fairly large party of unarmed Indian peasantry being attacked by the ordinary little *Bandas,* or Rhesus Monkeys, in the Punjab.] Also, it is just possible that the same or a related type of Cynopithecoid may be found in the Karakoram, and one of them could be the creature that a Mr. W. W. Wood and companions saw in 1944. He specifically states that this jumped "from side to side" or zigzagged. This is a most typical method of progression of many if not all monkeys when in a hurry on the ground, and especially on downgrades, but one which they adopt even on perfectly level, unencumbered areas. Also, please note that the locals with Mr. Wood definitely called the creature *banda* or "monkeys."

At this point I want to interject a very definite statement to my readers, to persons who may review this book, and to those of the scientific fraternities who might have gotten this far without having used the thing to throw at students or had

[274]

an apoplectic fit. This is that *I do not for one moment suggest that ABSMs are Giant Rhesus monkeys.*

What I *am* trying to say is that, in addition to the two very distinct forms of ABSM in this, the Himalayan South Tibet province—the *Meh-Teh* and the *Teh-lma*—there *could* be, first, a very large form of Coloboid Monkey in the coniferous montane forests, related to the Langurs and *Rhinopithecus;* and, second, a really giant form of *Lyssodes* or Stump-tailed Macaque, which *might* be the origin of *some* of the Tibetan (and *notably* the Tibetan) reports. The really giant *Dzu-Teh, Tok,* or *Gin-Sung,* of the eastern Eurasian Massif and the Indo-Chinese Block, definitely *is* an ABSM, and more than probably a full Hominid, but is known to the Nepalese only by hearsay from their Tibetan relatives. But there are still more complications in the Himalayan region. These are really of quite a different nature, and extend as far from ABSMery in one way as giant Cynopithecoids do in the other. This is the matter of Men.

This great province is not yet fully explored or known. When some soldiers employed by a person entitled the Rajah of Mustang, a sub-autonomous province of northwestern Nepal, killed an animal a few years ago that they did not know but which had been scaring villagers in their territory, it was declared to be a *yeti* (i.e. an ABSM). The beast was most adequately photographed (see Fig. 38) while still freshly killed, lying on a pristine white sheet. Later, it was carefully skinned with its extremities complete and was shipped with its boiled skull to Katmandu. It turned out to be a Sloth-Bear (*Melursus*). However, this is not the point. What is, is the fact that nobody had ever heard of Mustang; thought it was a kind of wild horse in our "West"; and that somebody was kidding. Even the wire-service representatives in Katmandu, capital of Nepal, could not get any clear answer as to whether there really was such a place, or to whom it belonged, even if only nominally. The same goes for most of the inner Himalaya and much more so for the Karakorams. There are some really delightful stories emanating from these parts, not the least extraordinary of which was solemnly put

[275]

out by two Canadian scientists named Jill Crossley-Batt and Dr. Irvine Baird of Montreal.

These two allegedly conducted ethnological studies there in the year 1921, and they stated that "In an isolated spot in the Himalayas, at 17,000 feet" they had discovered a "lost tribe of Chaldeans" who painted on goatskins with vegetable dyes, and who all lived to be 107 years old. Statements such as this just about floor me; more especially when some innocent is clobbered for remarking casually that he saw a funny fish in a net off Florida, or some such mild thing. Even the wildest moron playing hookey from a high school would be hard put to it to crowd more extremities into a single statement. Why Chaldeans; and, on what grounds? And who can tell that anybody lives to over 100, let alone a whole tribe; and why 107 years, precisely? The whole thing is a bit balmy but there it is, and we just have to try and cope with it.

This is, indeed, an exceptional case, but there has always been a great deal of mumbling about "lost races," "mystics," hermits, pilgrims, and outcasts in this area. True, quite a number of Hindu pilgrims do visit the Brahmaputra Gutter from India, and there are ascetics living all over the place high above the tree-line: also, there never was capital punishment in Tibet—that country being profoundly Buddhistic—and really annoying persons were always just thrown out of the community and told to fend for themselves. This, they have done for long periods, living until their clothes rotted away, while the law-abiding citizenry was absolutely forbidden to contact, aid, or have anything to do with these criminal outcasts. However, being Tibetans and Himalayans, and thus predominantly Mongoloid, these persons all started out with particularly hairless skins, so that they simply cannot be put forward as candidates for ABSMs. [Besides, they grow very long head-hair.] When, however, it comes to the Buddhist ascetics—the so-called Lung-Gompa—we meet quite a different condition.

These men deny normal life and take first to monasteries where they really study the supernatural, and in patterned stages, under persons with a tremendous fund of knowledge. What they learn is quite beyond us and, frankly, neither understandable nor even believed in by Westerners. However, they do in time seem to acquire some quite remarkable talents that smack of the magical. Dr. Julian Huxley has spoken seriously of their ability to melt a circle of 8 feet in diameter in 2-foot snow, simply by taking thought upon the matter; and others have described them as being able to teleport themselves; that is, to be transported instantaneously from one place to another; and, most certainly, to be able to send news in advance as quickly as by radio, though no radio exists. Of all of this I know nothing *factual* but of one fact I do have evidence. This is that the initiates to these disciplines do, at one stage of their training, go galloping about the countryside, stark naked, and in the worst of weather, and particularly at sunrise and sundown, for the good of their souls and the exorcism of sundry worldly hang-overs. They may then be a pretty eerie sight, charging through the rhododendron thickets and sometimes even howling a bit. But these chaps are almost as commonplace to the Himalayans as are mailmen to us: and they are not hairy, don't have separated second and first toes, don't eat raw mouse-hares or any other meat, and don't run around gibbering.

There is one rather delightful story about a *Hindu* pilgrim, however, which just goes to show what human beings can do. A certain Colonel Henniker of the British Army was crossing a 17,000-foot pass in Ladakh in 1930, in a blinding snowstorm, when he perceived a rather skinny fellow, clothed only in a loincloth, and using a staff, tramping stolidly Tibetward. Amazed, he hailed the man in English and received the astonishing and cheery reply "Good morning, Sir: and a Happy Christmas." [It was mid-July!] There may, in fact, be all manner of queer types wandering about in these appalling fastnesses; clothed or unclothed; fed or unfed; and everyone minding his own particular business.

It takes a great deal of patience and some ingenuity—as well as exceedingly good manners and taste—to get in with the local people and to be sufficiently accepted by them to hear what they really have to say. We of the West tend to adopt a lordly attitude to everybody else, and often in our own ignorance give away, by gesture alone, if nothing else, that we are mocking anything that we don't understand. The Himalayans are very wise, and perspicacious people.

But for all their wonderful qualities, it is not to the Sherpas and other Nepalese, nor to the people of the Himalaya as a whole that we must turn for some real pragmatic information about ABSMs. Rather, we should go to the Tibetans proper. Their whole attitude is utterly different, for they appear to have the whole thing down "pat," and, they just don't bother to argue the details. To them, there are three kinds of these creatures—called, as I have already said, by many names. They are not much interested in *Teh-lmas*, in that they dwell in the lower regions, of which there are none in their exalted land. *Meh-Teh* they know and treat as just another thing indigenous to the land, but of the hulking *Dzu-Teh* they take a really peeved notion. They say this vast creature is hard to handle and it raids yak herds; that they go in groups; they can get along in appalling climatic conditions; and they have all the ingenuity of humans, plus strength with which one is really almost unable to cope. That is why, they also say, they keep the skins of those which their compatriots slay, or mummify their bodies and put them away, but not so much out of respect but simply as "hereditary awful-warnings" to other men. Real Tibetans have spoken of all this to both Nepalese and to many foreigners in Nepal, and one much respected Lama named Punyabara even offered to bring back one of each of the three kinds, alive, if the Government would put up the money. My permissible comment is herewith terminated but perhaps I can afford to extend myself a little and make a few, more general comments at this point.

With all the above, how is it possible for anyone to state flatly that there is nothing in the Himalayan region to be investigated? This, I personally and simply cannot see. There have been those over the years who have endeavored to prove that nothing exists there; and many have tried by disproving or "showing up" one facet of the matter to show that all the rest is either myth, legend, or folklore. But, when you take each of these individual complaints, you find that none of

[278]

them jibes with all the others, while each of them in turn itself proves not to hold water. The ever-recurrent notion, for instance, that the tracks are made by local people wearing a particular type of loose footgear resembling a mukluk or moccasin—and which was recently again brought forward by one Michel Peissel in *Argosy* Magazine (December, 1960)— is obviously both absurd and impossible if only the advocates would just spend a few moments thinking logically about the matter. If this Mr. Peissel had considered the following facts for a moment, he would not have written as he did.

If these ABSM tracks—which, you will note, have baffled just about every really experienced mountaineer for over a century—were made by a local man wearing footgear such as he suggests, then, first, every one must have worn out the front half of both feet precisely, and in such a manner that neither shoe ever showed a single mark of where it ended or the bare toes protruded. Second, the men wearing these overshoes must all have been of an extremely rare type—if they ever existed—having the second toe larger than the first, and both of them, and *on both feet,* also widely separated from the rest of the toes. That there could be so many such freaks among the limited population of this one area is much too much to ask. Also, it is manifest that Mr. Peissel has never seen an imprint or a cast of the foot that made the medium-sized [or *Meh-Teh*] tracks. They are positively shocking when first seen, being absolutely enormous—and the gaps between the separated toes are enormous too, which could not happen physically if the whole was enlarged by melting and regelation. Almost the same goes for those, like Sir Edmund Hillary, who have attempted to debunk the scalps. Maybe these are made from the shoulder skins of a Serow (*Capricornis*), but were the makers not imitating something else they knew? And these things are, in any case, only playthings, like Christmas hats. Further, even if they are not genuine yeti scalps, what made the fresh foot-tracks?*

Let us not forget that the Kraken, the giant squid, was regarded as a fable for centuries until Prof. A. E. Verrill took a

* For a full account of this, see Appendix E.

small boat and went and got one alive off the coast of New-foundland. *Everybody,* except the North Atlantic fishermen had said that they did not, and *could not* exist, but reports of them persisted in coming in every year. I think people should pause, read the facts, and also consider a while, before making definitive statements about the ABSMs of the Himalayas, or of anywhere else, for that matter.

13. The Western Approaches

Despite all the current folderol, the real dividing line between the "West" and the "East" has always lain, and will always lie, along the eastern border of the U.S.S.R.

We have now reached the summit. Further, I have to admit, albeit with reluctance, that all my reportage up to this exalted point looks, both in retrospect and in view of what now faces us, pretty paltry. In fact, the old saw about straining at gnats intrudes itself on my attention, unwanted but persistent. It were as if I had up till now been squeezing a sponge of its last drop of information when what has already been said is reviewed in the light of what we now have to tackle. Whereas the reports even from such ABSMally rich areas as British Columbia may be counted on your fingers and toes, we now find ourselves confronted with literally thousands of them, spread over a thousand years in time, and throughout a triangular area with sides measuring approximately 5000, 4000, and 3000 miles in length. Moreover, these reports increase in number per annum on what looks suspiciously like geometrical progression so that the greater part of them are bunched up in the immediate past. Also it now transpires, the matter on hand has been pursued, and even scientifically pursued, in this area for over a century, though that pursuit has been plagued by all the same asininities and obstructions as elsewhere.

At this juncture a few words on the gruesome subject of geopolitics are called for. Most political boundaries are ridiculous. At one extreme we have gross misconceptions about "continents," rather fully discussed in Chapter 18; at the other, such

[281]

MAP XII. EASTERN EURASIA

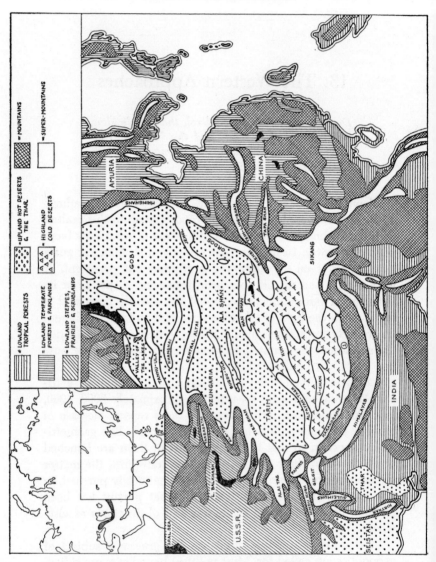

See facing page for legend

absurdities as the delineation of the North American state of Montana. In between these extremes man has further insisted on erecting quite arbitrary fences—such as that between the United States of Northern North America and the United States of Mexico—though these are sometimes called iron, bamboo, or "curtains" of other materials. Ridiculous terms like "the Near East" and "the Middle East," both of which lie in what is manifestly and geographically "*The West,*" add to the confusion; and then, to top it all off, we get purely political expressions such as "East" and "West," bits of both of which are now scattered all over the globe inside each other. Then, some buffoon (like Haushofer or Treitschke) must needs go and coin the phrase "The Heartland" but omit to define it. In some respects, such a concept is a splendid idea, as it implies a central blob which pumps away without cease or surcease, and, if applied to a certain area in central Eurasia, it makes a lot of sense ethnologically. Yet, the area that was finally pinned down for this happens always to have been one of the greatest ethnological blanks—this is the lowlands between the Urals and the great mountain barrier that cuts straight across Eurasia from southwest to northeast—while the "pumping" appears always to have gone on beyond that lofty barrier to the east.

MAP XII. EASTERN EURASIA

This most complex geographical setup in the world forms a vast triangle some 3,000 miles along its western face, which is the great barrier abutting onto the Russian steppes; some 2,500 miles along its southern curve, which runs from the Pamirs along the southern rim of the Tibetan Plateau to Sikang; and 3,000 miles from there along its eastern face to meet the Barrier in eastern Siberia. The whole of this is basically a high plateau, the central portion being a hot desert. Upon this, and all around its rim, lie immense mountain ranges. In the south, there is a super-upland, the plateau of Tibet, with even greater ranges upon it. It is an astonishing fact that the greatest of all mountain ranges in the world, that forms the southern rim of the Tibetan Plateau and contains the Everest Block, has no recognized name in English. The Tibetans and the Nepalese know this as the "Mother of All Mountains" or the *Muh-Dzhura rDzhung pBlhüm*. The Karakorams are the western end of this range.

If people insist on splitting themselves into two ethnological camps and calling these "West" and "East," they would be well advised to consider some ineradicable geographical facts. The most pertinent of these is this monstrous mountain barrier lying athwart Eurasia, since it has always formed, and will always form, the true dividing line between west and east. It lies along and constitutes the *eastern* boundary of the U.S.S.R.; and, if you want to be precise about the matter, it also forms by extension the southern boundary of that vast Union. Today also it forms the boundary between the Mongoloid-type peoples and the Caucasoid-type peoples; and I add the suffix "-type" most firmly because a not inconsiderable body of the peoples west of the barrier were original Mongoloids, and some on the east side originally Caucasoids but today (in the now almost classic expression of a certain comedian imitating a Chinese waiter): "So funny; all American look alike": so also do all Europeans, even the mongoloid Magyars. This great dividing line is of the utmost significance.

ABSMs are not found west or north of this line but they are reported from all along its edge and more or less all over the eastern area exclusive of the subcontinent of India and the eastern fringe of islands, as we have already noted. At the same time our information on ABSMs in this area, apart from the Himalayas and the Indo-Chinese Peninsula, comes almost exclusively from or through the Russians who are, of course, wholly in the *Western* area. This last fact may be rather puzzling to the general reader and somewhat aggravating to students of disciplines other than the purely geographic and biologic. Nonetheless, short of a major shift in the earth's crust, nothing—not even an all-out nuclear war—can alter the facts. Nature constructed our bed in the "West," and we might just as well make up our minds that we have got to lie in it together! [I cannot refrain from adding, purely as a student of plant and animal distribution, that we might also just as well give up any thoughts of trying to go and lie in any other peoples' beds; not only because, as in the case of eastern Eurasia, it is a bit crowded, but much more definitely because none of them are *our* environment. If we do so, we'll go Mon-

[284]

goloid or Negroid in time either by absorption or physical mutation, just as the Magyars have become Caucasoids in a few hundred years after landing up in our bailiwick.]

Considerations such as these are often regarded as what is euphemistically called political. They are not; they are purely biological. What is more, if such *facts* rather than a lot of (often mistaken) *ideas* were used to guide our policies and our activities, our species would get along much better. Early, primitive, and ancient man seems to have appreciated these facts if only instinctively, and acted accordingly. ABSMs seem to have had the clue since the first. Driven out of their original lowland forest homes they retreated into the montane forests, and particularly into those areas within those vegetational zones where Modern Man finds it hardest to get along. This is true "survival of the fittest": we might well emulate the forlorn ABSMs. The process happens also to make an otherwise appalling task a lot easier for this reporter. All I have to do is locate said particular regions, and the great mass of facts now to be presented then falls into a very fair semblance of order. The mess can be broken down into manageable parts—geographical units—and presented one at a time in logical sequence. To this I shall now proceed.

Let us assume that we have ended up at the northwestern end of the mighty Himalayas. This lands us in an area known as Gilgit which now lies in the north of Western Pakistan. [I apologize for this and a coming plethora of "political" definitions but there is nothing that a mere biologist can do about it.] At this point (see Map XII) you will note that we are very close to (on the right side, going west) the end of the almost as mighty Karakorams, which in turn constitute an extension of the "Mother of All Mountains," the *Muh-Dzhura rDzhung pBlhüm* of the Tibetans, and which we have called the southern Tibetan Rim. Ahead of us lies a most unpleasant complex of mountains known as the Pamirs or "The Roof of the World." These form a nodal point for all kinds of things in Eurasia—plants, people and other animals, languages, *and* ABSMs.

The Pamirs may be likened to a monstrous starfish with the appropriate five arms. These are vast strings of mountain ranges that go off in all directions—the Himalayas; the Karakorams; the Kunluns leading to the Altyn Tagh and Nan-Shans; the Altai Tagh leading to the Tien-Shans; and finally the Hindu-Kush going off to the west.* From this point we have first to follow the Hindu-Kush in order to get rid of a rather irksome business. This is that ABSMs have been reported from all along the extension of those mountains, which is to say along the Ala Dagh and Elburz in Iran to Azerbaijan and the Caucasus. There are those who regard the Caucasus as being in "Europe." As a matter of fact, nobody has ever made up their minds just where Europe does end in the east [vide: Europe: How Far? by W. H. Parker in The Geographical Journal, Vol. CXXVI, Part 3, pp. 278–297, September 1960]; and rather naturally, since it does not do so anywhere, being only a large peninsula at the western side of Eurasia. If this peninsula needs definition—and it does—it should be considered as lying west of the 30th meridian east which runs roughly from the White Sea to the Bosporus. The Caucasus area is profoundly in Eurasia.

The Hindu-Kush, Ala Dagh, and Elburz, together with the lower Caspian Sea, form the southern boundary of the Turkmen S.S.R. Between the Caspian and the Black Sea there are really two great mountain ranges with a lowland gutter between them. The southern is composed of Armenia and Azerbaijan in the U.S.S.R.; the north is the Caucasus. Both are still very rugged and over their upper reaches uninhabited by humans, and the latter appears to be a retreat of ABSMs. Their presence is fully accepted over both areas not only by mountain folk but also by inhabitants of the lowland villages and towns around their peripheries. As one of the Russian reports puts it, however, the younger generation put on a show of scoffing at the whole thing, probably in order to appear "modern," while the older people are most reluctant to speak about the creatures for deep-seated and most ancient "reli-

---

* Place names from now on are going to become as awful as political definitions. I have tried to confine myself to larger generalities that are shown on the map, and identify places that are not on that map by these generalities.

[286]

gious" reasons. This matter is made abundantly clear in Appendix A by Yonah ibn Aharon, who points out that there still remains a prehistoric animinism throughout this whole swath of Eurasia in which the souls of people enter the lower anthropoids, which latter are consequently held in such great reverence that even the mention of their names is most ill-advised. ABSMs, known in this area as *Kaptar* or *Kheeter*, seem to be regarded as the highest of all anthropoids and nearest of all "animals" to man.

There are dozens of reports on these *Kaptar* having been seen in recent years, as distinct from the endless older reports and myths, legends, and folklore. Many of these are very precise and really quite scientific in that they were reported by properly trained persons with the usual Russian mania for precision and suitable confirmation. This makes them the more instructive and convincing. I would that I could quote them in their original form but, alas, we don't read Russian so the best I can attempt is a paraphrasing of translations, using from time to time phraseology that may look quaint to us but which must be retained as expressing more exactly what the raconteur had in mind in his own language. Russian is almost as "mobile" a language as English and, when reporting in it, shades of meaning are most important. [Calling upon another language, to explain what I mean, the Spanish word *noticias* does not mean precisely "notices"; a good translator expresses it better as "advices upon (a subject) to everybody, by persons who are presumed to know what they are talking about" but with a distinct indication that the editor does not take full responsibility for same. This is a rather more precise form of our loose phrase "informed sources state."] One must bear in mind that the average Russian, especially when making a deposition or statement on which he may be called, often places more emphasis on the qualifying words than, perhaps, on the word itself. Thus "The Engineer X told me in Tomsk that when he was in Omsk . . ." has a very special meaning, and aids us in assessing what he finally records.

I could devote a whole book, let alone a full chapter to these reports of ABSMs from the Caucasus but, for obvious reasons can only give some examples. However, I will add the conclusions of the reporters since they are so very sane, orderly, and significant. The only other people who have published such sane statements on this subject that I know of are the Canadians. It is a pleasure to get back to fact without a gross overlay of preconceived ideas, prejudice, and doubt.

The main range of the Caucasus runs from the Black Sea coast about Krasnodar southeast to the peninsula on which the famous port of Baku is situated on the Caspian Sea. The range is divided into two blocks of higher mountains, the smaller in the northwest; the greater forming the boundary between the Dagestan A.S.S.R. and Chechen on the north and east side, and Georgia and what is called the Trans-Caucasian republics (Armenia and Russian Azerbaijan) on the south. There is a particularly wild area cutting across this block and known as the Tlyaratin, which embraces practically the whole basin of the River Jurmut and the upper parts of the Avarskoy Koysu which is a tributary of the Sulak, the main river of Dagestan.*

These mountainous regions are clothed in dense, montane, coniferous forests right up to the snow line and right down to the edges of the few villages that lie on the adjacent lowlands, and are, over wide stretches, really quite impenetrable. At the same time, the upper crags and rocky reaches are equally unapproachable except by well-organized professional mountaineering expeditions. Despite the most ancient civilization of the Caucasian region as a whole, and of the adjacent Armenian block to the south, huge areas remain quite unexplored. In these, large game reserves have been established, and these

---

* As far as possible I have endeavored to choose place names, such as Krasnodar and Baku, that can be readily found in standard atlases, and to use the traditional English spellings for these though these are almost invariably quite different from the official Russian and/or local spellings. Names that are not to be found on readily obtainable atlases or maps are spelt as given by the translators of the publications from which they were taken. In many cases in this and the following chapter I quote names that do not appear on any obtainable maps. These may well be altogether inaccurately spelt, having been rendered phonetically first into Russian and thence into English. The results may be quite horrible to the local citizens. For this I duly apologize, in the unlikely event that they ever read this book.

are populated by a very large and varied fauna including moose, some remaining Wisent or European Bison, Red Deer, mountain Sheep, Brown Bear, wolves, the great northern Lynx, and the Leopard. [The Snow Leopard's range does not extend west of the Hindu-Kush. However, Tiger occur in the Elburz Ranges even west of Teheran.]

Opposite the Georgians, on the northeast side of the main ranges, the hill folk are called Avars, those herders and hunters who have for centuries penetrated farther upward into these fastnesses than any others. Among them there is universal belief in and acceptance of the ABSMs they call the *Kaptar*. Surrounding peoples regard them with increasing skepticism as Folklore, Legend, or Myth in proportion to their distance from these unexplored uplands; which is the invariable rule as we shall see when we come to examine these matters (contracted to M, L, and F, in Chapter 17). The description they give of this creature is remarkably clear and quite invariable except for one set of facts. These concern the number of kinds of *Kaptar* that exist. The discussion on this point stems mostly from those who live farthest from the area where they are met with, and it has become enmeshed in a certain amount of straight myth, notably the curious notion that all of one kind are females. According to Russian investigators, however, those who so claim are the least likely to have firsthand knowledge of the matter, while they were quite unable to explain how this race of females reproduces and maintains itself. The notion of self-perpetuating, virgin birth, if I may so express the notion, has been widespread since time immemorial. It sounds absurd but, of course, it is not biologically impossible per se; at the same time, there is one very simple explanation for it. Even modestly civilized people sometimes separate the sexes in everyday living quarters, and my wife and I once spent some time with a tribal group of South Amerinds and had to reside in separate though adjacent villages. Then again, ABSMs seem to show a marked sexual dimorphism everywhere they are reported, this showing not only in size, but in color of fur or hair, while the young are said to look different again. Also, most ABSMs are stated to be solitary, only occasionally seen in pairs or with young in tow. The females, it seems, tend to associate in going to water, in food gathering, and so forth, while the males range widely. They are food gatherers rather than hunters and this we must not forget.

In the Caucasian region, the males seem to be encountered alone in the upper fastnesses whereas the females, which are readily recognizable it is said by the great and sometimes positively enormous development of their breasts (which, unlike any pongids, are pendant or hanging), show up at lower levels. Then, a Dwarf *Kaptar* is also spoken of, particularly on the southern face of the mountains, but as one Prof. V. K. Leontiev, who studied this business locally, with consummate discipline, observes, nothing is stated about these beings that obviates their being the young ones or "teen-agers," who also tend to band together and go off on their own. They are said to be smaller than the average man and to be clothed in reddish brown wool as opposed to the other two types—one of which, be it noted, is said always to be a male, while the other is always female; from which one may draw a rather obvious assumption one would have supposed—which are variously described as having dark gray, black, or silvered hair. This change of coat color, from gingery to gray-brown, to gray-black, and finally to white with age, is just as consistent with what is found among other Primates as is the change from shiny black in youth, as displayed by the sad little Jacko of the Fraser River, to brown and then grizzly. One must note that, with increasing age, those of us whose head-hair turns white will find that our axillary and pubic hairs do the same while those who have profuse chest hair will see that it also follows the head-hair in this respect. Thus a venerable male ABSM might be as white as the old chap who paced the truck in Oregon (at 35 mph, be it noted) and then popped into a lake. If Neanderthalers were hairy, they may well have had a fluffy wool, like that of a baby One-humped Camel as is so repeatedly stated by almost all the Eurasians who say they have met their local small ABSMs, and an "overcoat" of darker hairs like a muskrat and most other mammals of cooler climates, which develops with age, becomes profuse and dominant in the prime of life, and then goes silvery to pure white with age.

I cannot find any suggestion that there is more than one type of ABSM in this area, despite the fact that three quite distinct sets of names are applied to it there. The indigenous name is *Kaptar* and its derivatives and associates, but the Kirghiz "*Gul-i-aban*" group is also used among peoples of similar origin, while I find that the more distantly originating "*Almas*" stem also crops up in the form of "*Almasty*" and "*Albasty*." Some painstaking analysis of the origins of the reports of these names used in connection with the Caucasus area however brings to light the fact that the reports in which they were used were made by "foreigners" or at least by members of groups that are known to have moved in from the east. The Caucasus is an appalling mix-up; a sort of Grand Central Station for nomads, conquerors, emigrants, immigrants, wanderers, lost tribes, lost causes, and perhaps also indigenous evolution—hence the designation "Caucasoid" which actually means nothing. The oldest peoples in the area, which is to say those of whom we have no record of immigration, such as the Georgians and Avars, one and all adhere to the *Kaptar* designation for their local ABSMs—which, incidentally, have been perhaps facetiously called "Wind Men" by more frivolous outsiders!

That these manifestly original Caucasians—if not Caucasoids —are of one variety comes as rather a relief, especially at this juncture and before plunging into inner Asia, because there we are going to be beset by affirmations from all sides that there are not just two or three kinds in any one area, but that these are all quite different from others in other areas. I am not quite sure if we will be able to keep our heads through all that, and I am sure that I have not yet myself got it all straight, but in the meantime we may take what the Hollanders call a pause (but pronounce *powzer*) and try to come to grips with the *Kaptar*.

The clearest account of this creature is a firsthand one reported by none other than the Prof. V. K. Leontiev mentioned before and who is graced in one publication [No. 120, of the Third Publication of the Special Commission to Study the

Snowman of the U.S.S.R. Academy of Sciences, under the Direction of Prof. B. F. Porshnev and Dr. A. A. Shmakov] with the illuminating title of "Hunting Instructor of the Ministry of Hunting of Dagestan A.S.S.R." This is a man both of parts and of profound precision. I herewith paraphrase his account with due regard to that precision but with considerable compression.

It appears that in late July 1957, this gentleman with three associates was conducting an official investigation of a territory called the Gagan Sanctuary. On August 5 his companions returned to their headquarters as their work was finished, and Leontiev decided to make a few days' tour on his own. He was then at the head of the Jurmut River and spent two days there checking on some glaciers; he then trekked up a tributary stream. He notes that he came across leopard tracks on a patch of snow. After a rest overnight he continued onward and came across a set of quite different tracks on another patch of snow. He says that "you had the impression that this animal was walking on his toes—never getting very heavy on his heels . . . you could see that his big toe was unusually developed, but was it a toe or a claw? These footprints were deformed somewhat because of the snow being slightly in a melted condition."

The next night he camped under an overhanging rock but when preparing for sleep, "All of a sudden there came a strange cry. It stopped as suddenly as it started," he writes. "Then after a pause it repeated again; this time somewhere to the side of the original one. Then it was quiet. The cry was not repeated again. The cry was very loud. It wasn't like the yell of an animal—not any wild mammal or bird known to me could make such a sound, and yet it couldn't be a human being either. [And he is a professional wildlife conservator.] From where I was sitting to the origin of the cry was approximately 100 meters [110 yards], and at the time the cry was repeated, 200 meters. I just say approximately." The following day appears to have been a miserable one so that he decided to camp before dark at the head of the stream in a very dark gap. He ran out of matches and all the wood was wet but he just managed to keep the fire going long enough to brew tea; then, he chanced to look up at a neighboring snowfield to the south. Something moving thereon caught his eye and of this he wrote: "This creature was going across, ascending slightly the upper part, and away from me. At the moment I saw him he was approximately 50 to 60 meters away from me. It was sufficient to have only one glance of him to know that this was a *Kaptar*."

Leontiev goes on to state that it exactly resembled the descriptions he had obtained from all the locals adding, "He was walking on his feet, not touching the ground with his hands. His shoulders were unusually wide. His body was covered with long dark hair. He was about 2.2 meters [about 7 feet] tall." Realizing that this was a chance for the procurement of the most priceless scientific information but also realizing that he could neither catch nor, if he did, overcome the creature, Leontiev took careful aim and fired a shot at its feet. However, by this time the *Kaptar* was at extreme range for his rifle and he does not seem to have hit it for it turned to him and then with incredible speed waltzed about and ran up the slope with tremendous speed, cutting through the snowfield, reaching high rocks beyond and disappearing. Leontiev tried to follow but it was hopeless so he measured and sketched the footprints before it got dark. The next morning he re-examined these, made more sketches, and then spent the day searching around for the creature. Being out of food he had to leave the next day.

Altogether he estimates that he had the *Kaptar* in view for 5 to 7 minutes and pursued him for 9. He saw his back, left side and cheek; when he fired he had just a second's sight of the face for it was late evening, beginning to snow, and he could not see much detail. He then makes some most interesting remarks, to wit: "He was not too tall [7 foot would seem enormous to me, *Author*]; his shoulders were unusually wide; his arms were long—longer than a man's but shorter than a monkey's. His feet were *slightly bent* and very heavy [italics mine], and the whole body was covered with a dark gray fur. The length of the hair on the body was shorter than the hair or fur of a bear. He had especially long hair on his head. I had the impression that the hairs on the head were darker than on the body. I couldn't see anything of a tail. I couldn't see any ears. The head was massive, and when he turned to me, I saw for one second his face. It was somewhat like an elongated animal face, the general outline of the nose, lips, and forehead, or the chin or the eyes I couldn't see. I had the impression that his face, like his body was covered with hair. His back was slightly bent; he was stoop-shouldered. His general appearance was human-like. If you want to compare the *Kaptar* with some living creature the best comparison would be to think of him as a tall, massively built, wide-shouldered man, with a heavy growth on his face and the rest of his body."

Leontiev measured and sketched the *Kaptar's* footprints when only a few minutes old. Of them he says: "This footprint had a very strange formation. The whole print was about 25 centimeters [about 9 inches] long.* The general impression was of the toes pushed deeply into the snow. Also around the toes you could see some rough formation. The explanation is, of course, that he was walking with bent knees and like 'clawing' into the snow. The [outer] four toes did not come very close to each other, as in humans, but they were slightly spread out—about ½ of a centimeter to 1 centimeter. The width of the big toe was 3½ centimeters; in length, 9 centimeters. The length of the other toes about 5 centimeters. You had the impression that on all the toes there were very hard scar tissue formations—that the toes were widely separated and in between there was scar tissue formation. The entire print narrowed down toward the heel, and there were two parallel deep ridges like wrinkles.† You had the impression that it was not the whole step, and only the toes. This was not too unusual because when I looked at my own footprints I noticed that I put a little harder on the toes than on a heel and actually, that's the way the *Kaptar* would walk. The large toe was very far apart from the rest and it was very long. It seems when you look this over and study the print, the entire heel of the foot is covered with a thick growth of a tough hide interspersed with all kinds of little growths and heavy wrinkles.

* There would seem to be something wrong here. A 9-inch foot for a 7-ft. giant seems most improbable (*Author*).
† See the Shipton *Meh-Teh* prints.

There were no claws at all. This footprint has no resemblance to the footprints of any of the animals that I know. It doesn't look like a footprint of a bear, and, of course, is entirely different from a footprint of a human heel.

"The cry of the *Kaptar* is very strange and you cannot compare it with anything else. It consists of several repeating high-and-low pitched sounds, that remind you of the sound of a gigantic chord. There is certainly a kind of metallic quality about them. In the cry you can hear some plaintive note too. I, personally, did not experience any fear hearing this cry, but to me they seem to express the loneliness of a lost creature. I could not hear any coherent sounds, or perhaps I couldn't quite catch the fine shadings of the sound, just the way a human being pronounces them. The name *cry*, or terminology *cry* actually does not describe the sound that the *Kaptar* issues. This cry is peculiar, and so much of its own, that there would be many different ways of describing it and no particular way to give it *real* definition. At any rate, not any of the mammals or birds that I know have a cry similar to the *Kaptar*."

This is by no means the only close encounter with a *Kaptar* in modern times. First there are literally dozens of reports from locals including whole village populations who reported them about at various times, and sometimes for months and at low levels. Then also, one appears to have been captured in 1941 and physically examined by a lieutenant-colonel of the Medical Service of the Soviet Army, by the name of V. S. Karapetyan. I give this report verbatim as supplied to me, already translated, by the courtesy of the Russian Information Service. It goes as follows:

"From October to December of 1941 our infantry battalion was stationed some thirty kilometers from the town of Buinaksk [in the Dagestan A.S.S.R.]. One day the representatives of the local authorities asked me to examine a man caught in the surrounding mountains and brought to the district center. My medical advice was needed to establish whether or not this curious creature was a disguised spy.

"I entered a shed with two members of the local authorities. When I asked why I had to examine the man in a cold shed and not in a warm room, I was told that the prisoner could not be kept in a warm room. He had sweated in the house so profusely that they had had to keep him in the shed.

"I can still see the creature as it stood before me, a male, naked and bare-footed. And it was doubtlessly a man, because its entire shape was human. The chest, back, and shoulders, however, were covered with shaggy hair of a dark brown colour [it is noteworthy that all the local inhabitants had black hair]. This fur of his was much like that of a bear, and 2 to 3 centimeters long. The fur was thinner and softer below the chest. His wrists were crude and sparsely covered with hair. The palms of his hands and soles of his feet were free of hair. But the hair on his head reached to his shoulders partly covering his forehead. The hair on his head, moreover, felt very rough to the hand. He had no beard or moustache, though his face was completely covered with a light growth of hair. The hair around his mouth was also short and sparse.

"The man stood absolutely straight with his arms hanging, and his height was above the average—about 180 cm. He stood before me like a giant, his mighty chest thrust forward. His fingers were thick, strong, and exceptionally large. On the whole, he was considerably bigger than any of the local inhabitants.

"His eyes told me nothing. They were dull and empty—the eyes of an animal. And he seemed to me like an animal and nothing more.

"As I learned, he had accepted no food or drink since he was caught. He had asked for nothing and said nothing. When kept in a warm room he sweated profusely. While I was there, some water and then some food [bread] was brought up to his mouth; and someone offered him a hand, but there was no reaction. I gave the verbal conclusion that this was no disguised person, but a wild man of some kind. Then I returned to my unit and never heard of him again."

On the little map of Asia in a box at the left-hand upper corner of Map XII, you will see a small vermiform tongue of shading sticking out of the left-hand lower corner of the contained rectangle. This represents the extension of the Hindu-Kush Range, via the Ala-Dagh and the Elburz in Iran, to the Armenian highlands and the Caucasus in the west. This is the farthest west for ABSMs in the Old World unless some really very startling though admittedly vague reports that have just reached me from Sweden should have substance. The Scandinavian countries are hotbeds of myth, legend, and folklore regarding ABSM-like creatures of long ago but these new statements sound suspiciously like our own Northwestern ones. I must admit that this has quite unnerved me and I am not prepared to say any more until I have at least made some attempt to investigate. We may therefore turn east again and will follow that little wormlike strip back to the Roof of the World. Along the way, we pass through the Elburz Ranges.

These are quite surprising for their wildness and the existence therein of such obvious things as Tigers only a day's drive from Teheran. But then, I suppose it is really no more odd than Jaguars wandering about almost within sight of Los Angeles. Nonetheless, there is plenty of space here for lots of big as yet uncaught things and, by jingo, we get an alleged ABSM. This came to me from the indefatigable Bernard Heuvelmans, in the form of a plea for help since we are a sort of private "Bureau of Missing Persons" for the natural sciences, among other things. It transpired that a gentleman in New Jersey had written Bernard and stated: "When I was in the Army [in World War II], one man in my company was an engineer who had worked for an oil company in Persia. He and I talked together for hours and hours, as men do in the army, and I never detected him in a single lie, or what I thought was a lie, or even suspected that he exaggerated anything, but for one curious thing.

"He said that when he was working in Persia, some Persians brought around a 'gorilla' they had killed in the mountains. I was amazed that he should say such a thing. I assured him that there were no gorillas in Persia, or anywhere else outside of Africa. He said that it was as big as one, and surely looked like one. He saw it, and that was enough. I said that there were no anthropoid apes in Asia closer to Persia than the Malay Peninsula [sic]. He was indignant. Was I telling him that he didn't see it? Of course, he thought also that there were no gorillas outside of Africa—until he saw this one. He was a bit short-tempered about it, so I dropped the subject."

[I have not yet traced the gentleman concerned but his name is Daniel Dotson; his home state is Utah but he was in Washington, D.C. when he joined the Army. If anybody knows him, for the love of mike, please write me; and if you know where he is, don't wait on ceremony but extend to him my invitation to dinner forthwith. He can name the time and place.]

This is the only specifically Iranian (Persian) report that I have but there are others from the Iranian-Turkmen S.S.R. border, and more from the Iranian-Afghanistani border. The geography of this and the adjacent area, which I call that of the Pamirs generally, and to which we will now proceed, is so complicated both physically and politically that I have to resort to the accompanying little maps. Most of the material that

The borders of the U.S.S.R., Iran, Iraq, and Turkey, and the Caucasus. Dagestan is one of the Union of Socialist Soviet Republics.

The borders of the U.S.S.R., China, Pakistan, and Afghanistan. About these borders is the Pamir Range. B.A.A. is the Badakshan Autonomous Area.

immediately follows comes from Russian sources and I am simply following their breakdown of this into regions of their designation.* These have political or rather ethnic tabs on them such as Kirghiz, Uzbek, Tadzhik, Kazakh, or simply "Chinese" assigned to them. This is really most muddling for the boundaries of these groups are utterly bewildering and interlocking as the map shows, while all these peoples have been surging about for centuries, elbowing each other, and dozens more peoples, in and out of valleys and off plateaus, gradually getting themselves worked into a sort of political pudding. Also lots of them are still nomadic, while families

---

* These sources are first and foremost four Booklets issued by the Special Commission set up to study the Snowman Problem by the Academy of Sciences of the U.S.S.R., under the Direction of Prof. B. F. Porshnev and Dr. A. A. Shmakov. Bks. 1 and 2 were published in 1958; Nos. 3 and 4 in 1959 in Moscow. Secondly there are a number of articles kindly sent to me by Prof. Porshnev and a voluminous report made available by the Russian News Services.

and sometimes whole villages just up and move somewhere else. Then the tab "Chinese" in this case means simply that the place is on the Chinese side of the border, here principally Sinkiang, but also a whole host of other border provinces, autonomies, and such. Finally, a considerable percentage of the place names cited are not on any map; not even the most excellent, modern, Russian maps. This area must therefore be understood to encompass not only the Pamirs themselves, but the adjacent mountainous portions of Afghanistan, the Uzbek, Tadzhik, Kirghiz, and Kazakh S.S.R.'s and the Badakshan Autonomous Area [to be called simply the A.A.] unless otherwise stated. This of course runs off into the Karakorams to the east and the Ala-Tagh and Tien-Shan to the north. Most of the information from these regions was unearthed by the 1958 Expedition of the Russian Academy of Sciences to investigate the "Snowman" problem there.

In one of the Russian booklets cited, a map was included showing the distribution of myth, legend, and folklore about ABSMs in Eurasia; the areas from which reports of sightings, encounters, and tracks have been recorded within this century; and a dark globular blob covering this general Pamirs region. The legend states that this blob or blot was considered by the Soviet scientists to be the last remaining stronghold and the only remaining breeding ground of the Asiatic ABSMs. This is a very curious notion and not strictly in accord with either the published opinions of Prof. Porshnev himself [due to a certain very pertinent, and in my opinion, correct observation that he makes in one of his articles] nor with those of the Mongolian and Chinese scientists. In fact, I am of the mind that it was an idea imposed on the Commission by a sort of backhanded tradition stemming from the days before ABSMs were taken seriously even in Russia. Once again it was probably due to the old "*snow*man" bit; the everlasting reiteration that the creature or creatures *lived* in the perpetual upland snowfields, the obvious corollary to which was the biggest and most perpetual snowfields were the most logical places to look for them.

Professor Porshnev however states in what I can only describe as a stirring article in a magazine entitled *The Contemporary East:* "The expression Snowman is not supposed to mean a creature living among perpetual snow (or exclusively in the snow). Similar expressions are used in connection with some animals, like the Snow Leopard. It means only that this specimen (species) belongs to the fauna of the high mountain ranges. He appears on the snowfields or glaciers only while migrating. He lives, however, and finds his food *below* the snowline, among the rocks and alpine meadows, sometimes even in the subalpine zone, in the forests, as well as among the rocky sands of the desert and in reedy thickets. The alpine zone [i.e. Upper Montane coniferous forest] is known for its rich and lush vegetation and the variety of its animal life." Professor Porshnev is so exactly right.

Neither the Pamirs themselves nor the area generally are wholly snow-clad; as a matter of fact the whole is a vast hodgepodge of deep valleys, gorges, canyons, and intermediate ridges, and all the former are heavily forested up to considerable heights being at a rather low latitude. This may be called a wilderness area but it is not, strange as it may seem, anything so much like one as our own Northwest. There have always been people there, or barging through it, since most ancient times and today there are meteorological stations dotted all about it, while the extensive international boundaries that meander through it are not exactly left to the imagination or desires of the locals. Both the Russians and the Chinese have conducted rather thorough explorations into the area, while the Afghans live there, as do most of the Tadzhiks and Kirghiz, and quite a lot of other people.

Practically everybody who does live, or even camps there, is of a single mind about the existence among them, and all over the lot, of ABSMs. This is another case such as that of the Himalayas, the Great Gutter, and the Southern Tibet Rim, where the cases reported are just too numerous to detail as well as too consistent to be worth recording specifically. Such a procedure would be quite silly: rather like recording sightings of Mountain Lions from our Southwest. The bloody things are everywhere and seem always to have been; nobody locally paid much more attention to them than they did to other large wild fauna until outsiders started asking about them. Then they mostly clammed up; for two very different reasons, however.

First, the ancient animism mentioned above, is here even more deeply ingrained, but more shallowly covered by modern faiths such as Buddhism, and Islam than it is in the Caucasus, so that ABSMs being only just not men are regarded as ideal recipients for departed souls and should not be molested. This leads to taking special pains to steer foreigners away from them, while not mentioning their real names but referring to them vaguely, in generic terms. Secondly, to put the matter frankly, boiled ABSMs produced the most extremely potent and magical medicines for which really vast sums, in bar gold, were once paid in Russia, China, and especially in India. These medicines were known to the most ancient Chinese, to the Mongolians, the Tibetans, and to all Mongolic peoples all the way to Turkey. In the Pamirs area, the boiling, preparation, export, and marketing of these ABSM extracts (*moomuyam*) [called *mumer* by some] was carried on principally by Gypsies —referred to as the Luli or Asiatic Gypsies—who wandered all over the lot but mostly in directions exactly contrary to the normal annual migrations of the nomads for very obvious trade purposes. These Gypsies held a very peculiar and unique position in this part of the world. They were regarded as having sort of direct lines of communication both with God and the Devil, [and whole pantheons of other entities to boot] and so to be both able and sanctioned to tamper with most venerated things. Actually, like their Western congeners, they were consummate poachers, and since they could not be prevented from hunting anything, however sacred, and did not seem to suffer any dire consequences from doing so, they were assumed to have some special immunity or divine dispensation. At the same time, the whole concept of "Extract of ABSM" was probably a hang-over from most ancient ritual cannibalism, whereby token consumption of special parts of a powerful quarry or enemy [or even fellow citizen] was believed to transfer to you some of his powers. I witnessed just this process in the

Cameroons, West Africa, when an enormous male Gorilla was killed. The local Juju-chap begged bits of certain glands—and he knew his anatomy as well as any college demonstrator—and other parts of the body, made a brew out of these, and passed it around to all the hunters who took a token sip and smeared some on their gun barrels.

These two factors—the deep-seated reverence for ABSMs by the locals on the one hand, and their value as "medicine" on the other—have proved to be most potent ones in keeping information about the creatures from all outsiders. Personally, I suspect that there is something of the first attitude current among both the Northern and Southern Amerinds. This whole attitude in both its aspects comes to light in another way. This is the careful preservation of the heads and hands of ABSMs—and other Primates as well, it may be noted. The head, dried whole, has special significance, not for ingestion, like the extract, but as an object with its own medicinal qualities, and like any other sacred reliquary is kept hidden. This custom is pre-Buddhist but has been incorporated into Lamaist practice. The hands have another significance. They are kept as mere talismans, not having any deep religious significance, but rather because the hands of Primates (and men) have always seemed a marvel to Mongolian peoples, being literally the key to the success of both. There are mummified or desiccated hands kept in monasteries and by private individuals of communities all over Eastern Eurasia, from the Great Barrier, east. A few in Nepal have been shown to foreigners as we have related; others have been shown to Mongolian and Chinese scientists; and there are a few reports of them recorded in the Russian publications.

This is not the only aspect of ABSMery that presents a completely different face once we pass east, up and on to the great highlands of the Middle Mongoloid peoples. Here is the true heartland, not only of the greater part of modern humanity, but of culture also, for learning was apparently thriving there when even the Greeks were yet occupied in little else but bashing the Minoans' and each others' heads in, while we in the far west were running about clothed in blue paint and chip-

[303]

ping flints. The ancient repositories of knowledge and of documents lie sprawled up the great "basin" that forms the center of these eastern uplands, between the Great Barrier on the west, the southern Rim of Tibet on the south, and the escarpment on the east that fronts onto the lowlands of Manchuria and China proper. Around the periphery, along the Himalayas, through the Pamirs, and northeast up the Great Barrier to the region of Lake Baikal, there is only a secondhand knowledge of this ancient erudition or of its records of such matters as ABSMs; this knowledge moreover is often vague and distorted. The peoples of the western Pamirs were mountaineers, hunters, shepherds, and agricultural peasants; those of the Barrier itself mostly nomadic herders, who moved back and forth along the steppes that fringe the Barrier to the west and north, and stretch west to the Caspian. They were not literate and they did not support centuries-old libraries in monasteries, as did the inner Mongols. The Chinese on the other side of the uplands were settled agriculturists and at an early date took to city dwelling and the formation of city-states. They too developed an advanced "learning" but, despite the fact that "China" has for centuries nominally spread west to the Pamirs and to the inside of the Great Barrier, it absorbed more culture from those inner regions than it exported to them, while China proper was itself constantly overrun by Mongols coming notably down from the north through Manchuria.

When we get onto the great plateau, or rather into its great basin we will meet for the first time straight talk about ABSMs, rather than rumors, hearsay, and the somewhat dumfounded disbelief that we have encountered everywhere else, even among the most erudite. Educated Mongolians, using that term in its widest and proper sense to include all the peoples from the Siberia border to Nepal, and from Sinkiang to the Chinese escarpment, have a wealth of historical record about ABSMs, and are brought up to the notion that they still exist, *in several distinct forms,* all over their country, in isolated pockets, and all around its periphery in an almost unbroken line. Modern scientists of the Mongolian Peoples' Republic are fully aware of this and are beginning to restudy, reappraise, and make known to the world this store of knowledge, but they have as yet only just scratched the surface. The matter of ABSMs is really a rather abstruse item in their fund of knowledge. Mongolians are very practical people and although they have for millennia delved into every aspect of life, ethnology per se was one of the last of their interests. Wildlife was important, and medicine very much so, and it is in these literatures that amazing facts about ABSMs are found, as we shall see in the next chapter.

14. The Eastern Horizon

Anything marvelous like the compass or gun-powder that came to the West was once immediately said to have come "from China." This is doubtful on two counts.

Russian scientists appear to have been just as stunned as those of the West about a decade ago when they were confronted with the new turn in ABSMal events that took place after the Shipton foot-tracks uproar. Further, despite the fact that their jurisdiction has for a long time marched with the Great Barrier, and the expansion of their country was in the past centuries eastward, just as ours was westward, while their interests in inner Mongolia were multifarious, they don't seem to have known much more about this huge, truly mysterious subcontinent in upland eastern Eurasia than any other outsiders did. While the Westerners, led by the British, had been nibbling away at its southern border via India, and the Americans had shown some interest via China, neither had really even penetrated the great triangle. Some travelers considered rather intrepid had crossed it, and a few naturalists had accomplished bizarre tasks like unearthing nests of fossil dinosaur eggs therein, but the accumulated *lack* of knowledge about it mounted steadily. This is not to say that Europeans had not been traveling through it since very early times, for some became immortalized for their accomplishments like Marco Polo and the great Russian explorer Prjewalski in the last century. There were also lesser known but equally intrepid explorers, such as one Johann Schiltberger of Bavaria in the years 1396 to 1427. In modern times there have been men of exceptional perspicacity such as J. Nicholas Roerich and the

Englishman Peter Fleming. The list is of course almost endless, and in all this Russians have played a most prominent part. Yet, despite the fact that a very high percentage of these travelers throughout the ages seem to have mentioned ABSMs, and not just casually, the concept of the continued existence over enormous areas of some of our primitive ancestors, of sub-men, and possibly even of sub-hominids simply did not penetrate the collective mind of Russian scholarship any more than it did that of Westerners. The Russian expedition to the Pamirs went out every bit as unprepared as the Western expeditions to the Himalayas, with all the same preconceived notions and misconceptions, and it came back just about as mystified and empty-handed.

However, the Russians had in the meantime made special investigations in the Caucasus, and they had sent another party to the northern face of the Everest Block—which ended in a sad disaster—and they had offered their Chinese and Mongolian colleagues co-operation in investigating the whole matter. This intelligent approach was prompted in part by the growing tumult in the popular press and in scientific circles in the West about the "Abominable Snowman," and in part by certain historic discoveries of what may be called a purely bureaucratic nature by Prof. Porshnev. These prove to be a sad commentary on just about everything, but somehow make one feel a little better about some things. They demonstrate that we are not the only dumb clucks, or the only ones to let our scientific hierarchy obliterate any signs of novel thinking or unexpected discovery. I would like to tell this story in full not only for the sake of humanity but because its various facets point up just about everything that has been wrong with the study of ABSMery during the past century. Space does not permit and all I can give is the bare bones.

There lives in Moscow today a scientist by the name of V. A. Khakhlov who in 1913 submitted a full and detailed report on the east Asiatic ABSMs to the Russian Imperial Academy of Sciences. This priceless material was shelved, he was denied funds to continue his field investigations, and he was frankly told to shut up. Professor Porshnev happened to stumble on these reports in 1959 and sought out Dr. Khakhlov. He writes of his first interview: "Here he sits in front of me, this white-haired man, an emeritus scholar, a Professor of comparative animal anatomy, a scientist who made valuable contributions in the field of zoology. He talks about the discoveries he was about to make while he was a young man; his talk is enthusiastic and bitter at the same time. He is bitter not only at the general attitude taken by the pre-revolutionary Academy of Sciences, but at the action of his former advisory professor, P. P. Suschkin. In 1928, Suschkin came out with a startling, at the time, hypothesis; namely, that the region where the change from a monkey (sic) to a man took place was on the high plateaus of Asia . . . but not one word about the extensive contributions made by him [Khakhlov] or about his reports of the existence of 'Wild Men' in Central Asia." Need I say more?

Nor was Khakhlov the only enlightened scholar and enthusiastic field worker who was rapped over the knuckles and threatened with limbo at that time for the same reason. There was a young man named B. B. Baraidin who in 1905-07 specialized in Eastern folklore. He was given a commission to travel through Mongolia to Tibet on assignment from the Russian Geographical Society of [then] St. Petersburg. While doing so he encountered an ABSM at close range, while in company with many others in a caravan, and a young monk pursued the creature, which the locals called an *Almas*. Baraidin made a full report on this, but his boss, one S. F. Oldenburg, head of the geographical society and Secretary of the Academy of Sciences, ordered him to delete all mention of the matter from his report, stating that "no one will ever believe that, and it may prove embarrassing." At least, they were direct about it in Imperial Russia!

Young Baraidin had been befriended by a Mongolian scholar named Z. G. Jamtzarano, and when he told him of the incident, the latter was inspired to devote much time to pursuing ABSMs. This he did with the help of two assistants named A. D. Simukov and a Dr. Rinchen. The latter is now a Professor at the University at Ulan Bator, Mongolian D.R., still most actively engaged in pursuing the matter, and has given a great deal of information to the Russian Special Commission. It was these "discoveries" in the attics of Russian science that did more than all the firsthand reports of tracks and encounters along the entire length of the Great Barrier to aid Prof. Porshnev in mounting a proper investigation of ABSMery in Russian territory. Yet there remained a great skepticism, right up till the time of departure of the expeditions and investigators. In the previous year (1957) one A. J. Pronin, a hydrographer from Leningrad University, had made the world press with a story that he had observed an ABSM twice, for a brief time but at some distance on the Fedchenko Glacier in the Pamirs. This had at first been proclaimed by, but then just as violently decried in the Russian press, to such an extent that the inevitable debunking—which as usual amounted to nothing more than some "expert" saying that he did not believe him—was seized on by everybody as final and absolute proof that all ABSMs have never been anything but myths. I am sincerely sorry for Mr. Pronin, but I must say that this also makes me feel a little better: for it is manifest that our press also is not the only muddleheaded group or the only one that jumps to grovel in abject compliance with the least pontification on the part of an "expert." Reviewing press reports on the 1958 expedition to the Pamirs, I find an almost similar story—first considerable enthusiasm and even pride in this open-minded and truly scientific endeavor, then a sad retrogression to the age-old bolt-hole . . . "Sorry to have to do this, dear Readers, but I'm afraid we have finally to bury the poor Snowman. He turns out to be just a myth after all. We hate to see a good myth die; we need them in this day and age, but . . . etc, etc, etc." I have a whole file on these periodical requiems on ABSMs, the latest, as of writing, a highly facetious lead editorial in the *Christian Science Monitor* anent Hillary's scalps. They make amusing reading but are a sad commentary on intelligence in whatever part of the world.

Nonetheless, although this expedition to the Pamirs did not bring back a pickled ABSM, it did bring to light a wealth of most fascinating reports. Not a few of these were from resident Russians. One of the most notable only came to light later in a communication to Prof. Porshnev, who remarks of it in one of his articles: "Not only the 'authority of official science' acts as a hindrance to obtaining more information about the 'snow men.' There are other obstacles as well, which incidentally still remain: lack of co-ordination in gathering data is the most important [or most detrimental of all]. Investigators working in different regions are not aware that similar data is being collected in another area and, lacking this most basic tool of science—comparison—they are unable to accomplish anything. As an example we can cite a recent communication from a geologist by the name of B. M. Zdorick. He writes that much to his regret while he was in the Pamirs in 1926-38 he had no information about the Himalayan ABSMs [called *yeti*], and just could not understand all the stories he was told about furry men, *or even what he had seen himself* (italics mine, Author)."

"In 1934 Zdorick accompanied by his guide was making his way through a narrow path among a growth of wild oats on a little alpine plateau at about 8000 feet altitude between the Darwaz Ridge and the eastern reaches of the Peter the First Range. Unexpectedly the path leveled off and one could see how the grass was trampled on, the ground giving evidence that someone was digging around. There were splotches of blood on the path and remains of a gopher's skin. Just a little way from Zdorick and his guide, on a mound of freshly upturned earth, was a creature, asleep on his belly, fully stretched out. He was about a meter and a half in length (approximately 4 feet 10 inches). The head and the forward limbs could not be seen because they were hidden by a growth of wild oats. The legs, however, could be seen. They had black naked soles, and were too long and graceful to have belonged to a bear; his back was also too flat to be a bear's. The whole body of this animal was covered with fur, more like the fur of a yak, than the rich fur of a bear. The color of the fur was a grayish-brown, somewhat more prominent brown than a bear's. One could see the sides of the creature moving rhythmically in his sleep. The fear that took possession of the guide transmitted itself to Zdorick and they both turned around and ran for their life, scrambling and falling in the tall, wild grass.

"On the following day Zdorick learned from the local residents, who were much alarmed by the news, that he came across a sleeping '*dev*.' The local residents used another word in naming the creature, and Zdorick had the impression that they were using the word '*dev*' just for him, so that he could understand better. The local residents ventured the information that in valleys of Talbar and Saffedar there were a few families of these '*devs*'—men, women and children. They were considered like beasts, and no supernatural power was ascribed to them. They cause no harm to the people, or their stock, but meeting them is considered a bad omen.

"The geologist was very much surprised to hear that the '*dev*' was listed as an animal, and not a supernatural creature. He was told that the '*dev*' looked like a short stocky man, walking on two hind legs, and that his head and body were covered with short grayish fur. In the Sanglakh region the '*dev*' is seen very rarely, but they do roam about, either singly, or in pairs—male and female. No one had seen any young ones, but last summer they caught a grown one at the flour mill, where he evidently was eating either flour, or grain. This was at the eastern foothills of Sanglakh, only a few kilometers from Tutkaul. The captive was chained for about two months by the mill and was fed raw meat and flour pancakes. After that he broke his chain and escaped. They also pointed out a man who had a large scar on his head from a wound supposedly inflicted by the '*dev*.'"

The list of encounters with, let alone mere sightings of ABSMs throughout the Pamirs region generally (as defined above) are literally endless. The same can be said of the other major areas of the Mongolian upland triangle. These areas are as follows: first the super-upland plateau of Tibet with its three principal super mountain ranges, in the south the Rim with the Karakoram, in the middle the Kunluns that turn south to the head of Indo-China, and along the north the Altyn Tagh, that leads into the Nan Shans and on to the Tsin-Lings of China. Second, north of the Pamirs lie the Alai-Tagh, and from them stretch the Tien Shans to form the northern boundary of the Tarim Basin of Sinkiang. Next, north of these come the Grand Altai, forming the southern border of Mongolia proper. North of these are the Tannu-Ola and the mighty Khangai between Mongolia and Tannu Tuva. Still north again come the Sayan complexes and the Baikals, lying along the shore of the great lake of the same name. Then, in the Gobi Desert lie the Yablonovoi Mountains. Finally there are the Khingans running north to south between the Gobi and the eastern lowlands of Manchuria. There is some suspicion that ABSMery may have to be extended still farther north through the Stanovois, to the Dzhugdzhurs and Gidan Mountains which border the Sea of Okhotsk. There is also a most important triangle sandwiched in between the Nan Shans, the eastern end of the Tibet Rim and the upper end of the Indo-Chinese Peninsula, that has no collective name, but is filled with immense north to south ranges. This lies in Sikang, now incorporated into the Chinese Province of Szechwan. [All these subareas or natural provinces will be found on Map XII with the exception of the penultimate group which are in far eastern Siberia, and from which we have no definite ABSMery.] I cannot stress too forcibly the sheer volume of such reports and of those of foot-tracks, droppings, and other corollary evidence that have been found year after year all over all of these subareas within the great upland Mongolian Triangle. The full record of those that have been published—some 200, that have been properly investigated and assessed scientifically by competent specialists—will form the subject of another book. For now, I shall have to confine myself to a few samples and some further explanatory remarks about the country, vegetation, and general background against which they were recorded.

First, in the general Pamirs region, the Russian expedition brought to light half a dozen most recent and categoric reports. One was supplied by a man, described as "quite well-to-do," resident in Chesh Teb, who did a lot of hunting for pleasure. In 1939, in the spring, about 4 o'clock in the afternoon, while he was walking around he saw some man who actually jumped on him. "They started wrestling. This was a *Gul-Biavan*. The hunter was very strong and tall and heavy and once he was able to lasso a bear. Now, this hunter wrestled with *Gul-Biavan*. The *Gul-Biavan* was covered with short, soft wool and the man could not get hold of anything. On the face of this man there was also short wool and there was a terrible odor coming from him. Finally, the hunter was able to throw the *Gul-Biavan* to the ground, but at the same time he lost consciousness himself. The villagers came upon the man and brought him home. When he came to, it was late in the evening and he told how he met the *Gul-Biavan,* and the villagers told him that he was lying on the ground, and the ground around him bore evidence of this wrestling match."

In the same area intelligent local people made many sworn statements such as "A man in Roharv was traveling with two others through the Pass of Karategin and Vahio, when they saw a naked man covered with short black hair, who was slightly taller than an average ordinary man, and which had *a very strong smell*" [*italics mine*]. As elsewhere all over the world, this matter of a strong stink attached to ABSMs keeps cropping up throughout the east Eurasian cases. Then, there was the hunter, Andam Kerimov, from a place called Uskrog between Roharv and Bodaudi, who called the creature he encountered a *Voita* (just another local name for an ABSM). It was not much bigger than a man, was covered with hair but not much on its chest. It had a bare face, and ears sticking out, the nose was wide, and "over the nose and on the ears he did not have much hair." Rather pleasantly the report states that "At the time Andam met the *Voita* he was leading a goat but gave way to him—the *Voita*." A group of hunters named Alaer, Altibai, Matai, Beksagir, and Tastambek who were with the reporter and his father one Abdurahmanov Abdulhamid, when encamped for the night, heard "something treading lightly on the grass" and running out apparently with a light saw what they called a *Gul-Biavan* about 6 feet 6 inches tall covered with hair. "It had a powerful and unpleasant smell." This was in 1951.

Some of the most interesting information collected on east Eurasian ABSMs comes from Khakhlov's original inquiries at the beginning of this century, mentioned above. Khakhlov obtained most of this through that group of the Kazakh nation which had moved northeast and settled along the edge of the Great Barrier, north of Kirghiz territory and north of Lake Balkhash, in the area lying between the Abakan Mountains and Tannu Tuva. These people were actually foreigners to and were not acquainted with the uplands beyond the Barrier but they penetrated into it via certain lowland basins having entrances pointing to the west. The most notable of these is called Dzungaria which is an immense lowland pocket, into which the western steppes [i.e. prairies, to us] penetrate via two great valleys separated by the Tarbagatai Mountains. Patient inquiry by Khakhlov elucidated the fact that reports gathered by the Kazakhs from a wide area seemed all to come from Dzungaria. Khakhlov makes a point of noting that these reports came from herders, hunters, and those engaged in other pursuits strictly in that order numerically. His first most astonishing discovery, which has recently been much confirmed was that the ABSMs from that region had "been seen, captured, left footprints in sand, had an odor, resisted capture and yelled, and lived in captivity for a while."

"One witness, a Kazakh, stated that he was in the mountains of Iran-Kabirg and once, together with local herders, was taking care of a herd of horses at night. Toward dawn they saw some man prowling around and suspecting a thief, they jumped in the saddle taking along long poles with nooses which are used to catch horses, 'arkans' [lassos]. Because the 'man' was running awkwardly and not too fast, they succeeded in capturing him. While he was being captured, the 'man' was yelling, or rather screeching 'like a hare.' Looking the captured creature over, the herder explained to the visitor that this is a 'Wild Creature' not doing any harm to any one, and that he should be released.

"The 'wild man' was a male, below average height, covered with hair 'like a young camel.' He had long arms, far below his knees, stooped, with shoulders hunched forward; his chest was flat and narrow; the forehead sloping over the eyes with prominently arched brows. Lower jaw was massive without any chin; nose was small with large nostrils. The ears were large without any lobes, pointed back [like fox's]. On the back of his neck was a rise [like a hound's]. The skin on the forehead, elbows and knees hard and tough. When he was captured he was standing with his legs spread, slightly bent in the knees; when he was running he was spreading his feet wide apart awkwardly swinging his arms. The instep of the 'wild man' resembled a human, but at least twice the size with widely separated fingers [toes]; the large toe being shorter than that of humans, and widely separated from the others. The arm with long fingers was like a human arm, and yet different.

"When the 'wild man' at the insistence of the herders was allowed to go free, both Kazakhs followed him and discovered the place into which he had vanished: an indentation under a hanging rock strewn with high grass. The local residents offered additional information about these creatures: that they lived in pairs, seldom seen by people, and not at all dangerous to humans.

"A second witness found by Khakhlov stated that for several months he observed a 'wild man' in the regions of the River Manass, or Dam. This creature of female sex was sometimes chained to a small mill but was also allowed to go free. The general description was the same as of the male: hairy cover of the skin, stooped, narrow chest, shoulders were inclined forward, long arms; bent knees, flat insteps, spread out toes resembling a paw, the contact with the ground flat without the instep. The head is described in the same fashion—absence of a chin and a rise in the back.

"This creature seldom issued any sounds and usually was quiet and silent. Only when approached she bared her teeth and screeched. It [sic] had a peculiar way of lying down, or sleeping—like a camel, by squatting on the ground on its knees and elbows, resting the forehead on the ground, and resting the wrists on the back of the head [see p. 316]. This position accounts for the unusually hard skin of the elbows and knees—like camel's soles. When offered food, the female ate only raw meat, some vegetables and grain. She did not touch cooked

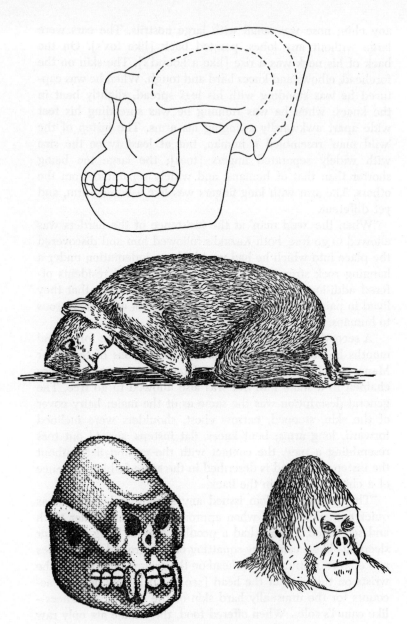

See facing page for legend

meat, or bread, although later she was getting used to bread. Sometimes she would catch and eat some insects. She would drink in animal fashion, by lapping water, or sometimes she would dip her arm in water and lick the water. When she was allowed to go free, she ran awkwardly swinging her long arms, to the nearby reeds where she disappeared."

Khakhlov notes that "This creature * has nothing in common with the *Jez-Termak* ('Copper-Nails'), or with the *Almas*." This is a most startling statement but one of the utmost significance, and also one that has been totally overlooked. We find on analyzing the reports from the general Pamirs area that, despite variations in coat color, all ABSMs there appear to be of one type. This is about man-size, and in many respects very manlike with, as is frequently mentioned, something like a primitive language or at least a vocal communication system more elaborate than anything known among animals. Its footprints, while having a very widely separated and extra-large big toe, are said to be human in form. This type—which incidentally seems to be the same as the Caucasian *Kaptar*—is most commonly called by one of the names in the *Guli-avan* group [due, of course, to the regionality of the languages in the area] and extends north into and along the Tien Shans, and east into the Kunluns and the Karakorams. Moreover, I do not know of any remarks to the effect that there are more than one kind of ABSM throughout these three regions. The larger and more bestial *Meh-Teh* type with *two* semi-apposed toes begins only east of the Karakorams along the Southern Tibetan Rim and, possibly, in the Himalayas; though—and this

* Known locally by the names grouped around the stems *Ksy* and *Giik*. The latter means "wild" as in *At-Giik*, Wild Horse.

(*Top*) Hypothetical skull of the *Ksy-Giik* type of Abominable Snowman as reconstructed by Russian scientists.

(*Center*) A drawing made by Prof. Khakhlov of the *Almas* type of Abominable Snowman from native descriptions.

(*Bottom, left*) An ancient mask from the great Mongolian plateau. (*Bottom, right*) Reconstruction of head and face of the creature on the mask, drawn by Russian scientists.

is exceedingly strange—I cannot actually find a single report of this type from any part of that range.

It would seem that Khakhlov's creatures from around Dzungaria are also of the *Guli-avan* type and that they extend north to the Grand Altai and beyond to the Sayans. The descriptions of those from the Nan Shans [which is to say the *northern* rim of the Tibetan super-uplands] seem also to be similar. This creature, which gives every indication of being a Neanderthaler-type of sub-man, and whose footprints *exactly* match those of a Neanderthaler discovered in a cave in Italy,* would seem to be a mountain dwelling form: in fact, they *are* montane forms, for not one single report of them from down on either the lowlands or even on the Mongolian Plateau itself have ever been recorded.

From the latter, which is to say Mongolia proper [with the Khangai Mountains that arise in its center] and the vast Gobi Desert, together with the Ala Shan, the Ordos, Turfan, and even possibly the lowlands of Dzungaria and the Tarim, we get reports only of the little *Almas*. These are much smaller and apparently even more human, and seem always to have been regarded simply as extremely primitive humans; hairy and without speech understandable to us, but having more or less all the human qualities such as suckling human infants and even, it has been alleged, "trading" with normal humans, in that they would leave skins at appointed places, and take away certain simple basic articles left there by the nomadic tribesmen in return. There is even a report of a scholar in a

* In the spring of 1948 the official caretakers of a cave known as "The Witch's Cave" near Toirano in Italy [all caves in Italy are government controlled] obtained permission to blast through what proved to be 11 feet of flowstone forming a blockage to one of the cave's passages, in the off-tourist season. They had seen bats flying in and out of a small hole leading into this and had rightly assumed that unexplored areas lay beyond. On breaking through, extensive passages with clean, smooth, wet, clay floors were found. On these there were enormous numbers of foot-tracks of large cave bears, of modern-appearing man, and of what are obvious Neanderthalers, together with many artifacts and even evidence of some kind of game played by throwing clay balls at a circle on the wall. The cave seems to have been finally abandoned and sealed by the flowstone curtain about 50,000 years ago. An outline tracing of one of the Neanderthaler footprints found therein is reproduced in Appendix B.

Mongolian monastery who was a half-breed *Almas*. This report comes from Prof. Rinchen, mentioned previously, and reads: "There was a lama in the Lamin-gegen monastery who was famous for his scholarship, and known under the name of— "a son of an *Almasska*." The father of this lama supposedly was captured by *Almas* and begot a boy with an *Almas* woman. Both father and son eventually managed to escape by joining a passing caravan. The boy was allowed to become a pupil in a monastery and achieved scholarly fame." The same informant, one Gendul from Khuremal of the Bainkhongor District, recently told [Dr. Rinchen] that in 1937 he saw in a monastery, Barun-Khure, an entire skin [or hide] of an *Almas* which was fastened to the ceiling of the temple. This skin was taken off by a cut along the back so that it remained practically intact and one could see that it had human-like legs and arms. The face was framed by long hair hanging from the head. The entire skin was covered with cabalistic signs and painted by the lamas. This *Almas* was supposedly killed in Gobi and brought as a gift to the monastery by a famous hunter, Mangal Durekchi."

Although I am getting somewhat ahead of my story, I would like to point out that the idea of a half-breed Neanderthaler becoming a great scholar is not to be scoffed at. Those sub-men had relatively large brains, while there is really no evidence that a large brain is necessary for a large intellect. One should take to heart the couplet that states:

> "Little brain, little wit.
> Big brain, not a bit"

and note that Anatole France's gray-matter capacity was only 1100 cubic centimeters while that of a proto-Neanderthaler so lowly as Rhodesian man was 1280!

Professor Rinchen, already mentioned, reports that a man by the name of Anukh, traveling in the South Gobi in 1934 with a companion "noticed in thick growth of saksaul grass a strange, two-legged creature that started running away from them. *'Dzagitmegen'!* [saksaul grandmother] decided both, and, making lassos out of rawhide, they started in pursuit of the creature. The fast Gobi camels had no difficulty in over-taking the creature whose body was covered with short wool. At the sight of people twirling the lassos, the creature issued such a piercing cry that the frightened camels would not budge any further and the creature was able to escape behind a rocky furrow."

Then, again, there is the story of "a caravan on the way to Kuku-khoto in Inner Mongolia. The caravan was traveling from the region of Uliasutan in Eastern Mongolia and approaching the southern borders of Khalkhi when it was decided to stop for rest near a place thickly grown with saksaul grass. When they were ready to start again the man who was sent to get together the camels could not be found anywhere. An old ex-perienced guide told the men that in this location may be some 'Dzagin Almas,' and advised that more than one man should go searching. After a while the three men who were sent out to search came to a cave and saw on the ground in front of it signs of struggle between two people—one having shoes, the other barefoot. The frightened men did not at-tempt to enter the cave and, recovering the camels, returned with their observations to the caravan, insisting that all should go to the help of their companion. The old guide again cau-tioned them against such an act. He stated that the *Almas* never kills people but, having captured one, will hole up for a while and will not come out of the cave. He suggested that they should wait till they came by on their way back and then attempt to free their comrade.

"And so on their way back they came to the same spot and, arming themselves with a gun, they decided to hide near the cave and wait till the *Almas* came out. They waited a whole day, and then toward sunset out of the cave emerged a two-legged creature covered with hair all over. A shot sounded and the creature fell dead. Reloading the gun, the men ran into the cave looking for their lost companion. They found him, but he seemed wild and listless. He never told anyone what occurred in the cave, he avoided talking to people, and in two months time he died."

Reverting to the creatures called *Jez-Termak** which is alleged to mean *Copper-Nails* [meaning, of course, "fingernails"] we should note that this name is applied to a larger, grosser, and more bestial type of ABSM than the Dzungarian *Ksy-Giik,* and is alleged to be found on the super-uplands of Tibet. It is said to be clothed in rather long, shaggy, very dark gray to black hair, and to have fingernails of copper. The implication in the folklore on this type and in all undocumented stories about it asserts that their nails are actually made of copper. This idea is illogical and as near impossible as anything could be but there could be a very simple and logical explanation for it. It is that the fingernails of some primates and notably adult gorillas are quite often bright copper-colored and even look burnished. The explanation seems to be that they are stained—as the material of which all nails and claws are made, known as keratin, so readily is—by the juices of certain fruits, barks, or berries on which they feed. I have collected monkeys of more than one species in Africa that displayed remarkable variations from the described coat-color and pattern by reason of bright red areas in the inguinal region; sometimes on the lower face, and even on the chest and inner arms. After preparing skins, it is a custom to wash them in warm soap and water before drying them for museum preservation. On doing this to these monkey pelts we were amazed to find that all this bright copper color washed out and left the fur pure white or yellow. Experiment elicited the fact that the ingestion of certain fruits, selected for us by local people, although looking green and otherwise quite innocent, produced this vivid red stain on the pelt of caged monkeys within a matter of days by dribbling from the mouth or anus; and, when said fruits were eliminated from their diets, the color persisted for weeks. Their fingernails also remained bright copper-colored until they grew out from the bottom up.

While everybody has for centuries been alluding to Tibet

* This term actually means "The Mountain one that scrambles using its hands," as is explained in the Glossary in Appendix A.

as the real center of ABSMery, there turns out to be surprisingly little of a concrete nature from that vast land. This may appear astonishing but should not be regarded as indicating that there are no ABSMs there. Quite the contrary, it appears to be the true "heartland" of the whole matter but, as I have been at pains to try and explain above, there is a deep-seated, proto-religious prejudice against speaking of the matter to foreigners while, at the same time, the number of foreigners to visit Tibet throughout the ages has really been extraordinarily small. However, almost all of those who have visited the country and written about their travels have mentioned ABSMs.

The American William W. Rockhill wrote in 1891 of hearing many stories from Tibetans in Pekin, China, and elsewhere of the "Hairy Mountain People" of their country but he discounted these reports as being merely cases of mistaken identity; namely, of bears. However, he goes on to say: "One evening, a Mongol told me of a journey he had once made to the lakes in the company of a Chinese trader who wished to buy rhubarb from the Tibetans, who annually visit their shores [i.e. the lakes of central Tibet]. They had seen innumerable herds of yaks, wild asses, antelopes, and *Gérésun Bamburshé*. This expression means literally "wild men"; and the speaker insisted that such they were; covered with long hair, standing erect, and making tracks like men's; but they could not speak."

W. M. McGovern in his book *To Lhasa in Disguise* (1924) says: "In nearly all parts of Tibet one finds the tradition of the existence of a primitive race of men, former inhabitants of the land who have been driven out of the plains by the Tibetans and who now dwell only in the passes and in inaccessible mountain crags. My own servant referred to them as snow men. They are said to be great hairy creatures, huge in size, and possessed of incredible physical strength. Although having a certain low cunning, they are deficient in intelligence, and it is the intellectual superiority of the Tibetans that has enabled them to oust the primitive snow men from the plains. It has been permitted to no white man to meet these snow men." He adds at another point that "Rockhill, who came

[322]

across this tradition in the extreme eastern part of Tibet [i.e. the land of the *Dzu-Teh*] suggests in his *Land of the Lamas* that the wild men may be nothing other than bears! For other parts of Tibet this explanation could scarcely be valid, since in many parts of Tibet, as, for example, in Kampa Dzong, Pari, and the surrounding districts, bears are unknown."

This is a rather significant statement in view of the constant insistence that all kinds of bears [though all varieties of the Eurasian Brown Bear] are found all over Tibet. Actually, it does not matter a "tuppenny-damn" whether they are so found or not, because the Tibetans know their local animals and their distribution much better than any outsiders do, have completely different names for bears and ABSMs with qualifying terms for the various kinds of each—none of which for one kind of creature could in any way be confused with those for the other—and would never for one moment confuse one with the other. Further, Rockhill himself states in another passage that "Lieutenant Lu Ming-Yang, when speaking of wild tribes to the north of the Horba country, assured me that men in a state of primitive savagery were to be found in Tibet. Some few years ago [that was before 1890, *Author*] there was a forest fire on the flank of Mount Ka-Lo, east of Kanze, and the flames drove a number of wild men out of the woods. These were seen by him; they were very hairy; their language was incomprehensible to Tibetans; and they wore most primitive garments made of skins."

This is a pretty startling statement, for Rockhill was a rather nasty kind of skeptic, given to debunking anything possible and imbued not only with the then current pomposity of all Western travelers, but with an almost modern craze for mediocrity and the disposal of all things that did not fit the accepted pattern. Having disposed of ABSMs proper as "bears standing on their hind legs" he then fell with a wallop into a report on primitive "humans," *with hairy bodies,* in a place where they "ought not to have been" [at least by Victorian estimation], and all on the say-so of a Chinese lieutenant. It is a strange commentary on the closed mind of even a famous traveler and at the same time an eye opener on our subject, for we must not forget that there is still ample room for whole hosts of mere "primitives" let alone ABSMs all over this vast Triangle. Further, we must not forget those forlorn people; still, at least until recently, maintaining a Bear-cult, called the "Hairy Ainu" of the Japanese island of Hokkaido. They seem to have been Caucasoids of an extremely early vintage and they certainly were hairy [and in some cases still are, despite their persecution!]. Hairy chaps, wearing skins, running out of burning Tibetan forests do not quite coincide with then [1890] or even current anthropological beliefs but this is no reason to gainsay their previous or continued existence. Apparently Tibetans took, and still take, all this quite complacently.

Just to make matters worse, this same Rockhill must add still another brief passage that states "Legends concerning wild men in Central Asia were current in the Middle Ages. King Haithon of Armenia, in the narrative of his journey to the courts of the Batu and Mangu Khans, in 1254-55 A.D., speaks of naked wild men inhabiting the desert southeast of the present Urumchi." Almost everybody who has traveled in Tibet both before and since the date of these statements have reported the same thing and some have said they have met the creatures. But, I stress again, these are all foreigners, since the Tibetans themselves just won't talk about the business. One story that has always fascinated me is that of a Kirghiz named Saikbaia Karalaein who told the Russian investigating commission about a Chinese family who had started wandering during the war and finally settled in Tibet. It appears that one of their women disappeared and they thought her to be dead. However "a year later she came back, and told them the story that she was taken, while gathering wood, by what she called a *Kish-Kiik*, or wild man. He was comparatively little different from an [ordinary] man but was covered with hair and could not speak. She also said that she was expecting a child by him. Hearing that, her husband killed her; and he was taken by the police. The woman also told where the wild man's den was. They went up there and actually saw wild men and women all covered with hair."

[324]

On several occasions Tibetans of higher education have said (see Chapter 12) or have been reported to have stated that they know of three distinct types of ABSMs in or around the periphery of their super-upland plateau, while, in addition, they speak of two "animals" with manlike or super-anthropoid characteristics. These latter they identify as, first a giant monkey and, secondly, the *Meh-Teh*. Of the other three "Man-Creatures" they are quite cognizant, affirming that they are first, the little dwarf *Teh-lma* of the lower valleys; second, the man-sized hairy one [i.e. the *Kaptar, Guli-avan,* or *Ksy-Giik* type]; and, third, something quite else. This is the mighty *Dzu-Teh* type known elsewhere as the *Gin-Sung;* a real giant, shaggy-coated, and able to stay for long periods in the ruggedest country; dangerous, a stock raider, but possessed of an almost exactly human-type foot. This, they and everybody else, agrees is not found along either the Himalayas or the Southern Tibetan Rim, nor even in the Nan Shans, but is confined to the unnamed triangle between these, upper Indo-China, and the Chinese escarpment. This same type seems to prevail also in the Tapa Shan, the Tsin-Lings, the ranges between Shensi and the Gobi, and again north along the mighty Khingans that separate the Gobi from Manchuria, and on into the Little Khingans that lie athwart the northern edge of that province. Also, as I said before there are indications that this type of ABSM may exist still farther north in the Stanovois, Dzhugdzhurs, Gidans, and even in the Anadyrs. This is of the utmost significance since it is only a skip from there to Alaska, while this *Dzu-Teh* type ABSM seems to conform very closely, if not be identical with, our *Sasquatches* and *Oh-Mahs*.

Here indeed is a strange situation to contemplate. We start out with a suggestion that there *might* possibly be some one kind of as yet uncaught and undescribed animal, probably an anthropoid [or a race of runaway human delinquents] in the Himalayas which somehow got colorfully called "Abominable Snowmen" and we end up with a whole galaxy of unknowns, spread over five continents, and concentrated in eastern, upland Eurasia, where, by the word of those people who know most about the subject locally, speak the local languages, and have devoted the most time to the matter, there appear to be no less than five very distinct types, each with its own characteristics and habitat; namely, (1) the Mountain Neanderthalers of the West, (2) the little *Almas;* also Neanderthalers, or mere primitives, of the hot deserts, (3) the bestial *Meh-Teh* of the Tibetan upper plateau, (4) the giant *Dzu-Teh* (*Gigantopithecus?*), *Tok,* or *Sasquatch*-type and (5) the tiny tropical, forest-dwelling *Teh-lma* of the southern valleys. This may sound fabulous but, the deeper you delve into the reports and the background, the more obvious and logical this becomes.

The distinctive nature of each of the five is perfectly in accord with the varying nature of other groups of mammals. Take for instance the wild sheep of this area. There are distinctive species and/or races in each of the great mountain blocks, while other hoofed animals replace these on the lowlands or comparative lowlands; one set on the *hot* deserts, another on the upland *cold* deserts. Then also, the actual geographical distribution is also perfectly consistent, in that one kind inhabits the far west (the Caucasus) and the western fringe of the plateau; another the comparative lowlands of the middle; a third the eastern mountainous edge; still another the Tibetan Plateau and its superimposed mountain ranges; and the last only the warmer valleys of the extreme southern periphery of the area. These divisions, furthermore, coincide with the distribution of both vegetation and vegetational types of growth. For instance, the arrangement of the latter going north from the Pamirs to the Sayans up the Great Barrier parallels (but is different botanically from) that going up the eastern escarpment from Indo-China to the Stanovois in Siberia. The whole picture, in fact, despite its enormous complexity [and our gross over-all lack of knowledge of the area] is perfectly logical and consistent with all natural facts and factors.

To reiterate—and I cannot help doing this and for some very real reasons—we should wipe away our sense of helplessness and hopelessness on taking our first look at Map XII, and just remember that this tremendous mishmash may be quite simply divided into five parts: the Great Barrier on the west; the central desert basins; the Great Barrier and escarpment on the east fronting Manchuria and China; the Tibetan super-uplands and their mountains; and, last, the fringe area of the Himalayas. This is eastern Eurasia in, as it were, a nutshell. The only things left over are the two enormous masses of uplands and mountains in Siberia, west and east of the Lena River respectively. These, however, do not at the moment concern us.

And so we find ourselves ending our world tour in an area that is only one stage removed from where we started. The animal life and much of the vegetation of far eastern Siberia is identical to that of our extreme northwest. What is more, as you go south from the Bering Strait on either side—down through Siberia on the Asian side, or through Alaska to the Yukon, and British Columbia on the American, you pass through the same succession of vegetational belts and moun-tainous zones at each latitude. Many large animals, like the Brown or Dish-faced Bears and the large Red Deer or Ameri-can Elk, have crossed from one to the other in comparatively recent times. The Amerinds seem to have done the same too; and the Arctic or Eskimo-type Mongoloids even later [unless they were on both sides all the time]. As I asked at the outset, what was there to prevent the Neo-Giants from doing so also, at some time? They are, of all the ABSMs, apparently the most rugged, surpassing in this respect the Neanderthaler *Kaptar-Guli-* (or *Gulb*) *-avans,* the desert-dwelling *Almas,* and the little warm-forest *Teh-lmas.* That the *Meh-Tehs* did not do like-wise seems to me fairly reasonable for it would appear that they are more apes than men and, like all of that ilk, are neither catholic in their tastes nor so readily adaptable as are the Hominids. Like the Gorillas in Africa and the Orangs in Indo-nesia they got into a special environmental niche and have re-mained stuck therein.

Turning to another aspect of the matter; the Pithecanthro-pines were manifestly lowland creatures and had plenty of space. Why should they go barging off into upper Siberia? Then, although there are Mousterian (Neanderthal) type stone implements scattered all over Manchuria and eastern Eurasia, the sub-men who made them—the Neanderthalers—appear to have been essentially a western species that spread from western Eurasia. The *Teh-lmas* and little *Sedapas* on the other hand need hardly be considered as candidates for emigration to North America; they are tropical types. It would seem to me that, if the *Dwendi* are just pigmy Amerinds, there are no ABSMs in the New World other than Neo-Giants except for the alleged *Shiru* of Colombia. What this might be I certainly don't know, and a great deal more than is at present on record about it will have to be established before anybody can hazard even an educated guess. But, life being what it is, I would not really be a bit surprised if it was the first ABSM to be collected in the flesh!

15. Some Obnoxious Items

Is it not true that there must be some physical evidence of anything physical? Is there any real evidence for ABSMs, and if so, does it prove anything?

You now have before you an over-all picture, and also some considerable separate details, of the statements made by all manner of people about ABSMs everywhere. It is an extraordinary galaxy of alleged facts. On the one hand, one could, I suppose, tear each individual one apart and suggest explanations of each of its parts. This is the procedure that both the general public, the newsmen, and the scientists have tried to do but, as far as I can see they have always fallen down on at least one, usually several, and often all of the separate aspects of these attempts, to say nothing of pure logic. On the other hand, looked at as a whole, all over-all suggestions put forward to explain the business as a whole, turn out to be equally illogical, often ridiculous, and usually demonstrably impossible. Take, for instance, the perfectly reasonable notion that the whole thing is a hoax.

This could well be so in any one place, for hoaxers are devilishly ingenious and conjurers are often really quite unbelievable. But, taken in the over-all of space and time—that is to say, as from about 500 years ago at the least, and all over most of five continents—the suggestion becomes a little ridiculous. If we insist, nonetheless, that the idea be pursued, we have to make the following assumptions.

Let us disregard everything except the matter of footprints and foot-tracks. These are of four basic types as reported and as copied in plaster of Paris—the pigmy with pointed heels;

the *Meh-Teh* type with huge second toe and separated big and second toes; the short, stubby man-sized Neanderthaler type; and the giant manlike with a double first subdigital pad. However, there are many variations of all three, though most notably of the first and third types and of these, particularly among the pigmies. These prints have turned up all over the world for, let us be ultra-conservative, and say at least a century. What is more, almost all of them have turned up in the most out-of-the-way places where they were least likely to be found—ahead of mountaineers who had lost their way, changed their minds, or who were breaking new ground; at the head of new roads; up uninhabited rivers deep in tropical jungles; and so forth. Sometimes they run on for miles.

This all being so, and it cannot be denied, if they were made by men for some reason, hoax or otherwise, those men must have been in association, world-wide, for centuries; have much skill; be reliably secret to a degree simply not known in other walks of life [especially the criminal]; and have a brilliant organization and tremendous sums of money behind them. One may perhaps also be permitted to observe that, being [as insisted upon by the skeptics] only human, they must have had an extremely powerful and coercive reason for making these ridiculous things. The notion that such a world-wide organization has existed, completely undetected for a century, seems, we must admit, to smack of the unreal. It is no good trying to explain one mystery by another even greater one. There is only one force that I can suggest that might foster such practices. This is some religious urge but I beg to leave this until later, for, to discuss it now, would be premature, while it would not be fully comprehensible until some other things have been said.

Still anent foot-tracks only, we then have to consider their being caused by men or other animals quite fortuitously and not by any specific intent. This is to say that, if made by men, they are due to strange foot deformities or to wearing foot-coverings of odd design. Nothing like the form of any of the four basic types of ABSM prints are known to be left by men. There is a recurrent theory that those left in snow are simply man's or animals' tracks which have been enlarged or deformed by melting and regelation. Not by any means have all ABSM tracks been found in snow. Quite the contrary: most of them have been found in muds and other soft soils. Melting cannot occur in mud. Another idea is that the tracks in snow are the result of animals loping along, putting two or more feet exactly into the same place or even—as recently suggested by Hillary—a whole group of animals, such as foxes, *stringing*, which means all following their leader and jumping exactly into the same spots as that leader.

Foxes do *string* in exceptional cases, and there are some animals that sometimes do place their hind feet almost exactly into the impressions made by their forefeet, especially some bears. However, in neither case can the resultant imprints ever possibly go on for mile after mile—especially in mud—without ever so much as a single apparent toe impression being out of place. The very idea is so preposterous as not to be even worth while considering. The ABSM tracks just go on and on with each right and left foot constantly and consistently reproducing itself exactly.

Still another idea is that the tracks in snow were left by men wearing partially worn-out footgear. In this case, however, the wearers must, in the first place (and for centuries to boot, and all over tens of thousands of square miles of territory), have possessed that extremely rare abnormality (a longer second than first toe) *on both feet*, as described above and in detail in Appendix B. Further, the footgear must always have worn out exactly and precisely, so that all toes on both feet were exposed, while the worn edges of the footwear never, ever, once, left any impression. This also is, of course, so manifestly absurd as not even to warrant further discussion. And so the whole of this wretched business goes. It does not matter which way you turn with regard to the tracks, but that you come up against a manifest absurdity.

As to stories, accounts, reports, and suchlike verbal statements there is really little we can do. They lack any kind of proof, and they fail to supply any kind of concrete evidence. The most one can do about them is to submit them to a crude statistical analysis to see if they display any pattern. They do; but it really only makes matters worse. Here you have a mass of illogicalities that appear to have a logical pattern; yet the pattern points to a further illogicality. Stories can only be repeated; while what we need is some concrete evidence—something physical that we can examine, try to analyze, and explain. Are, therefore, foot-tracks all that we have of a concrete nature? The answer to this is, of course, and as you must long ago have realized, no. There is, or is alleged to be, much other perfectly good physical evidence. There is also some cognitive, and also corollary evidence.

All evidence may, in fact, be broken down into six categories under three major heads—to wit, *Intrinsic, Cognate,* and *Corollary.* Under the first are actual physical items such as [alleged] whole mummies of ABSMs; dried heads or skulls; and parts, such as hands, on the one hand, and bits of skin and hairs, on the other, and including scalps, a bag said to have been made from a *yeti* skin, and some whole skins. Of the second category, we have, first, footprints and tracks, and secondly, excrement, while there are a few allegations of other possibly extraneous items such as beds, lean-tos, and primitive constructions in caves. Of the third category we have three quasi-concrete forms of alleged evidence: first, reputed calls and other sounds; second, stinks said to have been given off by the creatures; and, third, reports of things having been moved by them. These are of course almost as inconclusive (and illusive) as mere reports, in the absence of photographs or sound recordings. Finally, under the *corollary* class, we also have drawings and paintings, carvings and statuaries that are said to depict ABSMs.

The variety of all these items is paltry and the actual numbers of examples of each that we have are really quite fantastically small but it does seem incredible, at least to me, that this is all that has been produced over the ages for such a large series of alleged existing entities spread almost all over the world. I will admit that one hardly ever finds so much as a scrap of any dead wild animal anywhere but one would have thought that, even if no photos have been taken due to the creatures being nocturnal, at least one might have been shot, if only in self-defense. Of course, there are plenty of *stories* of them having been shot, but no parts seem ever to have been preserved. This is, perhaps, the most suspicious part of the whole affair—plain lack of concrete evidence. Further, most of what has now been produced has been shown either not to be of an ABSM, or definitely to be a relic of something else. Let me take these obnoxious items one at a time.

Starting with parts of the animals themselves, we actually have nothing but an allegation of a (nonexistent) skeleton from B. C. Then, there are reputed to be some complete mummies of ABSMs in several "Tibetan monasteries." A most erudite Tibetan, by the name of Tshamht bRug Dzün DahR dzhe Löh Bu, stated (in 1953) to Nepalese officials that he had inspected such corpses in monasteries at Riwoché in the Province of Kham, and at Sakya on the road between Shigatse and Katmandu. Then, a complete, dried head of a *Meh-Teh* is said to have been in the possession of the headman of the village of Chilunka, some 50 miles northeast of Katmandu for the past 25 years. Next, there are three mummified hands preserved in sundry of the Nepalese monasteries. These are desiccated skeletons of hands and wrists with some ligaments and dried flesh attached. One, kept at a place called Makalu, is attached to a forearm. A man says he has part of a skull of an *Oh-Mah* from California but he has not yet produced it. This completes the roster, except for a bag alleged to have been made of *yeti* skin, and three (or just possibly four) conical caps also allegedly made from their skins. Two whole skins were also produced.

This is a pretty paltry showing to begin with, but it actually boils down to practically nothing when critically examined. Let us so examine these items.

1. Neither any complete mummies, nor the dried head have actually been seen by anybody other than the respected Tibetan named above.
2. The two skins turned out to be and without any doubt:

[333]

first, of a Sloth-Bear (*Melursus*)—i.e. that from Mustang and mentioned above; and, second, that of a Blue Bear (*Ursus arctos pruinosus*). This was obtained in Bhutan by the 1960 Hillary Expedition.

3. The hand and wrist, with forearm, turned out to be that of a Snow-Leopard (*Panthera uncia*).

4. The hands and wrists (without forearm) are either two or three in number. There is a considerable mystery about these. All three have been photographed at Bhang-Bodzhei (i.e. Pangboche); one by several people; the other two only once as far as I have been able to ascertain. [All three are displayed among the photographs.] These I have numbered Figs. 2, 4, and 3. The first is the much photographed one; the second was published by Prof. Teizo Ogawa of Tokyo University; the third was photographed by I do not know whom.

Fig. 2 has rather broadly flattened metacarpals. The thumb is complete; the 2nd finger has only the basal phalange; the 3rd finger is complete; the 4th and 5th fingers are missing.

Fig. 4 has the thumb complete; the 2nd and 3rd fingers complete; the 4th with a small basal piece of the first phalange only; and a complete 5th finger.

Fig. 3 has a complete thumb; [possibly] two joints on the 2nd finger; a complete 3rd finger; and apparently no phalanges at all on the 4th and 5th.

With the exception of the photograph by Prof. Ogawa, the pictures available are extremely bad; taken from angles that distort the whole, and fail to bring out any of the details needed; and they are not so much generally useless as misleading. I have been unable to ascertain who took the only pictures of Fig. 3 that I have seen. They are overexposed. However, I have a notion that they are of the underside of Fig. 2, being held in bright light by some local helper. The only discrepancy between Figs. 2 and 3 is the [possible] extra phalange on digit 2.

Ignoring Fig. 3, therefore, we have two very old mummified and obviously hominid hands. The most notable is

Fig. 1, in which the metacarpals do, indeed, seem to be very wide and flattened.

5. Scalps are preserved at places usually written Pangboche, Namche-Bazar, and Khumjung. There may be another at Thyangboche.* As of the time of writing, that from Khumjung has been demonstrated by both blood and hair analysis to have been made from the shoulder-patch of a hoofed animal of the goat family known as the Himalayan Serow (*Capricornis sumatrensis thar*). Sir Edmund Hillary had one made for him in 1960 from the rump of a fresh Serow skin that he had shot. The hairs are identical. The hairs from the other two scalps seem to be also from the same animal.

6. The bag yielded hairs that are again microscopically and in general appearance identical to those of the scalps.

Thus, out of the entire roster of alleged bits of ABSMs we are left with two desiccated hands and wrists, one of which looks human, and one of which looks like that of a Neanderthaler—possibly, according to Prof. W. C. Osman Hill of London; almost definitely, according to Soviet scientists. Both these hands are extremely old.

Hairs from all of these specimens and from isolated tufts found on rocks, in bushes, on the ground, and associated with piles of excrement, have been microscopically examined. They show rather a bewildering array of characters. The identification of hairs is not nearly so easy as the layman might think. Hairs from different parts of the same animal look quite different, and we ourselves have five different kinds on our bodies at all times—head-hair, normal body hair, axillary hair, pubic hair, and some remaining lanugo or "fluff" like that on new-born babies. Even these look different, microscopically, at the tops [tips] and basal portions [bottoms]. Then, if you will just watch your dog around the year, you will see that he changes his coat twice, and that his winter pelt is quite different from that of his summer one. Also, many animals have patches of all kinds of strange and special hairs—like those

* The correct transliteration of these is Bhang-Bodzhei, Namdzhei-Bazaar, Khumh-Dzhungh, and Dhyangh-Bodzhei.

on the necks of moose, the rumps of some deer, and the quills of porcupines—quite apart from bristles, or facial or cranial *vibrissae,* which is to say whiskers and feelers. Also, almost all mammals are plentifully supplied with all manner of skin glands and many of these are surrounded by, or filled with, most extraordinary hairs. One of the four most valuable fixatives for our expensive perfumes comes from glands [called pods] on the insides of the legs of a certain kind of oriental deer. These grow the oddest bristles. Trichology, or the study of hairs, is an enormous subject as the late Dr. F. Martin Duncan of the London Zoo demonstrated by assembling the largest collection of mammalian hairs in existence during a lifetime but without anywhere approaching completion.

Blood analysis from specimens leached from old and dried skin or flesh samples is even more difficult but it is, if accomplished at all, considerably more precise. At least, you can say what it is *not.* Serological [or blood] comparisons have now been made between material obtained from various alleged bits of ABSMs, and compared with some Primates (i.e. monkeys), Man, rabbits, horse, dog, and some others. The results, unfortunately, have proved to be doubly inconclusive; first, in that none have matched and, secondly, because just the most likely types of known animals with which they should have been compared either have not been used or available. What is most needed in the case of, for instance, the *Meh-Teh* is a good comparison with the various mammals listed in Appendix D and, above all, with Mongoloid Man. As a matter of fact, none have actually been tried against either Gorilla, Orang, or any Macaque Monkey —the most obvious choices, one would have thought. Even more curiously, none of them have matched with any kind of goatlike animal (the *Capridae*) although the hairs from the same specimens match those of the Serows exactly. Thus, there is either some deliberate trickery here, or the scalps from which the serological specimens were taken are *not* Serows and the hairs only *look like* those of that animal. This, admittedly, does present rather a perplexing question.

[336]

Altogether, therefore, there is really practically nothing of a concrete nature even alleged to have come from or be of any ABSM that we can pin down. Matters are a little better with the next major category of physical evidence. These are the *Cognate*—i.e. the *ichnological,* which means the study of footprints and tracks—and *scatological,* or excrement. I will leave the former for further discussion (see Appendix B) after dealing with the scatological.

Specimens of excrement have been collected from various points in Nepal in the Himalayan area; allegedly from some points in eastern Eurasia (see Russian reports); and from the northern Californian area. Some specimens of the first and last have been most carefully analyzed in modern veterinary and medical laboratories and quite a deal of information about both their composition and the parasites in them collected. A lot can be learned about an individual animal from its excrement, as everybody knows from the common medical practice of stool examination. The study, as conducted scientifically, falls into two parts—first, that of the entire individual mass; second, that of its microscopic composition. Also, cultures are prepared from it, so that any contained organisms may be multiplied, examined, and identified. Also, the eggs of worms and other such comparatively large parasites are searched for and identified. All of these processes give us information about the animal that originated the specimen.

In the gross form the faeces alleged to be those of ABSMs, fall into two very clear-cut types—those from the Himalaya which are of large but not excessive man-size and are said to come from *Meh-Teh* and *Teh-lma;* and those of the *Oh-Mahs* from California. The only reliable examination of the former made in the field was made by Gerald Russell who had had many years of such field studies in Africa and the Orient while collecting mammals, reptiles, and amphibians for museums. He reported the form to be generally humanoid and the contents to be: "A quantity of pika (*Ochotona*) fur; a quantity of pika bones (approx, 20); one feather, probably from a partridge chick; some sections of grass, or other vegetable matter; one thorn; one large insect claw; three pika whiskers." Later, he examined also what appeared to be *Teh-lma* droppings near the river where he had found those creatures to be eating giant frogs. These contained bones of that animal and vegetable and insect remains in about equal proportions. Analyses of other *Meh-Teh* faeces have been made and variously reported but most of these stress the occurrence in them of remains of the little Lagomorph, the Pika or Whistling Hare (*Ochotona*). Further, Tom Slick was shown piles of the fresh entrails of these little animals on mountain screes where ABSM tracks were found. The locals asserted that the *yetis* hunted these little animals in their retreats between the loose stones, crushed them, partly ripped off their skins, tore out their entrails, as we might gut a fowl, and then ate the rest raw.

The Californian droppings are an altogether different matter, and I express myself this way advisedly. First, the individual piles of droppings are of enormous size, some (as that shown in Figures 10 and 11) being, as the ruler indicates, over 2 feet long. This was not an accumulation, all its parts being obviously of the same age. [Porcupines sometimes create toilets that they visit regularly and add to for long periods.] Their gross form is, moreover, of two distinct kinds—masses of fair (man-sized) faeces, and droppings of equal volume but of positively enormous man-shaped individual faeces. Sometimes these latter have a most extraordinary ropelike formation as if produced by a double bowel with interlocking spiral twists. Other samples have not, however, shown this twisting.

This presents one of the most positive bits of evidence for the existence of an ABSM, whatever it may be. Just about the only thing that can *not* be manufactured—at least to fool a medical man or a veterinarian—is faeces. Then, there is *no* large mammal in North America that can or does produce such droppings. The only alternates are large Ungulates or the larger Carnivores. The droppings of all the former are all pellet-like—from Moose to the smallest deer [and the Moose, incidentally, is not and never has been found in the Washington-Oregon-California coastal ranges, nor even in the Cascade-Sierra-Nevada Ranges] while that of the larger cats [here, only possibly the Puma] are most distinctive and do not, of course, contain mostly vegetable matter, as these *Oh-Mah* faeces do. The only remaining animals are the bears. Black Bear (*Euarctos*) are found in that region, and it is just conceivably possible that a few Brown or Dish-faced Bears (*Ursus*) might still be lingering there. Both these animals are omnivorous, but, as may be seen from the photographs, their droppings do not look at all like those of the local alleged ABSMs. There is, however, a matter that I urge most strongly should be considered along with these discoveries.

It appears that in certain circumstances human beings may give rise to just such faeces as depicted here. I have information on two such eventualities. The first is of Alaskan Eskimos who go on an almost exclusive diet of whale blubber in lean winters. This causes not just chronic constipation but a major blockage of the lower bowel which may result in retention for many weeks or months. Then, the family group goes in search of certain willows, the astringent bark of which they strip and eat. This acts as a very violent purgative. As a result of this, they finally manage to eliminate but not without great pain, splitting of the anus, and a great loss of blood. The sorry process was most graphically described to me in a letter from a U.S. Government agent in Alaska.

The other example of this medical obscurity that I have on record is that of what are called in China "Shensi-Babies." These are single, enormous, extremely solid faeces, eliminated by confirmed opium eaters, and sometimes by opium smokers, who have gone into prolonged periods of withdrawal due to narcotization; during which evacuation is ignored or actually physically impossible. Resultant faeces, when elimination does occur, are said to be, on occasion, as much as 2 feet long and 4 inches in diameter. It is just possible that some of the Amerindian peoples of our and Canada's northwest might have been periodically or occasionally subjected to some influences, odd diet, or narcotic that could cause like phenomena.

Quite a number of faeces have now been examined in properly equipped laboratories, and a few proper reports have been issued. However, the findings have not been pursued to their logical conclusions, and there has been a marked lack of any desire to issue positive pronouncements on them. I have seen such reports on *Oh-Mah* samples from northern California; of alleged *Meh-Tehs* from Nepal; and of the *Teh-lma* from the lower valleys of that area. The first appear to have been almost exclusively of vegetable matter; the second were of mixed content with pika hairs and bones included; and the third were basically vegetable matter in essence but included bits of insects. In two cases [one, a set of examinations made in a medical laboratory in Oregon of *Oh-Mah* faeces; the other run for Bernard Heuvelmans in the Brussels Institute] the eggs of certain parasitic worms were found. In both cases these were identified as belonging to the group known as the *Trichocephalidae,* and specifically of the genus *Trichuris.* This family of Nematode worms includes the "Hook-Worms." There is a species of *Trichuris—vide: T. trichura*—that is found in Man; other species come from a variety of other mammals. The size and proportionate measurements (width to length) of the eggs of each species are known and are quite distinctive. Those found in Brussels from the *Teh-lma* faeces appear to have conformed with the species that comes from sheep: those found in the *Oh-Mah* faeces were of three kinds. In a report on these, the specialist reporting stated that they could not be identified, however, due to their deterioration. Nonetheless, he got exact measurements of them and they could quite well have been identified, at least within certain limits. I am constrained to quote from this report:

"The largest egg is out of the range of human parasite ova, though Nematodes with such large eggs have been reported occasionally from various other primates." From this, the writer concluded that "The specimen (of faeces) is not human . . . is most probably primate . . . is most probably from a sheep or other herbivore."

This statement is equivalent to the British Museum's now famous dictum (see Chapter 19) that "Now you can see for yourself that this Abominable Snowman footprint is that of a bear . . . or a monkey." At this point I do refuse any longer to remain civil, though I still refrain from publishing the name of the expert who made the statement about the worm eggs. This is the kind of double-talk that one has to contend with, *ad nauseam,* in ABSMery; it is wholly unscientific; and, it is probably a deliberate evasion of the issue. The really alarming aspect of all this is that not a few samples of alleged ABSM droppings have now been collected and submitted to professional analytical laboratories but there does not appear to be any record of just what has been submitted to whom, what the latter found, or any proper carry-through of the analyses. There may be perfectly clear and valid evidence lying around in somebody's files showing that these faeces were produced by an anthropoid, if not specifically by a Pongid or a Hominid. If there is, we ought to hear about it—and in print—for the very simple reason that gross excremental masses of the size and nature of those from which the samples were taken could not have been dropped by any known mammals in the areas where they were found. Since this is so, if they contain species of parasitic worms found only in Man, anthropoids, or other Primates, it can mean only one thing—namely, that such a type of man, anthropoid, or other Primate lives where said droppings were collected.

Of what we call *Cognate* evidences of ABSMs, other than the scatological and ichnological; which is to be considered later and see also Appendix B, there are but a few isolated and not well authenticated items. Among these are "reports" or rumors that some Sherpas had found crude stoneworks in areas that they said were inhabited by *Meh-Tehs,* on the basis of droppings, animal refuse, and other items they said they found within them. This, in some measure, concurs with the lone story from British Columbia by the Amerinds of having found what they appeared to indicate they thought was a sort of incubation chamber constructed of crude piled stonework in a cave (see Chapter 3). Apart from this, we have the reports of a few central Eurasians, as given by the Russians, that the *Almas* dig holes in the ground and cover them with brush.

[341]

Of *corollary* evidence, we have really very little also. First, there are reports from many areas—central and eastern Eurasia, the Himalayas, Malaya, Sumatra, Mexico, Guatemala, South America, California, and British Columbia of extremely strange, high-pitched, long-drawn-out, gurgling whistles being associated with sightings of ABSMs and at other times when something unseen was heard moving about in the immediate vicinity. To these reports we may add the weird and unnerving sounds reported by the members of the first American Karakoram Expedition in 1938 (see Bibliography). I note also that Hillary, even after debunking the Khumjung scalp, and attempting to explain all the foot-tracks in snow by suggesting that they were made by foxes "stringing," chose the sounds reported said to be made by *"yetis"* as being one of the things that have *not* been explained. I find this rather odd as almost anybody can imitate any sound or can make up all kinds of weird calls. Mammals and birds, and even insects and reptiles, and especially amphibians make the most astonishing noises and variety of noises. A small spherical frog (*Rhinophrynus dorsalis*) known to the Mayas of Yucatan as the "Waw-Mooch" only makes a noise after a sudden rain but, although the animal is only about 2 inches long, this may be heard for over 2 miles. The tiny Demidorff's Galago (*Galagoides demidovii*), a minute Primate, that can sit in the palm of your hand, lets out screeches that make the whole forest ring for a mile, when a lusty man cannot make himself heard shouting at the top of his lungs, even when in sight of the persons whose attention he is trying to attract.

Personally, I lay little store by "noises," per se, but I must admit to having been profoundly shaken when the Amerindian couple, the Chapmans, in British Columbia gave out with exactly the same strange whistling call for their *Sasquatch,* that young Mr. Crew had given for me [and which I recorded on tape] in California, though neither party had ever heard of the other's existence. It is equally strange too, and it may be equally significant that, as far as I can make out from written descriptions, just the same very queer, very unhuman, and nonanimal-like (and invariably described as unearthly) calls have been attributed to ABSMs all over the world. [The awful roaring of the *Mapinguary,* the *Didi,* and others I lay no store by at all. All manner of most unlikely animals roar worse than any bull that ever lived in Bashan.]

Another possible corollary matter is that of the smell—or, rather, stink—of ABSMs. This has been remarked upon by Amerinds of both North and South America, Sumatrans, especially by the rubber tappers on the Malayan estates, and by Himalayans, Tibetans, and Mongolians. In fact, an overpowering "animal stink" is an almost regular attribute of close proximity to an alleged ABSM. This is a rather odd fact, but it makes some sense if these creatures really exist and are subhominids. One of the most terrible ordeals to have to undergo is to live with the nice little Pigmies of the Ituri Forests of the Congo Uele. They give off a smell that amounts to an overpowering stench and which is, to us, absolutely nauseating.* After many years of collecting wild animals and living with them, both in their native haunts and in captivity, I can tell quite a number of them down even to species, and blindfolded, simply by passing by their cage and sniffing. The greatest "stinkers," it has always seemed to me, are the Primates, and the larger ones in particular. I don't really mind it, but the smell of a large Mangabey is to me sensational, and I can tell before I enter a monkey house if they have a specimen of that genus housed there.

Whatever we may say about "stinks" has little meaning until somebody devises some method of bottling a smell and testing it against that of other living creatures. This is such an abstruse idea that it need no longer concern us and we may simply put the whole matter back into the same class as that of mere reports. I believe that the few other similar

* This they do not "mean" to do; and it is really no discredit to them. Negroes say that Whites smell like boiled, not too fresh, rabbit; some Whites say that Negroes smell awful, even to the fourth generation; and peoples who eat a great deal of "hot" or "piquant" foods may indeed have an extraordinarily powerful aroma, so that the old adage about Turkey-Buzzards refusing to eat Mexicans could actually have some basis in fact! Nor must we forget that what we call "sex" is possibly based, in its cruder forms, on smell, our most delicate sense. In fact, all animals stink.

facts that have been offered as evidence of ABSMs may be likewise treated. These amount to little in any case. First, there was a cairn raised by climbers on the top of a sacred mountain in the Himalayas that was destroyed; then, some boulders came down upon travelers at various places such as British Columbia, the Himalayas, and Manchuria. These could well have been set rolling by fleeing animals, by small vibrations set up by the travelers themselves, or by their conversation, especially where rocks are split at night by frost and may be teetering on a brink waiting only for the slightest imbalance to set them rolling. In California we have the reports of large oil drums having been actually toted across a road and thrown or rolled down slopes and of sections of culvert and other large objects like the wheel of a tractor-crawler being moved. But these, too, are simply reports. We have no physical evidence even in the form of photos of said objects before and after displacement. They are worthless as evidence of anything.

This leaves us—apart from the ever-recurrent tracks—with only depictions. What we need here is just one still photograph, however hazy, but we do not have even that. It is true that, however keen and agile a photographer may be, it is only a few times in a lifetime that he is at the right place, fully attending, with the right camera, film, lense, exposure, light, and everything else, and all pointing the right way at the right time to get a really worth-while news-shot. It does happen, but it is solely by luck. You could travel the Himalayas for two lifetimes with the best cameras ever invented at the ready and never even see an ABSM: and if you did, you still might be too scared, excited, or overcome even to press the necessary button. Bird photography is bad enough, but there you have either the nest or a feeding station to set the stage for you. Big game you can drive into in a jeep, and other rarer animals you can stalk, but the results you get diminish by some kind of geometrical recession compared to the rarity of the object sought. To get a filmstrip of an ABSM is really asking too much—and more especially since most, if not all of them, are alleged to be nocturnal.

To this end, trip-wires attached to infrared cameras and snooper-scopes have been advocated and tried. So far nothing has been obtained from either, except some excellent shots of startled deer and bears, though the wires have been tripped, broken, stepped over and apparently crawled under; and even, on one occasion, the camera was broken, while on another—and get this—it was opened, the film removed, and the camera itself replaced! That was no ABSM, unless our whole idea of human evolution is completely haywire. All of this latter took place in California! [And these are not the only suspicious happenings in ABSMery in that area.]

The only things "visual" that we are offered of ABSMs are a few very clear and precise drawings in old eastern Eurasian manuscripts, as we have related in the last chapter; some alleged "paintings" in monasteries in Nepal and Tibet, of which, I may say, I have never seen any photograph or reproduction; some sketches made under the direction of persons who have said that they have seen an ABSM; and a number of "artists' conceptions." These seem to me to be of values directly proportionate, in diminishing degree, according to the order listed. The ancient Mongolian texts really do show something and, being from very precise treatises on specific subjects [medical] and showing a large number of known animals most accurately and distinctly, they do seem to be worth-while of serious consideration. Monastic wall paintings might be fine if we could only have a look at them. Sketches made for those who allegedly have seen something don't say much, though it has to be borne in mind that police artists can, by questioning witnesses, finally produce drawings of wanted persons so accurate that they are immediately identifiable.

Mere artists' conceptions I lay very little store by, except those made by artists who are also zoologists, anatomists, and anthropologists—and such are far and few between indeed. Some of the grotesqueries produced in the name of science and especially of paleoanthropology and primatology, are simply fantastic. A lot of mere animal art is just as absurd—like Audubon's mammals which he twisted into all manner of im-

[345]

possible poses or stances in order to get them onto a piece of paper. Some of the "Apemen," "Cave Men," and "Our Ancestors" that have been published in serious works are an affront, and some of those that have appeared in higher-class magazines are absolutely laughable. You may remember one large series in color of some Stone Age people allegedly going about their daily lives which appeared a few years ago. In these elegant paintings all the men looked like ads for male muscle-building and most were clean-shaven and obviously of absolutely one hundred per cent pure Anglo-Saxon stock, while the women had figures like Hollywood starlets, but without certain mechanical aids, and long wavy hair. Their caves were swept clean; there was not so much as a scrap of bone in sight, and the firewood was all neatly sawn into handleable lengths! In one, there was even a herd of grade-A Jersey-type domestic cattle in the offing.

And so we are left only with the matter of tracks. And about these I have few words to say at this point. First, Ichnology, or the study of tracks of all kinds, forms a science of its own, with a sound methodology, and a very high degree of competence. It is divided into three major divisions: (1) that of animals, (2) that of people, and (3) police work generally, or that of everything that can leave a track. There is a strange thing about ABSM enthusiasts of both schools—pro and con— and this is that they simply do not seem ever to have realized that the most detailed studies have been made of all manner of man and animal tracks, and that photographs of many and detailed scale drawings of most have been published and are readily available. It seems, also, that the skeptics have not ever really looked at the published photographs of ABSM tracks or of the extant plaster casts of them. Had they done so, an enormous amount of verbiage and published mileage would automatically have been eliminated.

As an example, people are still suggesting that *Meh-Teh* tracks are made by bears. Bernard Heuvelmans' book, *On the Track of Unknown Animals,* has been available in English since 1958, and in it he shows in the simplest of terms the difference between bear tracks and hominid tracks, and how

[346]

one may invariably be distinguished from the other. There is nothing difficult about this: it is, in fact so simple that one would have thought that even the skeptics would have spotted it. [Simply stated, bears walk with their toes turned in and have their *outer* toe the biggest, whereas hominids walk either straight ahead or with their feet turned a bit outward, and they have their first or *inner* toes the biggest.]

The tracks of all bears and almost all other Himalayan, Tibetan, North and South American, and most other mammals are now known and on record. At the same time, the police forces of the whole world have for over a century been studying intensively all manner of tracks left by everything that moves and especially of people. If you only knew how much they can deduce from a single heel imprint, you would think a few more times before breaking and entering even your own home when you have forgotten your keys. Then again, as Tschernezky has shown, criminologists have made a special study of human feet—and come up with some extremely odd ones; such as those illustrating his paper (see Bibliography). Engineers—and especially road engineers—can work out to the last pound the weight of anything that leaves an imprint on any kind of soil or other compressible surface. Thus, with a large enough set of scale-drawings of animal imprints and tracks before you on the one side, of human footprints on the other, and some proper ones of ABSMs in the middle you don't have to wade through all the tripe that has been written upon this matter. All you have to do is take a good look.

The more technical details I have assembled in Appendix B. There are displayed the prints allegedly left by the various ABSMs, each duly tagged and, following these, you will find those of all the animals that have been brought up in this discussion with the exception of some absolute absurdities like one-legged giant birds and so forth. Then, there are the human feet and imprints from normal babies to grown giants; dwarfs, midgets, and abnormalities. Alongside these are those of the Pongids or apes. It has been said that one picture is worth a thousand words: these I think are worth a volume apiece.

In fact, without going into a lot of detail, technical or otherwise, it is quite plain that none of the ABSMs are either those of any known animal or any known type of human being. It but remains for me at this point, therefore, to draw your attention to a few salient and outstanding facts about these ABSM prints.

The thing to observe in the *Sasquatch-Oh-Mah-Sisemite-Mapinguary* type is that it apparently walks straight ahead with its feet turning neither inward nor outward. Therefore, it must bend or flex between the foot (the metacarpals) and the toes (digits) along a line at right angles to the line of travel. This gives us a point of reference to begin our study. If this is so, what at first looks like the "ball of the foot" is really a subsidiary pad at the base of the big toe [that, in all Hominids, unlike the other toes has only two joints]. The real ball of the foot is behind this so that, it is, despite its enormous size, really very short and broad. It has, in this example, what is called an Index of only 1.61—i.e. the number of times the width goes into the length. Further, the big toe is enormous. Then again, it will be noted from the photograph of this same print that there is a very pronounced and sharp ridge of clay running right across under the angle formed by the toes as they curve downward to obtain purchase. This is an invariable feature of the *Oh-Mah* prints. Now, even with our kind of *short* toes, mud would squeeze up between them in leaving a print of this nature. With these very long toes it should leave an imprint like that of a long-toed monkey. As it does not, something must have stopped it and piled it up. This can only be a webbing that runs up to the base of the terminal joints on all the toes.

The *Meh-Teh* or classical "Abominable Snowman" prints of the Himalayas, at first sight look just about man-sized but, when you handle a plaster cast of one, you get a profound shock. The thing is positively enormous and in some respects rivals the *Oh-Mah* prints which, though longer, look almost delicate and which are certainly in comparison most "refined." These things, as may be seen from the depiction of an impression of one alongside that of an ordinary human footprint are grotesque, and bestial. They also show features that, though not at all apelike in fact, digress from the human pattern most widely. They have an enormous big toe; but they also have an even more enormous second toe; and both are widely separated from the other three little toes, and they curl curiously inward toward them. This thing is not human at all.

The *Ksy-Giik–Almas* of Eurasia are notable for the size of their great toes and also for that of their "little toes," both of which are wide. The whole foot, moreover, is very short and broad and splays out in front. Otherwise it is human enough. I would just ask you to look at the outline of an imprint left by Neanderthal Man in the cave of Toraino in Italy. I do not think that I need to say any more on this score, except to remind of the Russian scientists' identification of one of the mummified hands.

The Pigmy type—*Agogwe–Sedapa–Teh-lma*—show a rather wider variety of form, but most display the peculiarity of a pointed heel, combined with small size, compactness, and more or less equality of toe length. This is the easiest print to fake and it is the nearest to some animals, but it has its oddities. Actually, I do not think we have enough accurate tracings or photos of them to assess, and the only plaster casts that seem to have survived are not worth-while. [The best proved to be those of Malayan Sun-Bears!]

Whatever is making the so-called ABSM prints thus comes in at least four forms. Moreover, these four forms have persisted for centuries. If this is all the work of a secret society it has four national chapters, but each of these would appear to be allowed to operate in the territory of others, for often three types will appear in one area, and in several there are two. The method of indentation of the prints also is most ingenious, for it has been estimated by road engineers that some in North America—which had no "impact-ridge" around them, as they would have if they had been *stamped* into the ground, but had distinct "pressure-cracks" all around all of them, which can be caused only by a steady push downward —have been calculated to have needed a minimum of 800-lbs each to be made! Also, if a device is used, it must stride along, not roll, for it can surmount inclines that no man can, can step over things, go around things, alter its stride on either side or both sides, pivot, flex, dig in with its toes going up and its heels going down, and do a lot of other things that no machine built could do unless it stood about 50 feet tall, and was so loaded with gadgetry that it would weigh tons. Yet, whatever does make the "Bigfeet" can go under an 8-foot tangle of branches without doing more than break off the little dead twigs.

Thus, of actual physical evidence for the ABSMs we have possibly one or two desiccated human-looking hands, a few piles of excrement, and, now, some hundreds of miles [in the aggregate] of tracks. We are right back where we started—with lots of reports but practically no facts. Is there anything else or anywhere else that we can try for information? There is. Two leads seem promising. Let us turn to these and see what we can unearth.

16. Our Revered Ancestors

Unless you believe in spontaneous creation, we, like everything else, must have had an origin. And this goes for ABSMs as well.

We now seem to have done several things, which had better be straightened out before we go further. (1) We have hinted at just how many of the "skeptics" have blundered into absurdities, contradictions, and frauds. (2) We have destroyed the value of most of the physical evidence, by subjecting it to some proper examination. (3) We have, as a result of the above, "canceled" ourselves out rather neatly, and now need a fresh line of approach, in order to see if any valid avenues still remain to be examined.

Positive arguments put forward against the existence of ABSMs really fall under four heads. These are that the whole thing is (a) lies, (b) hallucinations, (c) hoaxes, or (d) repeated cases of misidentification. That they could all be nothing more than straight lies is, I feel confident in saying, quite impossible. Too many people, fortuitously associated at the time, have actually found the foot-tracks, and all over the world. Moreover, it seems highly improbable that all the reputable people, so many of them well-known, listed in the previous chapters would all make up the same lie to describe the same creature and then say they saw it. For similar reasons, we may dispose also of the hallucination theory. First of all, there is a considerable doubt about the very existence of "mass hallucination" in which several people think they see the same thing at the same time. Then, hallucinations don't leave foot-tracks, hairs, or excrement; make wild yells, or move cairns. This brings us to the hoax theory.

MAP XIII. THE OLD WORLD

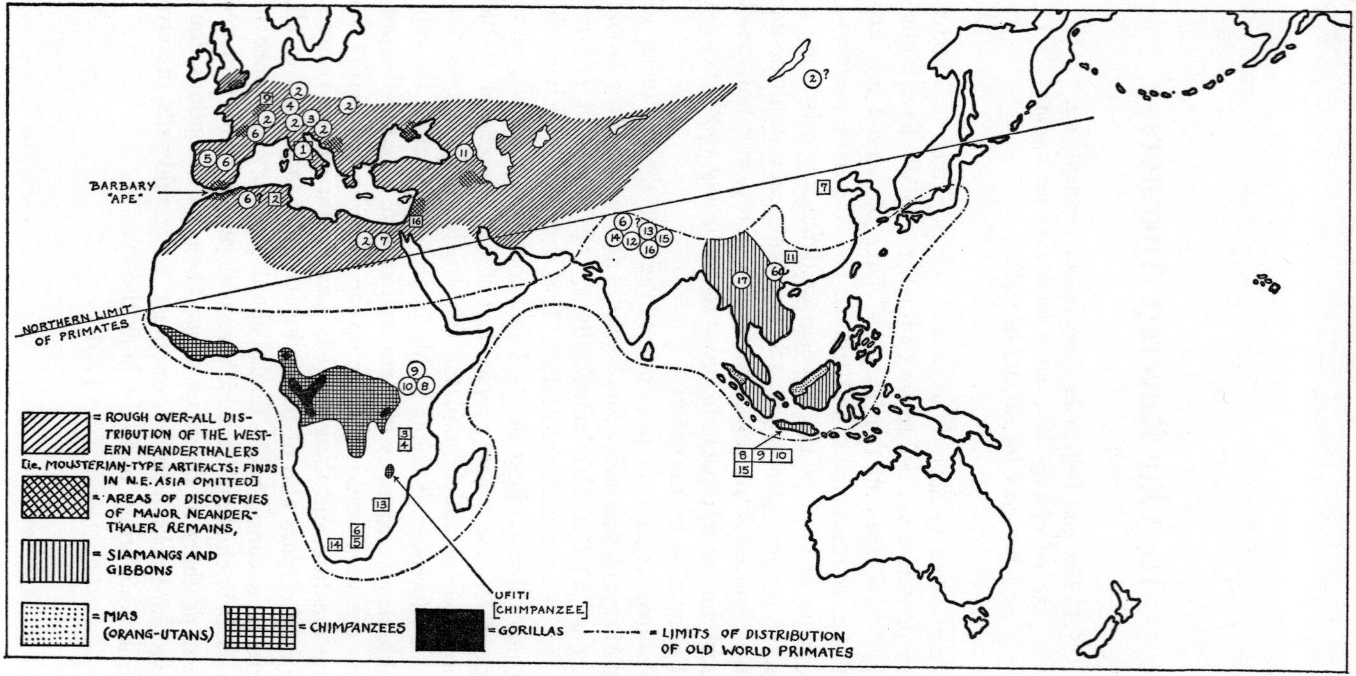

See facing page for legend

NORTHERN LIMIT OF PRIMATES

BARBARY "APE"

= ROUGH OVER-ALL DIS-
TRIBUTION OF THE WEST-
ERN NEANDERTHALERS
(i.e. MOUSTERIAN-TYPE ARTIFACTS; FINDS
IN N.E. ASIA OMITTED)

= AREAS OF DISCOVERIES
OF MAJOR NEANDER-
THALER REMAINS,

= SIAMANGS AND
GIBBONS

= MIAS (ORANG-UTANS)

= CHIMPANZEES

= GORILLAS

= LIMITS OF DISTRIBUTION
OF OLD WORLD PRIMATES

 [FITI [CHIMPANZEE]

This is the most difficult one because it is *possible,* however improbable it may appear to be, and whether you can suggest any way in which it could be done or not. As to the latter, I personally don't have the slightest idea, or any reasonable suggestions to make, though in some special cases I think I could duplicate some of the observed results by the exercise of a lot of energy, time, and money. However, I will repeat once more, magicians and professional conjurers can do the most amazing things that sometimes seem, to the rest of us uninitiated, to be quite impossible and even illogical, while

---

MAP XIII. THE OLD WORLD

Showing the over-all limits of distribution of the living Pongids and the location of finds of the remains of fossil Pongids and Hominids. There are those among anthropologists today who maintain that the entire tropical and the whole of both the north and the south temperate belts of the Old World were inhabited in succession by, first, sub-hominids, then Australopithecine forms, then Pithecanthropines, then Neanderthalers, and finally (either contemporaneous with the last or following them) by Modern Man. Whether or not the Neanderthaler type preceded Modern Man does not alter the fact that the type of stone implements that the former made is found at lower levels all across Eurasia; a western form, reaching to the Great Barrier; and an eastern, beyond that essential divide on the great upland plateau of Mongolia. The distribution of reports of ABSMs coincides closely with that of fossil Hominids and Pongids in eastern Eurasia, Orientalia, and Ethiopia.

NUMBERS

1 = Oreopithecus
2 = Atlanthropus
3 = Zinjanthropus
4 = Africanthropus
5 = Australopithecus
6 = Plesianthropus
7 = Pithecanthropus pekinensis
8 = Pithecanthropus robustus
9 = Pithecanthropus erectus
10 = Meganthropus
11 = Gigantopithecus
12 = Homo heidelbergensis
13 = Homo rhodesiensis
14 = Homo saldhanensis
15 = Homo soloensis
16 = [Recent Find]

NUMBERS

① = Oreopithecus
② = Pliopithecus
③ = Austriacopithecus
④ = Paidopithex
⑤ = Hispanopithecus
⑥ = Dryopithecus
⑥ⓐ = Dryopithecus keiyuanensis
⑦ = Propliopithecus
⑧ = Limnopithecus
⑨ = Proconsul
⑩ = Xenopithecus
⑪ = Udnabopithecus
⑫ = Sivapithecus
⑬ = Hylopithecus
⑭ = Sugrivapithecus
⑮ = Bramapithecus
⑯ = Ramapithecus
⑰ = Pondaungia

[353]

hoaxers and funsters have gone to the most extraordinary lengths to pull their stunts. One of the classic examples was the famous Würzburg "Fossils." These were a number of little clay tablets inscribed with crude drawings of animals and ancient Hebrew and other scripts which some students planted in a quarry where very ancient fossils were being brought to light by their professor. The nature of fossils was in dispute in those days, the general opinion being that the Almighty had put them into the rocks to test man's faith in the Biblical tale of creation. Another classic hoax *may be* [and I say this advisedly for reasons that we will see in a moment] the allegation of faking of the lower jaw of the very famous Piltdown Man. Anybody can comprehend how such as these were done once one knows that they *are* hoaxes but it is sometimes hard if not impossible for us to see how conjuring tricks are accomplished. However, while I haven't the foggiest notion *how* such tricks as "abominable snowman" tracks might be made in the circumstances among which they have been found, I *do* have a suggestion to offer a bit later on as to *why* they should be. For the nonce, however, let us just say that the hoax theory is extremely abstruse and has probably been adequately disproved, or, at least as of now, proved to be impossible. This leaves us with the business of mistaken identity.

I went over this briefly in the last chapter and can only add that, while in some cases a known local animal can be conjured up to possibly explain the alleged "sightings" of the creatures themselves, and even for the excrement and the hairs, there are no living animals known that can make any one of the four main types of footprints. Further, I would again stress the fact that the idea of some of them being made by four-footed beasts putting their hind feet into the imprints of their front ones, or more especially of a series of animals all jumping into the same hole for miles on end, is quite absurd and impossible.

This completes the roster of debunking explanations. Are there any positive suggestions as to what ABSMs might be?

There is one and it comes in three parts: to wit, that ABSMs are as yet uncaught and unidentified living creatures. There are three suggestions here: first, that all or some are unknown apes; second, that all or some are left-over relics of sub-men [i.e. what used to be called "ape-men" and "men-apes"]; or third, that all or some are remnants of very primitive humans. And, in view of everything else, this would certainly seem to be the best, most logical, and most probable suggestion; especially since the really extraordinary galaxy of other animals both small, large, and enormous which have come to light only in this century and right up to this decade [i.e. the new herd of Woodland Bison in Canada in 1960]. The question then immediately arises: What kinds of animals, sub-men, or primitives? Let us examine this straightforward question.

The first thing we have to do is to list the ABSMs and try to classify them according to whatever characters and characteristics they have or are alleged to have. To lead you through all the arguments by which I have arrived at the following general descriptions would take volumes, and be most irksome and dull. Most of the essential facts have already come out as we have reported the stories about them, and from what little physical evidence there has been left to us. The rest is technicalities, but each and all of the facts have been checked and the data on them is on file.

First, we should understand that the number of names for ABSMs (see Appendix A) has nothing to do with the number of different kinds of these creatures. There are literally hundreds of names for ABSMs still in use today, and hundreds more in over half the languages on earth and in many more that have now passed from common usage. Second, the number of individual localities where they have been reported is again not any guide to the number of kinds there may be. Like other animals, ABSMs seem to have wide distributions, some much wider than others, while some [and perhaps distinct species, or sub-species] appear to have very restricted distributions. Third, this distribution is not in any way as haphazard as it at first appears to be, while apparent inconsistencies and complete illogicalities in it are not only perfectly logical if one particular aspect of geography is taken into consideration—that is the geography of vegetative forms (see Map XVI; and the explanation in Chapter 18)—but actually go far to confirming the validity of the whole business. A fourth point we should bear in mind is that size has nothing much to do with the matter, for the distinction between the pigmy, man-sized, *Meh-Teh*, and giant forms is blurred in any case, while there may be large, medium, and small races, sub-species, or species of any genus of animals—and even in the same locality. This assessment is therefore based on one major and several subsidiary criteria.

The basis is the geography of vegetational types—desert, scrublands, savannahs and prairies, orchard and parklands, woods and forests, and most especially of montane forests on uplands and mountains. The supporting data are, first, the degree of "humanity" or "humanoidness" of the individual creatures as reported or alleged; second, the over-all extent to which their bodies are human; third, the degree in which their footprints approach those of man; and fourth, to some extent, how they are said to behave. They are listed below in accordance with these principles, those at the top being the most manlike, those at the bottom the least manlike, but it should be clearly understood that this does not mean that the latter are any more apelike. This is another matter that will be tackled in a minute. ABSMs then, seem to go like this:

I. *SUB-HUMANS* (East Eurasian and Oriental). Of about standard man size; hairy or partially hairy; head-hair differentiated from body hair; occasional use of very primitive tools such as sticks, bark cloth, clubs, hand stones; wary but not unfriendly; strong odor; some form of vocal communication but no true speech; good rock-climbers and swimmers; crepuscular and diurnal, possibly nocturnal also; may "trade."
    (1) Proto-Malayans, as appeared on rubber estates 1953.
    (2) Yunnan Hairy Primitives, as reported by Chinese.

(3) *Ksy-Giiks,* of Central Eurasia; possibly a Neanderthaler.

(4) *The Almas,* of eastern Eurasia; a small kind of (3).

II. *PROTO-PIGMIES* (Orient, Africa, and possibly Central
and Northwest South America). Smaller than average
humans, to tiny; clothed in thick black or red fur but
with differentiated head-hair that usually forms a mane.
Go about in pairs or family groups; wary but inquisitive;
apparently a very primitive form of language; toes sub-
equal and heels small or pointed; good tree-climbers and
swimmers; tropical forests down to seashores and swamps;
omnivorous, insect, fish, and small animal eaters plus
fruits, leaves; very nervous.

(1) *Dwendis,* of Central America, possibly only dwarf
Mayas.

(2) *Shiru,* of Colombia, S.A.

(3) *Sedapas,* of Sumatra.

(4) *Sehites,* of West Africa.

(5) *Agogwes,* of East Africa.

(6) *Teh-lmas,* of valley forests of the Himalayas.

III. *NEO-GIANTS* (Indo-China, East Eurasia, North and
South America). Taller than average man by at least a
foot or two; much bulkier, with enormous barrel torso
and no neck; head small, practically no forehead; heavy
brow-ridge and continuous upcurled fringe of hair right
across same; head-hair not differentiated from body hair
and all comparatively short; dark gray to black when
young, turning reddish or ocher-brown, and getting silvered in old age; face light when young, black when
adult; prognathous face and very wide mouth but no lip
eversion; eyes small, round, very dark and directed
straight forward; feet very humanoid but for double pad
under first toes, and indication of complete webbing to
base of last joints; has no language but a high-pitched
whistling call; nocturnal; does not have any tools; mostly
vegetarian, but takes some large animals and cracks
bones; retiring and very alert, wily, and afraid of man

but will attack if cornered, molested, or scared. Indication that they try to kidnap human females for breeding purposes. Food collectors; make beds in open or in caves. Drink by sucking.

(1) *The Dzu-Teh* (*Gin-Sung, Tok, Kung-Lu*), of Indo-China and Szechwan.

(2) *The Sasquatch* (*Oh-Mah, Sisemite,* etc.), of North and Central America.

(3) *The Mapinguary* (and *Didi*), of South America.

IV. *SUB-HOMINIDS* (south central Eurasia—i.e. Nan Shans, Himalayas, and the Karakorams). In every way the least human. Somewhat larger than man-sized and much more sturdy, with short legs and long arms; clothed in long rather shaggy fur or hair, same length all over and not differentiated. Naked face and other parts jet black; bull-neck and small conical head with heavy brow-ridges; fanged canine teeth; can drop hands to ground and stand on knuckles like gorilla; habitat upper montane forests, but descends into valleys in bad weather and digs for food under upland snowfields; color, dark brown; nocturnal and somewhat inquisitive; usually flees but may make simulated attacks if scared, and carry them through if the person gives ground and is alone; temperamental and bestial when aroused, being destructive like an ape; foot extremely un- or non-humanoid—second toe longer and larger than first, and both these separated and semi-opposed to the remaining three which are very small and webbed; heel very wide and foot almost square and very large. Omnivorous but with a preference for insects, snails, and small animals; will kill larger game. Lone hunter and food collector; wide traveler like all carnivores.

(1) *Meh-Teh* (and by other names), of the Himalayas.

(2) *Golub-yavan* (and other rather similar names), of the Kunluns, Nan Shans, and Tsin-Lings.

This completes the roster and calls for some comment. First, I have omitted anything that might exist in the Colombian Massif of the Andes except the little *Shiru* which seems in every way to agree with the No. II class above—namely the Proto-Pigmies. Should it so prove to be, then the *Dwendi* might probably go into the same group. At the moment, and for reasons that I go into more fully in the next chapter, it is my opinion that the latter are just groups of Mayas or related peoples, some of whom are really almost pigmies (see Fig. 56, the photograph of a Mayan mother, standing beside me, holding one of my godchildren).

I have also omitted the *Muhalu* and the Tano River giant of Africa as, in view of Mr. Cordier's report, and the nature of the former's footprint, the thing is definitely an ape. Left out also, are Dr. Moore's tailed creatures which as I have already said I personally think were large monkeys. This leaves but one form in doubt. This is the little *Teh-lma* of the lower valley forests of the Himalayas. Of these there are two conflicting and diametrically opposed opinions, which cannot be reconciled. One party claims that they are giant Macaques or Rhesus-type Monkeys, such as I discussed in the last chapter under their proper name of *Lyssodes*. This is fair enough and well taken. However, the tracks left by these *Teh-lmas* were found, copied, and examined by none other than Gerald Russell who is, in my opinion, just about the one man, apart from Mr. Cordier, who could really interpret footprints; and he states categorically that those of the *Teh-lmas* are definitely humanoid, and he demonstrates this with plaster casts. Also, he says, the creatures always *run* on their hind legs, which simply is not a simian [or monkey] characteristic. I think therefore that the *Teh-lmas* must be classed in the Proto-Pigmy group. The only other doubts are whether there really is any difference between the North American, Central, and South American Neo-Giants; and between the *Meh-Teh* of the Himalayas and the *Golub-yavans* of the ranges immediately north of Tibet. In both cases the descriptions of the two lots seem to be identical: they could, in each case, simply be races. Therefore, after disposing of the "animals"—mostly apes but some may be monkeys—we are left with eight or possibly twelve types.

To reiterate, these are: Four very primitive sub-humans; four proto-pigmies; two or three neo-giants; and one or two really "abominable" and bestial creatures. It is of course possible that the Proto-Malayans and the Yunnan primitives could be two forms of the same; that the *Ksy-giiks* and the *Almas* are only a size difference of another; that the *Dwendis* are fully human; that *Teh-lmas* and *Sedapas* are only racial forms of the same creature; that all four giants are but one form, for reasons of their distribution that we will see later; and that, as we have just said, the two "abominable" ones are the same. Such a further combination, or "lumping," gives us a fairly manageable list and perhaps a more believable one. It also coincides with geographical and other requirements. It goes like this:

I. *SUB-HUMANS*
  (1) Indo-Chinese-Malay, and south Chinese.
  (2) East Eurasian (*Ksy-giik–Almas*).
II. *PROTO-PIGMIES*
  (1) Oriental (*Sedapa–Teh-lmas*).
  (2) African (*Sehite–Agogwes*).
  (3) American (*Dwendi–Shirus*).
III. *NEO-GIANTS*
  (1) Oriental (*Dzu-Teh–Tok–Gin-Sungs*).
  (2) American (*Sasquatch–Oh-Mah–Didi–Mapinguarys*).
IV. *SUB-HOMINIDS*
  (1) Tibet and Himalayas (*Meh-Teh–Golub–yavans*).

This is still rather a "tall order" but there it is; and, we can't just sit back and deplore it. Something has to at least be suggested. The next questions, therefore, are: if there are all these creatures still running about waiting to be found, what exactly may they be? Also, do we have any ready candidates on our own family tree that we *do* know to have existed and to which we might assign any of them?

Here, for almost the first time, we are on surer ground, for we do indeed have plenty of candidates and, moreover, all in most convenient locations, and in many ways looking just right. All these, what is more, are on *our* particular branch of the family tree, and on rather convenient places thereupon to boot. This calls for said family tree, but even before we look at this it might be worth-while turning to Appendix C and taking a look at the much more extensive and general "Tribal Vine" as I call it, of the major group of mammals to which we belong, and which is known as that of the Primates, or Top Ones. On this you will be able to see at a glance just who your relations are and also just how widely separated you are from the less pleasant ones, and particularly from the Pongids or Apes.

I include the apes in the accompanying tree because there is all this endless talk about our being descended from them [which we are not] and also because of the wide use of the terms Ape-Man and Man-Ape, both of which now have to be abandoned; for, however non-human a *Meh-Teh* may be, neither it nor anything else can be halfway between Man and Ape.

Here, there appear for the first time on our canvas a number of new characters. These need introduction.

Since the publication of Charles Darwin's *Descent of Man*—not any longer perhaps the incorrect title it once seemed—anthropologists have been digging away all over the earth trying to find our ancestors. The procedure has had its ups and downs; its sudden great discoveries, and its patient piecing together of chance fragments; it has had its hoaxes, false leads, and other alarms and excursions; and sometimes its executors have gone a bit balmy; but, by and large, it has really made the most remarkable progress. Much of the story has been oft-told, but there is a crying need for a straightforward over-all account that brings matters right up to date. It is an enormously complex story and there remain in it both many blanks, great and small, and some appalling muddles. The worst of the latter, currently and rather surprisingly, concerns Modern Man (*Homo sapiens*) himself, and most especially in his earlier forms. The archaeologists have pushed him back in time to terrifying lengths on the grounds that he along with a few submen of the Neanderthaler type were the only toolmakers, but then the paleo-physical anthropologists [which is to say the searchers after fossil men's anatomy] suddenly popped up with two horribly nonhuman-looking types of creatures, both of which seem to have made fairly good tools. These are called the Australopithecines of South Africa, and the related Zinjanthropines of East Africa. Also, another group of sub-hominids called the Pithecanthro-pines of Indonesia and north China, proved in the latter area not only to have made quite usable tools but to have used fire in the latter. This has considerably upset our original ideas about toolmakers.

While all this was going on, other archaeologists searching for artifacts, as is their profession, and anthropologists search-ing for old human bones, and also the zoologists searching for extinct animal remains, and paleo-climatologists, and paleo-oceanographers, and glaciologists, and a whole bunch of others, even to geomorphologists and people concerned with wider matters like the IGY, kept turning up what appears to be evidence of Modern Men in ever more ancient [or earlier] deposits and strata. So, we have two sorts of floods of knowledge coming from opposing directions—one working back from the present, the other working forward from about a million years ago—not just meeting head on, but overriding and infiltrating each other. While the existence of modern-type Man himself has been pushed far back, the continuing existence of sub-humans and even of sub-hominid creatures has crept steadily forward in time.

Despite this, we find ourselves today no more advanced with the problem of Man per se than we were at the begin-ning, while we are actually in a greater muddle about both his beginnings, past distribution, and affinities now than we ever were. There are other complications too. The nice old idea that the Neanderthalers were a sort of Model-T Man, from which we arose but which itself later died out, has also gone all haywire. First, we now have bones of quite obvious modern-type men from strata just as early, if not earlier than the first Neanderthalers, and the Neanderthalers turn out to have been much more modern-man-like when they began than when they finally died away. In fact, they progressed backward as it were, getting ever coarser in appearance and structure. Then, there has been the distressing affair of Pilt-down Man.

This character, in the form of several pieces of a cranium and most of half a lower jaw was said to have been discov-ered in a gravel pit in the south of England by a man named Dawson in the year 1911. These were shown to Prof. Sir Arthur Smith Woodward, who declared them to constitute the remains of a new and very primitive form of sub-man with the brain of a human and the face and teeth of an ape. The fragments of the skull were assembled in various differ-ent ways by various experts; the mandible was completed in theory by extrapolation; and a single canine tooth was fitted into the general scheme so that a pretty fair assemblage was created upon which tendons, muscles, and skin were in due course modeled, ending in some very fine "artists' concep-tions" of the original owner of the bits. And so it went till 1953 when investigations made in the Department of Anatomy at Oxford and of Geology at the British Museum using new and elaborate methods of dating materials, indicated to some research workers that the lower jaw was a fake, and made from that of a modern chimpanzee by coloring with chemicals, artificial abrasion, and the filing of its teeth to match the human pattern. The single upper canine tooth, which is rather doglike, was declared also to be that of a modern chimpanzee, and also to have been tampered with. This "disclosure" made a great splash in the press. Unfortunately it now transpires that just about every aspect of it is as phony as Piltdown Man himself is alleged to be.

First, even these researchers admitted readily that the bits of the *skull* (cranium) are very old indeed. They are also very odd, being enormously thick but showing, by their curva-ture that they belonged to a very big brain-box. Comparison of the grains of rock still in their interstices would seem to indicate that they came from an exceedingly old strata for any hominid—no less than the Red Crag Beds of East Anglia, which is actually far "worse" than anything claimed for them by Messrs. Dawson and Woodward who said they came from a comparatively late Pleistocene river gravel—a mere difference of a million years! Next, the fragment of lower jaw is not, by its shape, that of a chimpanzee. It could possibly be that of a young orang-utan but it has one feature [called the simian shelf] more in conformity with some extinct apes than with any living one.

The final examinations made of the jaw [the cranium had been admitted by everybody to give good evidence of being hominid, human, and about 50,000 years old, *wherever* it came from] were made by a man who ought, above all others, to know what he was talking about. His name is Dr. Alvan Marston, a dental surgeon and a trained anthropologist and, furthermore, the discoverer of the famous Swanscombe Skull. He read a paper on his findings to the Royal Society of Medicine in London.

To this most august body he showed radiographs of the teeth "in which it was possible to see that the pulp chamber, or nerve canal, is filled with grains of ironstone and sand. This points to the fact that it was a young animal, which had not finished growing, and in whose tooth the pulp canal was still empty. In the Piltdown tooth, the entrance to this cavity is blocked with a piece of stone which has become cemented in, as stones are cemented into stalagmites in caves. This shows that it could not have come from a recent ape. Moreover, the crown is of a sort that is never found in existing species. It is found in the fossil *Proconsul*. The palatal surface of the root is flatter, too, than in existing types. This suggests a smaller mouth, and this (in turn) is borne out by the poorly developed *simian shelf*, such as those of certain fossil apes."

Dealing with suggestions that the teeth of a modern ape had been taken and deliberately ground into a shape more in keeping with human shape by some unscrupulous person, Mr. Marston said (*ex* Leonard Bertin, Science Correspondent of the *Daily Telegraph*, London):

I went into this matter very fully in a paper in 1952, after studying the matter for several years, and I can say that neither the canine nor the molar teeth have been mutilated, much less by Mr. Dawson (who discovered the Piltdown skull), who knew nothing of dentistry.

This is a very important matter to ABSMery for it points up two facts: first, that a very modern type of man [i.e. the cranium] was around Europe some 50,000 years ago; and second, that mandibles of most ancient apes have been disinterred [even if only in the Mediterranean area from where Dawson, the alleged discoverer of the Piltdown Man seems to have obtained many of his other fossils] for a long time, and *can have* most extremely hominid or humanoid-appearing molar teeth. These things we must bear in mind. Thus, both the Piltdown cranium and jaw are extremely ancient. However, it does seem to be true that they don't belong together and that they were never deposited at Piltdown, but probably were transported there by Mr. Dawson along with some phony bone tools and a few other odd bones. The gravel beds in which they were *said* to have been found have been extensively dug and sifted and not so much as one bone of anything has ever been found in them.

I go into this not only because it is a pertinent example of a hoax, plus the almost total unreliability of supposed "experts" [on some occasions, at least], but also because it shows the limitations of the much vaunted modern dating techniques, the manner in which the press can be completely misled, the lack of knowledge of one speciality by persons trained in another, and a galaxy of other obscenities that plague the whole gamut of the sciences. In this case, we have the added importance [to us] of evidence at the same time of a really very modern-type of man which, if some experts are now finally correct, could antedate quite a number of the so-called sub-humans, and sub-hominids.*

In our search for candidates for living ABSMs, therefore, we need not go dashing off into the remote past looking for bandy-legged, long-armed, brainless, gibbering peoples, before considering very carefully the large choice of manlike ones that have been around for a few thousand years and, maybe, even since before the four recent crustal shifts or ice-advances.

* By "sub-human" I mean Hominids that are not evolved into a form we can call *Homo sapiens:* by "sub-hominid" I mean species of Hominids of *genera* other than *Homo.*

We do not actually have a real definition of a true Man as opposed either to a sub-man or a sub-hominid. Anatomically, we may be able to draw a fairly fine line, saying that this, that, and the other cranial characters are typically of *Homo sapiens,* whereas others are not. However, I could name two prominent anthropologists who claim that they themselves are almost perfect Neanderthalers—i.e. living examples of sub-men! The reasons for their claims are perfectly valid as far as their bone structure, and posture goes. Also, I may say, both of them and especially one who is a North European are almost completely hairy all over: a most startling sight on a white sand beach in summer! When it comes to features other than osteological, such as skin color, hairiness, shape and size of teeth, gait, length of arms, thumb manipulation, toe agility, and so forth, we simply have no established criteria. We have been wrestling with what we call "race" for so long we have completely overlooked many much more important points about living human beings. Skin color really has practically no significance whatsoever, and it may change throughout life; as witness the number of Congolese babies born bright pink. Head-hair does show some classifiable features; so also do some oddities like "pepper-corn" hair growth as found among the Bushmen-Hottentots, the Mongolian-fold on the upper eyelids of Mongoloids, the "larkspur" heel of some of the Negroids, and so forth. These are special adaptations and they have nothing to do with basic hominid taxonomy.

The fact is, we cannot draw a line between "men" and "sub-men" and in many parts of the world today all manner of intermediate forms—both individuals, tribes, and whole races—still exist. It is only within the last few years that anthropologists have seriously suggested that the "Blackfellows" of Australia are really a separate sub-species of *Homo sapiens,* if not a distinct full species, having all manner of characteristics that most of the rest of us don't have—such as a different heat-regulating system, and other features. Then again, the yellow-skinned, glabrous Bushmen, with their steatopygy [or fat bottoms], the strange form of the male penis which is often permanently semi-erect, and the odd development of the female labia minora into huge flaps that may fall even to the knees, and which are known as "Hottentot Aprons," it seems obvious, really stand quite apart. Just because their head-hair is very tightly spiraled, and they have greatly everted lips, it used to be thought that they were sort of "primitive Negroes." This is quite absurd as they do not have any single feature that is typically Negroid, nor do they share any of their own odd ones with that race.

Likewise the Negrillos of Africa and the Negritos of the Orient, or Pigmies, as we call them, were until recently also thought to be a sort of offshoot of the great Negroid stock. But they too have practically nothing in common with the true Negroes. Apart from their tiny stature [as opposed to the exceptional tall stature of Negroids] their lower leg is shorter than their upper, they have reddish skins, they are covered with a yellow down sometimes developing on the limbs into quite thick hair; their blood type is quite different, and they have many other odd features, all of which are quite contrary to those of the Negroids. So also are they to those of any other race—Bushman, Australoid, Caucasoid, or Mongoloid. Then there were once the Tasmanians. These seem to have been an extreme and almost pigmy form of the Australoids and really to have been almost another species. They are extinct.

The Negroid so-called "race" is apparently the newest, and it is the least pongid-like of all. [Apes have no lips, the straightest of hair, the shortest legs and longest arms, and a host of other features that are the exact opposite of those of the Negroes.] The most pongid-like are the Caucasoids which have non-everted lips, straight hair, and so forth. The Mongoloids are really very different from both. Their absence of body hair and very thick long straight head-hair, round in section, is very odd; so also are the proportions of the parts of their limbs, with small hands and feet, short lower limbs and long upper. It is also curious that, despite their enormous fecundity, the Mongoloids become "lost" in crossing with the Caucasoids and sometimes in one generation, whereas they vanish completely at the first cross with Negroids. It has been observed—and by entirely "unprejudiced" people—that it takes nine crossings with Caucasoids for a Negroid to lose all his special features. The Negro in fact is a strongly dominant type and also a very new one who does not actually enter into our picture at all. Nor does the Mongoloid unless, as was once suggested, he developed quite separately from the Pithecanthropines. Rather is it with the Pigmies, Bushmen, and Caucasoids that our story is concerned.

Even if we don't know where "sub-man" ends and "man" begins we do know that, quite apart from myth, legend, and folklore, there was once [and in some cases still seems to be] a group of not-quite-humans spread all over a vast area from Morocco to the Pacific, and from the southern border of Eurasia [which, incidentally seems to have remained the domain of the surviving Neanderthalers] to central Africa, southern Arabia, Ceylon, the East Indies, New Guinea, and the greater islands immediately beyond. Everywhere we go throughout this vast swath of the earth's surface we find traces of peoples so primitive that they are variously alleged to have been hairy, to have had tails [a mere profligacy, as we have explained], to dwell in trees, have had no proper language, be cannibals, lack fire and even tools, and generally to be "Those who lived in the land when our ancestors first came from . . ." Osman Hill has brought to light some exceedingly interesting facts about one of these races called the *Nittaewo* in Ceylon.

These little, mostly Pigmy, primitives that seem once to have inhabited the whole of the tropical belt of the old world, provide us with most suitable candidates for our Proto-Pigmy Class of ABSMs—the *Sehite–Agogwes* of Africa, and the *Sedapa–Teh-lmas* of the Orient. These little ones are alleged to be really very human in many respects and their footprints are as human as they can be. The facts that they are hairy and gibber do not, as we have seen, necessarily put them into any bestial class nor even out of the human. They could just be leftovers; the "Devil-*Sakai*" that can really use the trees as highways. If there really are such Proto-Pigmies in the New World, represented by the *Dwendis* and the *Shirus,* they must have traveled around the long way by the Bering Straits land-bridge at an early date, and become isolated. These two little ABSMs would certainly seem to be pigmy primitives, rather than sub-hominids or even tiny races of sub-men.

We come now to the odoriferous characters who invaded the plantations of Malaya in 1953 and who appear to have sent their females to solicit young Chinese girls. These seem in almost every way to be thoroughly human despite their odor, nasty teeth, and excessive hairiness. There is no mention of them being *covered* with fur; rather, that they all had great mustaches and long head-hair, and very hairy limbs: They were also said to have light skins. All of this points clearly to a human type and even Caucasoid at that, primitive maybe, but still not even a sub-man. The same goes for the hairy primitives of inner Yunnan, reported by the Chinese. There is no implication that these were sub-men or bestial; just completely wild "people" without speech, and which could even be tamed and which would then show what appeared to be pleasure at accomplishing simple tasks and in the use of clothes. In fact, I feel rather strongly that these two types—which, incidentally you may note are the only two for which there are no recognized specific and distinguishing names—are simply very primitive peoples that have somehow managed to keep out of sight until things like the British bombings of the Communists in Malaya and the Communist stirring-up of country life in China brought them to light.

For the northern types—that is of Eurasia, in particular—we must wait until we look into myth, legend, and folklore in the next chapter, though, be it noted, that was the land of the Neanderthalers and everything about the *Ksy-Giiks* and *Almas* and all the others reported from that continent seems to point solidly to their being just such creatures.

The two remaining types of ABSMs, the Neo-Giants and the *Meh-Tehs,* present us with problems altogether different from any that we have so far encountered. Here, we come to the real core of the matter. These are the *Dzu-Teh, Tok, Gin-Sung,*

Sasquatch, Oh-Mah, Sisemite, Didi, Mapinguary type on the one hand, and the *Meh-Teh, Golub-yavan* on the other. We may well call these the "Inevitable No-men."

What could the Neo-Giants be and why should they have the apparently extraordinary distribution that they are alleged to have? At first both questions sound unanswerable but both are really amenable to very simple suggestions. Some years ago (1937) one Dr. von Koenigswald was searching through bottles of old fossil bones and teeth in a Chinese apothecary's store in Hong Kong when he came across a human molar tooth that was at least ten times in volume that of any ever grown by a man. And thus started the affair of what has been named *Gigantopithecus,* an enormous something, that once inhabited south China and left its bones in limestone caves. The controversy about this creature has been extensive and intense. Dr. Koenigswald's associate, Prof. Weidenreich, named the tooth *Gigantopithecus,* which means the giant "monkey" or by license "ape," rather than *Gigantothropus* or the Giant *Man,* because he was a very conservative and ultra-cautious soul. However, even before further remains of the brute had been found, other leading scholars stated that it was misnamed and was definitely a Hominid. [I had the privilege of examining the tooth all one afternoon in the American Museum of Natural History, and comparing it with the molars of all manner of men, current and fossil, and with apes, and for what my opinion is worth, it is certainly most strongly hominid.]

The tooth remained a ghastly enigma until 1956 when a Chinese farmer by the name of Chin Hsiu-Huai dug guano out of a cave in a mountain named Luntsai in Szechwan and spread it on his field. In this was found a part of a jaw with teeth of the same kind. Dr. Pei Wen-Chung, doyen of Chinese anthropologists, set up a prolonged search and found some fifty more teeth and, allegedly, a number of limb bones of the creature. He said that these indicated that it was a 12-foot tall, bipedal, carnivorous [sic] ape, than which there could hardly be a longer list of *non sequiturs.* Its teeth are utterly human, not just humanoid or hominid; if it walked erect, it was not an ape—not at that size and weight; and if it was carnivorous [which its teeth do not at all indicate] it was, again, not an ape as that seems to be just about the only distinguishing thing about the diet of that group—they are all profoundly herbivorous, though gibbons will take insects.

The other question debated about this brute has been whether [if it is *not* an ape but a Hominid], it belongs with the Pithecanthropines of North China and Java—to wit: *Sinanthropus, Pithecanthropus,* and the giant *Meganthropus.* This is not really very important to us but the manifest fact that it was a Hominid and not a Pongid is so, and leads to certain potent observations. If it was really that size, or even over *six* feet tall, it must have been a terrestrial creature, and if it was an ape it would have walked on all fours like the gorilla. Nothing that size can travel by treetops. If it was not an ape, it started out with the hominid type of foot, which is what is called plantigrade, and neither it nor its ancestors ever needed to develop a specialized great toe, which was opposed and worked like a thumb. Thus, this creature, primitive as it may have been, probably had a very human type of foot on which to support its immense bulk. Whatever it was, it lived in what is now southern China.

Now let us look at Map X. This area is a part of Orientalia, and is today subtropical. The mountains that surround it are those of the Indo-Chinese Massif and of the Szechwan Block. These areas are the lands of the *Dzu-Tehs, Toks, Kung-Lus,* and *Gin-Sungs*—the huge, furred "bear-men" or "men-bears" of ancient Chinese, Mongolian, and Tibetan legend and of current ABSM sighters. But then comes another thing. What else lives in and previously lived in this area? This is the land of the Metasequoia, of the raccoons called pandas, of certain curious little insectivorous mammals, of several odd amphibians, and of numerous invertebrates including a lot of most rare and odd parasitic forms. And where else, if anywhere, are any of these or their only relatives found today? In the northwestern part of North America!

There is still a continuous causeway of mountains from Szechwan all the way [to the west of China proper] to and through Manchuria to eastern Siberia. Because of increasing altitude toward the south (see Chapter 18), this is clothed in the same type of montane forest all the way. The same kinds of forest start again on the other side of the paltry Bering Strait, in Alaska, and continue on down in an almost unbroken chain to Tierra del Fuego at the very bottom end of South America. Moreover, sometime during the recent ice-advances and retreats, all manner of Siberian animals crossed over to the New World—like the Brown Bears, the Moose, the Elk, and others; and finally, the Amerinds, and then the Eskimos, did so too. Why on earth, should or could not a large sub-hominid also have done so, and simply by following the richly stocked montane forests all the way? That low temperatures could have prevented or even dissuaded them from doing so is just not valid, for, if the *Dzu-Tehs* are their living representatives, they can travel in snow without any trouble, and crossing the Bering Straits [even without a land-bridge due to alterations in sea level or elevation of the land], is no problem, for you can always walk across the ice in winter. It looks, therefore, very much as if Bernard Heuvelmans might have been right when he suggested that the largest type of ABSM in northern Orientalia could be a descendant of the *Gigantopithecus*, and the bolder his suggestion seems now, when it is realized that at *that* time (1952) the consensus was that that creature was an ape.

There remains then the *Meh-Teh–Golub-yavan* group of creatures, the original "Abominable Snowmen" which, as it now turns out seem to be the least "human" of all. Their distribution is odd but may be fully rationalized once again by referring to a map on which both topography and vegetation are shown (see Maps XI and XVI). The creature is obviously an inhabitant of the upper montane forests, but of the temperate zones; not of the tropical, such as occur on the Indo-Chinese Massif. As is explained in Chapter 18 the various vegetational *belts* that girdle the earth are repeated upward on mountains as *zones* and in the same succession as found at sea level, traveling from the equator to either pole. Further, in this arrangement, 600 feet of altitude is equivalent to one degree of latitude. Now, it so happens that the whole of central eastern Eurasia rises steadily to its southern rim [or, alternatively, tilts down northward to the great depressions of the Tarim to the Gobi]; and it also so happens that this tilt is just enough to create identical conditions for vegetation on the upper slopes of the enormous Pamirs-Kunlun-Nan Shan string of mountain ranges which run along the northern rim of the Tibetan Plateau, *and* along the mighty Himalayas to the south. The Pamirs themselves are too high for this type of vegetation, but it is continuous around their eastern face, so that one can travel in the same type of forest all the way from northern Assam west to those uplands, then north, and finally east all the way back to the Tsin-Lings in central China. This great *U*, lying on its side, is just the alleged distribution of these creatures. By this point, you will notice that when we speak of ABSMs, we are really referring to their alleged foot-tracks. Everything else about them stems from mere reports. Our sole problem here is, then, what could leave footprints of the nature attributed to these *Meh-Teh–Golub-yavans*.

These prints are really very odd indeed. Nothing at all like them is known in any hominid *or pongid*, either living or extinct; the outstanding difference between the two being that the big toe of apes is enormous and widely opposed, while that of all known hominids, though larger than any of the other toes, is not much separated from them and lies parallel to, and is bound to them. The *Meh-Teh* prints are in some respects intermediate, in that the big toe *is* considerably opposed; but then, so also is the enormous second toe.

The opposition of the big toe of the Pongids is an extreme speciality and was obviously developed by a tree-climbing animal, and, once developed, it has persisted [i.e. been unable to be gotten rid of]. In those apes—and notably the gorillas—which due to their weight have had to come to the ground and stay on it, and would much better have a foot like ours, it still persists. There is, however, the question of rock-climbing, and there are monkeys that have brought this activity to a high art, notably the baboons and macaques. However, these retain the fully opposed big toe and do not in any known example show any signs of having so developed the second toe. Thus, these *Meh-Tehs* must be a special evolutionary development of their own, at present without known ancestors. Just because the Pithecanthropines are known once to have existed in the Malaysia-Indochinese-Chinese swath of provinces; and just because the Himalayas are nearby and shown on all our atlases as being "in the same continent," the suggestion has often been made that these ABSMs may be descendants of those sub-hominids. We do not have the skeleton of a foot of the Javanese Pithecanthropines but we do have some foot bones of the north Chinese ones (known previously as *Sinanthropus*), and they are quite human and do not show even any tendency to the extreme oddities of the *Meh-Teh* feet, which are quite non-human. Dr. W. Tschernezky has discussed these feet fully in a paper in *Nature* (Vol. 186, No. 4723, May, 1960) and he therein shows, that despite these extraordinary big- and second-toe arrangements, it is fully plantigrade. Hence it is neither pongid nor hominid. What could it be?

I know of no answer to this question, and the only reasonable suggestions are that it is either (1) a very primitive hominid that for some reason developed that kind of foot, or (2) a very advanced pongid that did so after coming to the ground at a very early time. Frankly, in view of the "character" attributed to these ABSMs and their alleged actions I personally think that they are more pongid. Also, it would seem to be somewhat more in accord with what we know of the processes of morphological evolution to suppose a further adaptation of a foot with an already opposed big toe by changes in the second toe, rather than for a human-type foot to develop not just one but *two* opposed toes. Thus, I would place this type of ABSM as it is shown on the family tree; namely, as an early offshoot of the Pongids.

17. In the Beginning . . .

It's a funny thing, but all histories start by saying "In the Beginning . . ." and then proceed to describe all sorts of things that happened before.

In almost every book that I have written, I have found myself, sooner or later, disposed to interject a remark which, above all others that I ever heard, left the greatest impression on me. This was made to me by a V.I.P. in a distant and unvisited part of West Africa—a Paramount Chief. Having fixed me almost to a point of hypnotization with his enormously wise and expressive eyes, he stated solemnly: "The best place to begin all stories is at the beginning." Then he shut up and waited. As Paramount Chief of the region he was endowed with the status of Chief Justice both as an executive of the Government of the Protectorate and as paramount native Justice of the Peace; he was ultimate arbiter of all law; and, in Africa, this means deciding upon the validity or otherwise of stories. The Chief knew human nature.

It would seem that this positively cosmic piece of advice must have been offered long before the dawn of history, for all peoples seem to have taken it to heart. There simply is not a history—religious or secular—that does not start with something like "In the beginning . . ." The Bible gets off to a flying start in this respect, beginning, in its original form, "In the beginning, the Lord created heaven and earth . . ." With this, hardly anybody, and not only Jews and Christians, disagrees. All the other great religious histories start in much the same way. However, whether historians begin like this or, in the more secular fields, with the beginning of their nation as *the* original [and everybody claims this prerogative] people, one

[375]

MAP XIV. THE WORLD

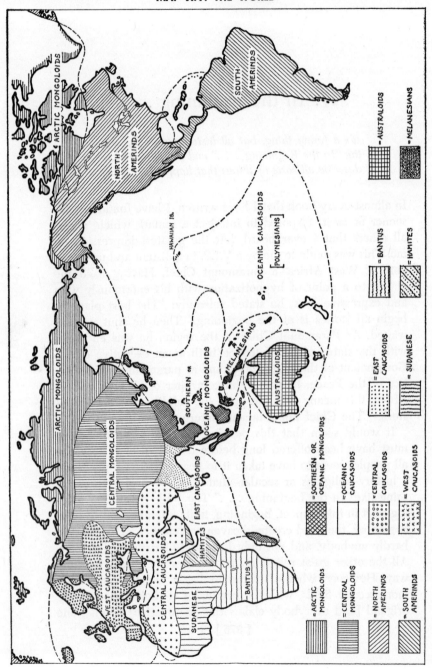

immediately or very shortly encounters a most odd circumstance. It transpires that, regardless of the fact that almost everybody is agreed that the Almighty started the whole works, there was an enormous length of time (or elapsed time) before *We* appear on the scene—usually by God's design, but sometimes just out of the blue. This is passing strange but you will find it if you dig back far enough into any statement made by any people about their origins.

Humanity seems universally to have attempted to adopt the Chief's admonition, but in the majority of cases to have run into a profound difficulty. The usual way around this was the declaration of spontaneous creation, either, as among those whom we consider to be the more advanced thinkers, by the One Power, or, as among those we say are of lower intellect, by a pantheon. This gets rid of the problem as to what went on before *time* started, as well as what was before this earth was formed. There is only one conflict in all this and that is between one group—and it is actually far the larger one, though this is seldom realized—that contends that neither time nor space have any beginning or possible ending, and another which contends that they did have a definite beginning and may have an absolute end. This latter party asserts that

---

MAP XIV. THE WORLD

Today, the entire land surface of the earth, apart from Antarctica, the Greenland icecap, and a number of islands, mostly in the polar regions, is officially "inhabited" by Modern Man. Actually, he lives on considerably less than a tenth of the total land surface. Modern Man is divided into two very distinct groups—the Primitives and the Non-Primitives. The former consist of the Australoids of Australia and, in part, Melanesia; the Bushmen of Southwest Africa*; the little Negrillo Pigmies of central Africa*; and the Negrito Pigmies of the Andaman Islands, the Malay Peninsula, and the Philippines.* There are three divisions of the Non-Primitives—the Mongoloid, the Caucasoid, and the Negroid. Of the first, there are five subdivisions; of the second, three; and of the third, two, but with another group—the Hamitic—derived from intermixture with one of the Caucasoid groups. Currently, the West Caucasoids and the Sudanese Negroids have greatly extended their range, notably to the Americas. [* Not on map due to scale.]

before the beginning there was but God. These are abstruse matters of the higher philosophy but, as we shall see, they have a most profound bearing on our very pragmatic concerns.

The Bible in a way attempts a rational confluence of these two otherwise opposed points of view, by stating first that everything is indeed a willful act of God, but that it was, as concerns this earth and us, only an interlude in eternity. It then gets down to specific facts about this latter as, really, quite a separate subject. Thus it has two beginnings; the first cosmic; the second, terrestrial. But then again, unlike almost all other religious histories, it starts off still a third time with the beginning of *Man*—and by this is meant what we call *Homo sapiens;* not just the Hominids. On this last business, it is rather specific.

Now, this sort of thing crops up also in just about all historical "beginnings," even down to those given by the most primitive and ancient peoples who have ever left a record of their ideas. It is really a most odd fact and one which seems, to us, utterly illogical; namely that, if asked how it all started, everybody invariably says that thus or thus characters, who are usually claimed as the tellers' ancestors, came to the land and smote the horrible creatures which were already there, eliminating them and thus starting EVERYTHING. Sometimes the whole business gets really complicated as is so very pertinently explained in the Bible. For this I turn to a good friend of mine —a brilliant young scholar; the Rabbi Yonah N. ibn Aharon, B.D., S.T.M.—who has provided me with the following properly classical rendering of the appropriate passages from the Book, together with those pertinent commentaries that only true scholarship can provide. He says:

The earliest Biblical reference to genetic variation within the human family is in the sixth chapter of Genesis, where we read: "And it was when Man began to multiply on the face of the Earth and daughters were born unto them; and the Sons of Those from upon High (Jerusalem text: those who are worshipped) saw the daughters of Man, for they were good-looking, and they took unto themselves wives from among such as they might care to choose. And Yehovah said, My power can never benefit the perverted ones who have made themselves human; and their days shall (number) one hundred twenty years. In those days there were (already) nephilim in the land; (it was) thereafter that the Sons of Those from upon High came unto the daughters of Man, and there was born to them those Giborim who were ever after considered to be a divine people (lit.—the people of the Name)."

Nephilim is often translated "giants," but the commentators tell us that they were so called because men would fall (nophel) on their faces with fright at the sight of them. (cf. Ibn Ezra, B.K.) The giborim, who are later on referred to as giborei tsayid, are reputed to have been "as tall as a tree," in contrast to the shambling nephilim (cf. Yoma). Giborei tsayid means "The Mighty Ones of the Hunt."

Our main concern must, however, rest with the creatures who terrified the Israelites during the Exodus from Egypt and their period of wandering in the Sinai desert. These were the Sheidim—the Destroyers—who had been known to the Patriarchs (Abraham, Isaac and Jacob) as the Seirim—the Hairy Ones.

The best scriptural description of their characteristics may be drawn by inference from the account given in Genesis 27 of the manner in which Rebecca connived to win the Rights of the First-born for her youngest son Jacob, as against the prerogatives of his twin brother Esau, who is described as "coming forth first from the womb 'reddish with a great mat of hair' " (as thick as a wig, as Rashi puts it). Esau grew up as a hunter, very much ashamed of his deep red covering, which earned him the nickname Edom, or Reddy, as the vernacular might have it. The implication of the text at every turn is that, although ostensibly Jacob's twin brother, Esau was no true son of Rebecca. It would appear, at least that she felt that way about it, because she forced Jacob to seek the blessing of his father, Isaac, whose weak sight made possible the success of the ruse. To this end, Rebecca, slew two young goat-kids (lambs, according to some sources), and clad Jacob, whom the text describes as her "smaller" (not "younger" as one would expect) son, across the back of his neck and around the wrists in their hide (Genesis 27:19 ff.). Thus prepared, Jacob went into the presence of Isaac, his father. It is from this interview that we learn most about the Hairy Ones, insofar as specifically Judaic sources are concerned.

Jacob was successful in passing himself off for Esau on three counts: (1) his apparent hairiness, which we have just accounted for; (2) his voice, and (3) the odor of his clothing. Jacob's speech, when he entered his father's chamber and identified himself as Esau, is thought by many not fully to have convinced the wily Isaac, but this impression is based on a misreading of the text. Isaac is quoted as saying "the hands are the hands of Esau, but the voice is the voice of Jacob." In reality, the passage reads, "the hands are the hands of Esau, and the voice of Jacob is higher pitched (than that of the one who stands before me)"—kal kol (as the passage is actually written) instead of kol kol, the popularly accepted reconstruction.—Indeed, the commentators bear out this interpretation when they tell us that Jacob counterfeited the voice of Esau by growling from deep within his throat (cf. Klei yakar, etc.). The outdoorsman odor of his clothing was achieved through more obvious means, for Rebecca simply appropriated the cast off garments left over from Esau's last visit. The stench, we are told, was akin to that of "a field of rotten potatoes." *

A composite physiological picture of the Hairy Ones shows us that they were Hominids, and, indeed, were close enough to modern man to be capable of intermarrying with other human races. They had long arms, and a mass of reddish hair covering their bodies, darker about the head than below; they attained a height of about 4½ feet (par for those days, I'm told) "like the stone hounds of Aram." Beneath the shaggy exterior, the bone structure was clearly evident, particularly on the legs, which were short and very straight, the elbow, neck, and heel joints being unusually large.

The habitat of this race is definitely known to have been restricted to the Sinai Peninsula; there are similar beings known to have lived in the South of Egypt, but the history of the Hairy Ones (as the R'aya M'himna remarks) was apparently bound up closely with that of the Israelites, who had to cross their territory to get to Egypt; and, inasmuch as the Israelites were relatively poor farmers, they were dependent on the Egyptian trade for their sustenance in years of famine, until the time of the Babylonian invasions. This limited habitat will, at any rate, account for the fact that the Hairy Ones never harassed the Egyptian homeland. But, just as they threatened war against Jacob after their candidate had been defeated in the election of a Nomad chieftain (which was, after all, the political role of the Patriarchy), they had no intention of allowing the Israelites to dwell in peace after the latter had made good their escape from Pharaoh. So intense was their onslaught that, within days of the Exodus, the Israelites were forced to dig trenches, and cover them with branches so as to protect themselves from the stone-throwing barrage of the Hairy Ones. One tactic which the latter favored was to cover themselves with pieces of foliage, or sand, so as to camouflage their whereabouts; the Israelites hoped to do the same for their dwelling-places (succoth). The Bible is

* Curiously, this crops up repeatedly—in Canada, the Himalayas, and in Central Eastern Eurasia. However, the potato was unknown in the old world prior to A.D. 1500.

understandably quiet about this humiliating state of affairs in the passages which relate to the miraculous deliverance from the Egyptians; and, we may add, the "air-drop" of a special food concentrate called manna, that kept the people from starving after their supply of Egyptian roast lamb had run out on them, leaving them to the mercy of the Hairy Ones, to whom they were ready to offer their children in return for foodstuffs (S. Retsinuatha IV, ¶16c).

As soon as things settled back to normal, however, the God of Israel decided to assert his power against the Hairy Ones. His real reason for doing this is that, in the days before He ever thought of speaking to Moses, they had rejected him as their deity, in favor of a certain Azazel. Thus, no sooner was the Holy Sanctuary completed, than Yehovah gave orders to Aaron, the High Priest, who, as we read in Leviticus 16:7, "took two of the Hairy Ones, and caused them to stand before the Lord. And Aaron put tags on the Hairy Ones, one tag for Yehovah and one tag for Azazel. And Aaron made an offering of the Hairy One which he had tagged for Yehovah; and the one which had been tagged for Azazel remained alive in the presence of Yehovah, who sent it off to Azazel (who lived in) the desert, that it might atone (for the sins of Israel, according to the commentators)." Lest the casual reader mistake the *seirim* of this passage for the "hairy goats" (seirei 'izim) of the verses immediately preceding it—as did the famous but unhappily incompetent modern scholar S. L. Gordon—Leviticus Rabba, the most ancient of the exegetical works of Judaism, spells it out for him: "These *seirim* are none other than the Destroyers, the sons of Esau." The similar meaning of seirim and sheidim is among the ten best cases of authenticated synonymity in the Hebrew language.

Thus, it may be understood that the "scapegoats" so glibly mentioned in the English translations, and in countless works on anthropology and psychology, were, in fact, manlike creatures, with considerable biological, if not spiritual and intellectual affinity to the rest of mankind, and that Aaron, acting under orders from Above, committed an act which, to our mortal eyes, looks like something very close to human sacrifice. The commentators are agreed in admitting this, but, as is the case with Jacob's deception of Isaac, they do not debate the morality of Aaron's obedience to the will of the Most High. The sacrificial rite was symbolic of Israel's complete subjugation to its God, and a repudiation of the power of the worshippers of Azazel, of whom the people had become so much afraid that they were sacrificing their produce to that deity at Yehovah's expense. Israel must turn out the Beast within, and sever itself from the sons of Esau, born of the same woman (Zohar).

Perhaps the most poignant affirmation of this need, and the best scriptural justification of the views expressed in this essay, is to be found in the lines of Moses' great poem, the Ha'azinu, as given in Deuteronomy 32:13, where he says of the "generation of Egypt," meaning his contemporaries:

"They have sacrificed to the Destroyers,
    (who come) not from on High;
Those from on High, they have failed to recognize,
For these New Ones have come from close by.
Your fathers (O, Israel!) did not thus abominate themselves!"

And the word for abominate, in the Hebrew text, contains the same root letters as the word Seirim—an unmistakable play on words, and a fitting end to this discussion.

To which I would add that it is even more of a coincidence [perhaps] that the newspaperman, Henry Newman, should, three millennia later, have by pure mistake named certain similar creatures by an antonym in our language—for an object abominated is abominable.

I give this most curious, and in some ways wondrous sidelight prominence here for several reasons. First, because it is from the Bible, the very tenet of our faith. Second, because it is, as far as I can see, an absolutely unassailable example of the only modern credo that tries seriously to cope with what otherwise may be, and often has [I believe erroneously] been considered to be evidence of an uncompromisable conflict between our science and our religion. Myths, Legends, and Folklore [hereafter called MLF] may be laughed at or cast aside as the hallucinated maunderings of uneducated and often unthinking man; or, as in some quarters today, they may be elevated to a position of gross reality just as if they were about the only things that we can really rely on. Neither attitude is realistic or warranted. A lot of myths are straight history; a lot of history is pure myth. In the field of religion, and not just theology, there has always been a very widespread misunderstanding of the simple and obvious fact that most *religious* histories are clearly bipartite—being, on the one hand, philosophical; and, on the other, entirely pragmatic. The Bible, as

[382]

I have tried to point out above, is thus duplex in content; and, moreover, on the purely pragmatic level it is really three quite separate secular histories—that of the Universe, of the Earth, and of Modern Man. The Philosophical ground-floor is not, of course, our concern here: nor are the origin or history of the universe or this planet. What we are concerned with is the origin of man. On this matter Darwin had neither the first nor the last word.

I personally consider these pragmatic and secular parts of the Bible to be straight history written by sundry groups of the first peoples to use their brains, the first to try to investigate their environment, and the first to make some logical sense out of it. The Semitic peoples, starting, if we may use language as the criterion, with the Chaldeans, were certainly contemporary with and probably preceded—even if they did not actually play a part in the origination of—both the Harappa-Mohenjo-daro culture of the Indus Valley and that of the Egyptians of the Nile. They certainly seem to have been the first to leave written records. However, in their day, the greater part of the world was a pretty wild place and many things that have long since vanished were then still at that time right in their own back yard. By this I mean that they had primitive races [perhaps hairy] living at their very doorsteps, just as New Yorkers today still have mink, raccoons, and opossums wandering about the parks immediately over the rivers that separate Manhattan from Long Island and the mainland. Thus, anything they have to say about such leftovers then still extant, is thoroughly worth-while studying and analyzing.

From the preceding statements extracted from the Book of Genesis, I can but infer that in the peninsula of Sinai [seenigh] there still lived at the time of the Exodus (1317 B.C.), a not inconsiderable number of hairy fellows of hominid genetical background, even up to throwing stones and breeding with the Israelites but who were at the same time advanced enough to have some primitive form of religion with a "god." I refuse to disbelieve these passages in the Old Testament: *ergo*, I must accept them as historic and thus that these types existed. If they did so, it is, once again, no good just sitting back and saying "I don't believe it," or "So what?" or even just sitting back. It behooves us to get to work and at least speculate what they might have been, and why; and what happened to them. The same may be said for exactly similar types of creatures that appear, just as matter-of-factly, in early passages of almost every other secular history. And they do so appear.

I do not propose to go into the details of MLF. Not only is it not specifically our province but it is, except to specialists, incredibly boring; in fact almost as boring as having to wade through the names and perquisites of gods in multiple pantheons. Also, with respect to ABSMs or ABSM-like creatures, the whole business becomes unutterably monotonous for, from all over the world, the stories told are nothing but almost word-for-word repetitions of the modern reports that I have already given aplenty—giant, funny, or pigmy foot-tracks; tiny, man-sized, or giant hairy people; high-pitched whistles or gibberings; abducting of young human females usually followed by their release; and an almost invariable smiting or eradication of such types "In the beginning." The whole dreary business is a bore but it does still have very great significance, for it means that almost everywhere [apart from Australasia, Oceania, and Antarctica, as far as we know] sub-humans if not sub-hominids inhabited the whole earth prior to the arrival of the first *Homo sapiens* persons who proceeded to oust them or at least take over their territory.

In ferreting out *noticias*—as the Spanish so aptly put it—of the existence of these sub-humans in all the welter of written, transcribed, spoken, and remembered MLF, one does, however, have to be extremely careful to observe one basic fact. This is the very clear distinction made by most peoples—though little so by Caucasoids of the West during the past few hundred years—between three types of *Beings;* exclusive of the all-pervading Spirit, or God. These are: (1) *Divine Entities,* being representatives of God, gods, demi-gods, or disembodied noncorporeal personalities of another world but which may appear in this one and influence it. These are entities in their own right that, while being able to assume human form or "enter into" humans, do not change their own identities. (2) *Disembodied Spirits* of various kinds. These may be the souls of people, dead or alive, mass-produced ancestors, spirits of animals, plants, stones, or anything else, either collective [generic] or individual, together with all manner of *lares* and *penates*. To most peoples these are just as real as living people, animals, or plants. (3) Unknown or as yet undiscovered but live, corporeal things.

ABSMs have always fallen very clearly and distinctly into the third class. Nowhere in the world is there any doubt about this. If asked, the "benighted natives" will usually say something like the Nepalese at Pangboche when asked by Stonor about the *Meh-Teh* alleged to have been seen the year before. The answer he got was "How could they [i.e. any of Nos. 1 or 2] leave footprints?"

I have a fancy that a somewhat extensive galaxy of alleged creatures in the folklore of Western Europe is of this same most pragmatic nature. If you come to look into what was said about Fairies, Pixies, Trolls, Titans, Vampires, Ghouls, Gnomes, Imps, Bogies, Brownies, Elves, Leprechauns, Satyrs, Ogres, and Fauns [as diametrically opposed to "ghosts," "specters," "apparitions," "spirits," "phantoms," "wraiths," "spooks," "banshees," "lemures," or "lorelei," which were definitely of Class 2], you will find that they may all be summed up by the classic line from the somewhat bawdy old English song that begins "There *are* fairies at the bottom of *our* garden."

Creatures, usually hairy, generally malignant, only rarely benevolent, but perfectly capable of breeding, as well as communicating with human beings, form the basis of these tales. And *note*, they come in four convenient sizes. The same may be said for all similar types known by whatever other languages all over Europe, North Africa, and a great part of what is today Russia. There seem, indeed, to have been "in the beginning" ABSMs of just the usual four types—pigmy; man-sized [and specifically of the Neanderthaler kind]; giant; and the bestial *Meh-Teh* with its abominable feet [cloven?] and pointed head.

Anthropologists have shown that most Australopithecines were tiny things like the modern Pigmies, while their cousin *Zinjanthropus* for all its enormous jaw development and molar teeth [its canines and incisors were tiny] was itself a little creature. There were undoubtedly "Little People" all over the place in ancient times. As to their having been "Giants in those days" we don't really know about western Eurasia, or Europe as we call it. We have a huge jaw from Germany [named *Homo heidelbergensis*] and the preposterously thick cranium of the thing alleged to have been found at Piltdown. In the Orient we have both *Meganthropus* from Java, along with *Pithecanthropus robustus* from the same area, and *Gigantopithecus* from southern China. Thus there were definitely "giants" available in southeast Asia and these could quite well have crossed over to the New World, along with hosts of other large animals before or during the Pleistocene ice-advances, and then have filtered on down to the Matto Grosso and the Guiana Massif. If there ever was a giant in Ethiopian Africa, it *could* just possibly have been of this stock; but we also have other very rugged-looking types there in so-called *Homo rhodesiensis,* fragments from Algeria (*Atlanthropus*), and from Tanganyika (another *Meganthropus*) which might have provided ABSM material on that continent. We discussed the candidates for the man-sized types in the last chapter—primitive modern men in Indo-China, and Neanderthalers in Eurasia. The matter of the bestial *Meh-Teh* type has also been investigated. This leaves us with a few vague rumors from Africa, North America, and the Indo-Chinese Block.

The masses of reports from Eurasia, ranging, as we have seen from the Caucasus to Manchuria, appear to have a distinct cohesion though to be of more than one *specific* type. It is interesting to note that this was the land of the Neanderthalers per se [Rhodesian and Solo Man only look somewhat Neanderthaloid], and the descriptions of the ABSMs seen there in no way conflict with our findings on that branch of the human stock. Likewise, the MLF that pertain to such concrete entities [i.e. Class 3 above] from this whole block of territory

[386]

provide us with as good a reconstruction of Neanderthal sub-man as any anthropological institution has yet concocted.

In other words, modern and historical reports of ABSMs; the findings of palaeoanthropology from bones and artifacts; and MLF, all converge and literally combine—yet on a precise regional basis, perfectly in accord with both ethnography and phytogeography. This is the clincher; so let us examine these two aspects of the matter.

As currently defined the major branches of the human (*Homo sapiens* subsp.) species are distributed as illustrated on the world map on page 376 (Map XIV). They fall into four pretty clearly defined lumps; one with five major divisions; one with four subdivisions; and the others with two. Spotted about, but very sparsely, are also the remnants of two other basic and more primitive groups both now nearly extinct—namely, the Bushmen and the Pigmies. [Of course, there are also the recent wanderings of the western Caucasoids and Negroids but these I have ignored as not being in any way germane to our story.] From this map one may see more or less how the world was about the time of Columbus, and before the expansion of Europe had really gotten under way. Armed with this, one may then proceed to consider ABSMs in MLF, and in point of current fact. Both classes of data fall exactly into a single pattern.

The status of both these folk-tales and current reports of ABSMs in the Americas is too confined and obvious to need much comment. One needs the more detailed vegetational map (see Map III) to elucidate the regional features. The northern tree-line clearly divides the Asiatic or Arctic Mongoloids [i.e. the Eskimos] from the North Amerinds. ABSM reports from this continent fall into two classes: those of giants right across the top and then down the western mountains; second, the much vaguer mumblings about "the little red men of the bottomlands" from the Mississippi drainage.

There is undoubtedly a great deal more ABSMery about South America but the reports of it are scattered through the voluminous local presses of its many countries, while our knowledge of the beliefs of and the factual information possessed by the indigenous Amerindian peoples is sadly limited as yet. Just as in Central America however, cases have apparently gotten on to police-blotters with some regularity, and several reliable travelers have made reports. Among these are three that Bernard Heuvelmans has sent me.

In 1956, the geologist Audio L. Pich found on the Argentinian side of the Andes at a height of over 16,000 feet enormous human-like tracks, with prints about 17 inches long. In 1957, the Brazilian newspaper *Ultima Hora* of Rio de Janeiro stated that similar footprints had been found in La Salta Province of Argentina, and went on to say that a newspaperman found the people of a village named Tolor Grande in a turmoil, due to eerie calls at night emanating from the Curu-Curu Mountains. These are said to be the habitat of a dread creature called the *Ukumar-zupai*. In 1958, *La Gazeta* of Santiago, Chile, of May 6, published a report of an "ape-man" seen 50 miles from a place called Rengo by a party of campers. Several other witnesses are also quoted, and one Carlos Manuel Soto swore out an affidavit on May 13, which includes the statement that "I saw an enormous man covered with hair in the Cordilleras." It was also stated that the local police had investigated.

Turning now to the Old World, we find quite a different situation. Let me first dispose of Ethiopian Africa, something that I have really already done in that I tried to point out in Chapter 9 that, apart from the vague Tano giant and the *Muhalu*, which seems pretty definitely to be a pongid [just as the natives have always asserted], there is nothing to report but the widespread notion that pigmy races were once much more widely distributed; and still are so today, while some of these are so very primitive indeed as to be hairy. Most African peoples have a large and splendid pantheon of gods, and they also almost universally believe forcefully in another whole world of disembodied spirits of all kinds, but they make the clearest distinction between both of these and mere unknown animals, of which they still speak aplenty. If any of these were hominids, Africans would be the first people to say so. They don't.

This leaves us with the continents of Eurasia and Orientalia for, as we have said, there is nothing to report from Australia, Melanesia, or Polynesian Oceania.

But here comes a rather ticklish matter. Map XIV which displays the distribution of modern men prior to the expansion of Europe has one most astonishing feature. This is the almost exact coincidence of the distribution of the Caucasoids and Mongoloids with that of the true continents (see Map XV) and with certain major boundaries between Vegetational Belts (see Map XVI). The coincidence would seem impossible did we not know that Man, being an animal, is just as confined by the limits of the environment in which he evolved as is any other animal; while the major factor in *any* environment is the *form* of its vegetation. However, this is not the ticklish matter.

Both the Mongoloids of the Old World and the Caucasoids are subdivided into three major lots, though all of course merge to some extent. But, if you dig back into the origins of all three branches of the Caucasoids, you will almost without exception find that they are known to have, said to have, or believed to have originally come from central Asia. There appear to be remnants of some really original Europeans in the Basques; of the Middle Easterners, in such isolated spots as the Canary Islands, the Atlas Mountains, and Abyssinia; and of the Easterners or Indians, in the southern part of that peninsula; but everywhere we look, we find a residue of Mongoloid penetrations or immigrations going back for millennia. The Semitic peoples alone would appear to have stayed where they originally evolved and to have rebuffed these Mongoloid hordes; an aspect of history that is of the utmost significance. Since the Caucasoid seems to be rather strongly dominant to the Mongoloid type, that type soon disappears physically when it slops over its own precise borders—*vide*, the purely Caucasoid appearance of the Slavs and of the still later Magyars today. However, while they may appear to disappear physically, their MLF usually linger on and become rooted in the lands they conquered or swamped. This has been most particularly the case in Europe; much less so in the Middle East; and surprisingly little in India, despite the many great Mongol invasions thereto and their long periods of dominance there.

In studying the traditions of Eurasia, we must therefore regard the area in two parts—the first that of Europe and central Asia to Korea; the second, that of the Middle East or Semitic world. Likewise, when we come to Orientalia, we have to make an absolutely clear distinction between Caucasoid India on the one hand, and the lands of the South or Oceanic Mongoloids on the other. There is, then, the added complication presented by the fact that the Northern or Arctic Mongoloids of the Old World also have clear traditions of ABSMs along with beliefs in a great number of such related creatures of the past, which they share with those of the Central Mongoloids *and* the Europeans. Throughout this whole vast area there is an almost universal "belief," amounting to a true folk memory—and which may in many cases almost be accepted as historical fact—of previous, now extinct, inhabitants of the land, who were sub-human.

Also, there is really no clear line drawn between these historical traditions and reality as we have pieced it together from archaeological and anthropological diggings and delvings. Nor is there any clear demarcation between sub-men and full men, in that lots of peoples seem never to have quite decided whether interbreeding was permissible or even possible. Since primitive *man* would presumably try to breed with anything sufficiently like himself on purely biological and instinctive grounds, the line may never really have existed in the first place, and therefore there may always have been crossbreeds [such as have been found in caves in Palestine between Modern Man and Neanderthalers] and thus of all manner of degrees of "man-ness" and "sub-man-ness."

It is interesting to note that MLF about such [and thus about what we call ABSMs] have everywhere shrunk back progressively through time from the initial centers of civilization. To put it crudely: they disappeared progressively—first as accepted fact to become folklore locally but still fact over the border; then they became a legend locally, and folklore over the border, but remained fact "in far countries"; finally, they became mere myths locally but, going outward, first a legend, then folklore, and finally something only rumored as still existing in very far-off lands. This is only logical, for the earlier inhabitants of the land were either exterminated, absorbed, or driven out; and, as the centers of more advanced culture began to merge, the poor sub-hominids, then the submen, and finally even primitive true men had to keep moving out until they got into isolated pockets—mostly forested mountains where they were finally hunted out—or withdrew into the great uninhabited and unusable uplands. Anthropological history is absolutely clear and precise on this process, and the whole history of MLF marches along beside it throughout Eurasia.

The situation in the Middle East, that is from Mauretania to the Pamirs and south to the borders of Ethiopia and Orientalia (see Map XIV—the Central Caucasoids) was somewhat similar, but appears to have taken place on an earlier time scale, and to have been more rapid. The reasons for this are twofold. First, the "Modern Men" of this natural province appear to have been the first to become civilized and organized; but, second and much more important, this whole area suddenly dried up climatically just about the time civilization began, and it has continued to do so ever since. In fact, this desiccation may have been the primary *cause* of the development of civilization as a whole in the first place, for it must have acted as a tremendous spur to human efforts to survive. Sub-men and really primitive peoples seem to have been disposed of in very short order in this province on both these accounts and also because there were no great forested mountain blocks or uninhabited uplands for them to retreat to, either in it or for long stretches around it. The Sinai Peninsula was one of the few that there were, and we have seen what was there in the passages quoted from the Bible earlier in the chapter. Yet again if you look at this same map you will perceive the very significant fact that ABSMs are reported from the Caucasus, the mountains of northern Persia, and the Pamirs; while very strong traditions and even some historical records (see Pliny) of their previous presence lingered on in Morocco till Roman times and in extreme southern Arabia till much later. The truth of the matter is that the primitives, sub-men, and others of the Middle East had nowhere to retreat to but deserts where they could not live; they could not cross the Mediterranean on the one hand or the Arabian Sea on the other to get to forested Europe or India; and when they went south [if they did] toward Ethiopian Africa, they ran head on into a large and most vigorous population who would just not admit them—the Negro peoples.

Today we find an immense amount of evidence—see Chapters 13 and 14—that not a few real primitives and/or sub-men (i.e. in both cases colloquially, ABSMs) seem to have managed to survive in the vast unused mountain blocks that cover the lands of the Central Mongoloids throughout central Asia; and it is possible that they may still be spread over the even less-known and practically unpenetrated uplands of the North Mongoloids in Siberia. However, the Russians only absorbed this immense subcontinent in the 19th century and they simply have not even yet been able to explore it fully; any more than we have parts of Alaska, the Yukon, British Columbia, and the Canadian Northwest Territories. This is a land of continuous—and particularly difficult—coniferous forest, actually forming the largest continuous plant growth on the surface of the earth. We may expect many surprises from there, more especially as there is plenty of already known ML *and* F among its inhabitants that is most pertinent.

The situation in Orientalia is similarly obvious; or, at least, it should be by this time. To take India first: as the human population grew—and it started to do so enormously at a very early date—the primitives and others had to get out. Here they had a fairly wide choice. First and most obvious were the forested uplands of the southern part of India itself and the island of Ceylon beyond; second, they had the mighty

Himalayan ranges hard by; and third, they had the Indo-Chinese Massif to retire to. Now, it seems that they went in all these directions for the legendary and recorded history of India is full of references to primitives in the southern Peninsula and in the mountains of Ceylon, until quite recent times [see especially, Bernard Heuvelmans' book for those in the latter]. Then, there are still some very primitive peoples in those areas; and these, like the Senoi of Malaya, in turn have traditions about even more primitive and often hairy people who preceded them. With characteristic pragmatism they do *not*, however, report them as still existing. Hence, no ABSMs, in India proper, today that is.

The Himalayas may be regarded as being "in India" and they certainly are in Orientalia. We have already heard quite enough of current ABSMs in that province, but we should add that MLF and all the rest about them there is, and always has been, rife throughout the entire country. Further, there are all manner of odds and ends of peoples still living in complete isolation in the area; as witness the so-called "Chaldeans" of Messrs. Jill Crossley-Batt and Dr. Irvine Baird, and the incredible "Jungli Admis" of McIntyre, whose account of which goes as follows:

There are some curious specimens of humanity to be found dwelling among the forests about the Chilpa, called "Razees," compared with whom the villagers are quite civilised. These villagers described [them as] "Junglee Admi" (i.e. wild men of the woods), as they termed them to me, and as being almost on a par with the beasts of the wilds they inhabit, subsisting on what they can secure with their bows and arrows, and by snaring.

My old friend Colonel Fisher, senior Assistant Commissioner of Kumaon, gave me the following short account of these interesting barbarians. "They were the original inhabitants of the country about there, but the persecutions to which they were subjected by the Kumaon Rajas, and especially by their neighbours the Goorkhas, were so cruel, that they abandoned their hamlets and retired into the wildest and least inhabited parts of the country, and lived on wild roots, fruits, and fish, and game, and lost all recollection even of their language. I was told by the Rajwar of Askote, they themselves have entirely disappeared from Kumaon, though there may be a few yet on the banks of the Sarda in our territory, or the thick jungles on the Nepal side of the river. The last time I saw a man and woman of the tribe was at Askote in 1866, and they were caught for my special benefit. We gave them a few rupees, but they seemed to value them as much as apes! They would eat anything given to them; and both the man and the woman wore long hair down the back, and used leaves stitched together for clothing." From this, the condition of these remnants of an almost lost race appears to have been still much the same as, we may suppose, was that of Adam and Eve after the fall.

In this area tradition, rumor, ethnology, and proved reality all come together into one inextricable web of history, so that one cannot really draw any hard and fast lines between them. One has to steer a very steady course, bearing in mind, the very dangerous rocks of theology, mysticism, and our Classes 1 and 2, of [believed-in] noncorporate *Entities*. The area is also positively crawling with outcasts, hermits, religious initiates and now with displaced Tibetans, Communists, mountaineers, and goodness only knows what other types. Also, there are five kinds of bears in that country, several species of large monkeys, and at least two kinds of alleged ABSMs. However, in the minds and opinions *of the locals,* as the ethnologists quoted above so clearly state, there is really no confusion whatsoever about all this vis-à-vis the ABSMs. They have it quite clearly in mind which class of entity or creature is which, and the *Dzu-Teh* of Tibet, and their own *Meh-Tehs* and *Teh-lmas*—and the old tales and belief about them—form a distinct and clear-cut class of their own.

So we are left with the fringe lands of the South or Oceanic Mongoloids or what is often called Southeast Asia. This, as may be seen from Map XIV, coincides exactly with the remaining part of Orientalia but for its two overseas extensions to the island of Madagascar to the west, and out into the Pacific to the east to encompass the Micronesians. Among these peoples are the Japanese, the south Koreans, the Chinese proper [as opposed to the Manchus], the Indo-Chinese, which is to say Vietnamese, Laotians, Siamese, Cochins, Burmese and all their associated peoples, and the Malays and assorted Indonesians including the Filipinos. [The division between these and the Melanesians and Australoids is not precisely along Wallace's Line, but somewhat east of it.]

Here again we encounter all the same confluences and confusions between the findings of the physical palaeoanthropologists, the ethnographers, the philologists, native myth, legend, folklore, tradition, history, and current ABSMery In fact, this is par excellence just as we said above "the great mix-up." Although it is the homeland of the South or Oceanic Mongoloids, it has also been for untold millennia a sort of doorway between the West and the East. Just about everybody [apart from the Amerinds] have at one time or another streamed through here, either one way or the other, and most of them seem to have left some remnants of themselves as well as of their cultures, their beliefs, and their traditions, scattered all over the place.

First, there were undoubtedly sub-hominids here in the form of the Pithecanthropines; then sub-humans in the form of the Neanderthal-like Solo so-called men of Java; then a race of pigmies that everybody says were hairy and lacked proper speech; then what may be called Modern Pigmies—*vide* the Semang of Malaya, and others on the Andaman Islands and in the Philippines; next, small dark and possibly primitive Caucasoids of the last Vedda, then Dravidian types; next, another lot of peoples who have been called glibly the Oceanic Negroids, who have ended up as the Melanesians. Meantime, the true Australoids, or "Blackfellows" of Australia, seem once to have dwelt thereabouts before moving down into their southern land and becoming isolated. Next, came three quite distinct lots of Mongoloids, ending with the modern Malayans, Indonesians, and Siamese. And just to completely confound the issue, a group of very advanced Caucasoids passed right through from west to east, and on into the Pacific to form the Polynesians. There are moreover, as I say, traces of all of these passings to be found all over Malaya and Indonesia. Moreover, there is still ample room for primitives and ABSMs all over this continent, i.e. Orientalia, apart from Java where the population is too great. The Indo-Chinese Massif is an enormous unknown and mountainous forest country; the peninsula of Malaya has great wildernesses; and Sumatra and Borneo are, in a manner, still empty. Even in overcrowded China there are large mountainous areas that the Peoples' Republic has not yet got around to organizing. Here it is not, however, worth-while even starting to discuss MLF. The sheer volume of these is too great; and that which deals with ABSMs seems to have no ending, and this is concentrated in some areas, but singularly lacking in others [such as Borneo], as I have pointed out. Moreover, it trails off in all directions, both in time and space, into living primitives—like the poor so-called "Hairy Ainu" of Hokkaido—and into types that even omniscient Chinese Communist officials seem unable to classify.

It should by now be fairly obvious that these abominable MLF stories cannot be ignored and may often add very considerably to our knowledge of both the past and the present status of ABSMery. Were there but one single case in all of it that did not jibe with the established precepts of vegetational distribution, general geography, anthropology, ethnography, *or* modern reports, we might have to think again and reappraise it—if not doubt it: but there is not so much as *one* single inconsistency in the lot. Thus, we must accept all of it as evidence for the previous existence of primitive races of modern men, of sub-men, and of sub-hominids. In doing this we have to remember only three sets of facts.

The first is that all men are xenophobes. The second is that almost all men believe almost as fervently and completely in a nonmaterial world as they do in the material one. Third, that the three great major divisions of Modern Man [who inhabit most of the earth today] *are* different. These differences are to some extent physical in their most basic aspects, such as spirally curled hair in the Negroids, partially curled or "wavy" hair among Caucasoids, and straight hair among the Mongoloids; but most of the other standard physical criteria break down—such as the degree of eversion of the lips, skin color, comparative length of limbs, and so forth. However, there is no doubt whatsoever that the three do differ—and radically—emotionally.

[396]

Emotion has nothing to do with intrinsic instinct or intelligence, aptitude, or ability: it concerns only the way in which things are done. In this, the three main types of modern men behave according to the nature of their original environment. The Mongoloids, developed on endless plains with nothing but a horizon to look at, are "contemplative"; the Negroids, developed in a land of violent colors, contrasts, changes, and multiple life, are "emotional"; the Caucasoids, having everywhere struggled upward amid a welter of physical problems, like wildly varying seasons, endless mountain ranges, rivers and arms of the sea to be crossed, ice ages, and so forth, are basically "mechanistic" or what they choose to call "practical." The Australoids, the Bushmen, and the little Pigmies are frankly quite beyond our ken. All three of the last seem to live and operate in a world so strange that we push most of its precepts into a vague realm that we have named parapsychology.

18. Some Basic Facts

Much of what we accept as fact turns out to be fallacy; and, many things that we would normally consider complete rubbish prove in the end to be quite true.

Having now met the ABSMs, you may well be a bit bewildered, and the over-all impression you have gained of them will still be that they are a pretty polyglot lot. Likenesses between two or more there can be but, on the whole, there does not appear to be any one feature common to all. This, however, is a gross misconception. As with so many things in life, the whole question has been presented popularly on completely false premises, or has, at least, been given a wrong twist, which is not only most misleading, but initially set us all off on the wrong trail. The causes of this are singular, and psychologically, very potent. They are basically, the name, "abominable *snow*-man," that has become attached to it. This is the one single, outstanding fact that has become attached to all ABSMs. Yet, *none of them live in snow*, or in any place that is either perpetually, or even for any substantial part of the year covered by snow. This, indeed, is a negative fact but it is of very great import because it has led to the misdirection of almost all our serious attempts to solve the problem.

There is much misinterpretation of all the evidence—notably in the department of footprints and tracks, which have always, as a result, been immediately assumed to have been made in snow. [Tracks and imprints made in snow can be most misleading in a number of ways and are open to all manner of interpretations. See Appendix B.] Those left in

mud have altogether other significance. The best and sometimes the only evidence we have of the existence of many
creatures is nonetheless most frequently seen on snow surfaces and, since snow does fall upon and lie on the ground in
many areas where ABSMs are reported, or near to those
places, it is in a way natural that the two should have become associated. However, the search would never have followed the course that it has, and it might well by now be
over, had it not been for the coining of the delightful but
nonetheless nonsensical term "snowman."*

There is, however, another feature that all ABSMs have in
common. This has remained just as obscure as the snow bit
has been prominent. It is that all of them are forest dwellers
and, it seems, basically inhabitants of mountain forests. Even
the Malay Peninsula types, the Sumatran *Sedapas,* the little
African *Agogwes,* and the Central American *Dwendis,* though
sometimes reported from sea level or even from coastal
beaches, are invariably hard by large mountain blocks or
substantial uplands that are not inhabited by humans and
which are mostly unmapped and unexplored. The only terrain where as yet unidentified creatures of larger sizes can
exist today is that which is forested, be it equatorial rain-
forest [i.e. jungle to most people], *taiga* or spruce forest, or
even the endless *mulga* of Australia. Among forests, moreover, the two most likely types to be inhabited by such creatures are swamps and mountains. Almost all of the "new"
animals that have come to light in this century have been
found in one or other of these sorts of forest; and with a considerable emphasis on the mountainous.

Given the misleading tag of "snowman," our whole search
has been further diverted by a really extraordinary sort of
mass blindness that must be basically psychological. While
tracks have on many occasions been found in snow and at
high altitude—there not otherwise being snow at the latitudes

---

* It is interesting to note—and the fact should be noted—that Linnaeus,
the founder of modern systematic nomenclature, actually gave a name
to an ABSM, which he called *Homo nocturnus;* i.e., The Man of the
Night.

MAP XV. THE WORLD

EURASIA

ORIENTALIA

AUSTRALIA

ETHIOPIA

ANTARCTICA

ERICA

COLUMBIA

[OCEANIA]

= ABSM
REPORTS

MYTH, LEGEND &
FOLKLORE OF ABSMs

concerned—and while some of these tracks have been followed for some distances, and continued in snow, an end has never yet been found to any one of them that do not either come from or return to forests. Nobody seems ever to have considered the fact that our own human tracks in snow, especially crossing high passes in such places as the Himalayas, in no way indicate that we *live in* places of perpetual snow. Quite apart from this, it would seem obvious that no creature, even of our size, could find enough food in any such place to reside there perpetually. *Mountain* snowfields in the tropics, sub-tropics, temperate, and sub-arctic belts are perpetual and, unlike the Arctic *lowlands* where the snow vanishes for some months each year, are not underlain by a mat of rich vegetation which may be dug or scratched for in winter. A limited number of larger animals, like the Muskox and Reindeer, can gain a living on the Arctic lowlands but none live on or can survive on ice, on snowfields, or on ice-caps such as those of Greenland and the Antarctic; *and none can live on the perpetual montane snowfields* of other areas. ABSMs may, and often do, it seems, live right up near the tree-line [mostly because that is their last retreat] and they naturally and customarily cross over the upper snowfields to

---

MAP XV. THE WORLD

The surface of the Earth is somewhat clearly divided between areas of two kinds. One, which we call the land but which includes certain peripheral areas at present under shallow seas, forms rafts of certain kinds of rocks of lighter density, some 40 miles thick. The other, which constitutes the ocean floors, is covered by a much thinner layer of these strata. The hydrosphere—or water capsule of our earth—finds its own level due to gravity. As a result, the first areas are subaerial, the latter subaqueous. The former are "land-masses"; the latter "oceans" (with adjacent seas). The first are not, however, the *Continents,* which are specific land areas, with associated promontories and islands, each of which has an unique history, structure, flora, and fauna. These are seven in number, with the islands in the South Pacific forming an additional unit. Current reports of, and myths, legends, and folklore pertaining to, ABSMs are now recorded from five of the Continents.

get from one valley forest to another and particularly when disturbed by loathsome mountaineers! Gerald Russell's observations on this are most pertinent.

Even the Russians were initially misled into sending all of their scientifically mounted expeditions up into the worlds of perpetual snow and bare rocks—in the Caucasus, in the Pamirs, on the Tibetan side of the Everest Block, and in the Sayans. Some of their first reports in 1958 displayed distinct surprise at their failure to locate any evidence of the creatures in these places while the map they issued showed their belief that ABSMs not only lived in, but had their only remaining breeding ground in such an area in the Pamirs. The truth of the matter is they, like everybody else, were more or less hypnotized by the silly expression "snowman." At the same time, neither they nor anybody else seems ever to have mapped the *world* distribution of reports of ABSMs, plus the MLF of ABSMs—though the Russians did attempt this for Eurasia. But, of much greater significance, is the fact that not even they correlated that map with the one factor that is of paramount importance in elucidating the distribution of *any and all* living things—as well as a host of other matters such as disease, civilization, industry, and so forth. This is the distribution of vegetational types.

The study of vegetation, though a department of botany, has very little to do with the details of that subject; more especially when it is the geographical aspects of it that we are investigating. Plant geography, or phytogeography as it is called, concerns the geographical distribution of species, genera, families, and/or larger groups of plants; the study of vegetation is concerned with the distribution of the various types and forms of *growth* of plant-associations. In the latter, the actual species, genera, families, or bigger groups of plants themselves really do not matter at all; it is the manner in which they grow— namely, to what height, how close together, in what form (as trees, shrubs, or herbs), and so forth. Thus, any one patch of vegetation in one area may have as its dominant tree a palm, but in another a broad-leafed, hardwood, deciduous tree, and in still another a pine, yet all three "forests"—if they be such—

may be of the same *vegetational type*. Further, an area of, let us say, orchard-bush, in one place may consist of one kind of stunted acacia tree, standing widely apart with, under and between these trees, tall grasses of two species; but, suddenly, in an adjacent area, the whole scene may change—often along a very precise line—to tall trees of the same species, but standing much closer together and having only short grass below; and this grass may be of a new species or one of the first two or both. In other words, the appearance of the vegetation has changed but the plant species have not. This is called a change of *facies*, a term borrowed from the geologists, who first coined it to define the appearance of strata of rocks in various places, which may be totally unlike: thus, in one place a shale, in another a sandstone, in a third a limestone may be found, and yet all still be of the same age and laid down, or deposited, under the same shallow sea.

The study of vegetation is a grossly neglected science and, although it has played a very lively part in geography, and especially in animal and plant geography, since long before mere plant distribution studies were initiated, it has never been given the place it warrants. In fact, there is not yet even a single textbook devoted to it, as opposed to general phyto-geography. And yet, the whole of plant, *and animal*, distribution is wholly dependent upon it; while, all the most important aspects of human life such as agriculture, much of industry, and even nationality, race, and all the larger characteristics and characters of human beings are wholly subservient to it. It is amazing that the last thing to be mapped in any country has been and still is its vegetational forms. Only one state in this Union has done this—California—while there are states, like Texas, that have vast afforestation, soil bank, and other programs which are entirely dependent upon detailed knowledge of the distribution of vegetational types and *facies*, but that do not yet have a single map of any such.

Actually, the most important map of *any* piece of land is a vegetation map. Even topography [showing altitude] is really of minor importance. To zoology, and such matters as stock raising, it is not only just essential, but so vital that it really amounts to the *only* feature of the land concerned that is needed. As a very broad example, it may be noted that nearly all our cattle are of the wrong breeds for the vegetational belts and zones in which we now keep them and try to raise them. Take the Hereford breed of cattle, for instance. These were developed in Herefordshire in England in the middle of the southern North Temperate Deciduous Forest Belt, yet we try to raise them on prairies, scrub belts, and near deserts and then we wonder why this stock deteriorates and needs constant infusion of new blood from the old (original) country. The cattle that should roam our Western ranges, and by the millions, are either Masai or Ankole bighorns from Africa, or Sindi Humped Cattle from Pakistan. And so it goes with almost every animal and plant that man tries to rear; as well as to man himself. Hollanders from the coastal marshes of Europe are never going to thrive in upland Colorado; Spaniards from the windy *desiertos* of upland Spain are not even going to survive in the Canadian boreal pine forests or in the Florida everglades. They either die out, or move out.

The mapping of vegetation is thus the sole most important task for the terrestrial geographer. Details of rainfall, topography, soils, and all the rest are purely secondary and can come later. And colonizers, agriculturists, stock raisers, and others would be well advised to drop all other studies until that has been accomplished.

If, therefore, we want to attempt any sort of interpretation of the distribution of any living thing, the first task we have to perform is to ascertain the distribution of the vegetational types throughout the areas concerned; *and also* around the world, so that we may have some notion as to the significance of the purely local distribution. In our present case, it therefore becomes necessary even if only briefly to outline the basic principles of vegetational classification and geography. This is a tall enough order, but before we can attempt even this, there is another more basic matter that has to be straightened out.

It may seem almost impertinent to say that such a thing is necessary; for it is, alas, a sad commentary on the present state of our understanding; and it is a terrible indictment of our educational system that it should be so. The truth of the matter is that the very fundamentals of geography—all of which have been published for nigh on a century—are simply not known, generally, in one particular and most vital respect. This is the basic matter of the definition and delineation of the real continents.

An immense amount of rubbish has been talked throughout the ages, published in past centuries, and is still mouthed today about seven continents and "the seven seas." None of these expressions have any but the vaguest connection with reality. First, seas and oceans are completely different things, with different structures, histories, types of fauna, flora, and so forth. There are actually five *oceans*—the North and South Atlantics, the Indian, and the North and South Pacifics.* All the rest of the surface of the earth covered by salt water, is *sea*. There are six *land-masses*, that emerge from the seas— North and South America, Eurasia, Africa, Australia, and Antarctica. These land-masses are *not*, however, *continents*. Continents are intrinsic areas of land under air, just as oceans are really areas of "land" under water. They form distinct units, each having its own construction, history, fauna, flora, and so forth. What is more, the confines [edges] of these true continents do not, except in some exceptional details, coincide with those of the land-masses. This has been manifest for about a century but we still persist in calling the latter by the names for the former, and sometimes vice versa.

There are *seven* continents. These are outlined in detail on Map XV. They are, can be, or in some cases, might better be named as follows: (1) *Erica* [after the bawdy old Norseman who first located it for the "Western" world of historical man], which we now call "North America" and which stretches from the northern tip of Greenland to the Isthmus of Tehuantepec in southern Mexico, and from the western Aleutians to eastern Greenland. (2) *Columbia* [in memory of the Italian, Christopher Columbus] is the next area. This continent stretches from the Isthmus of Tehuantepec and the Florida Strait, between

* For the exact definition of these see my *Follow The Whale*, page xviii, Boston, Little, Brown, 1956.

the peninsula of Florida and Cuba and the Bahamas to the extreme tip of Tierra del Fuego. (3) *Antarctica,* which is almost two sub-continents divided by a long deep channel now filled with ice. (4) *Australia,* which is the austral or southernmost generally, and which includes a lot of partially sunken land running north to what has now been named "Wallace's Line." There remain three others and these are going to cause us somewhat more trouble.

Let us take (5) *Ethiopia* first. This was the original western name for the vegetated lands south of the Sahara Desert, and includes the whole of the African land-mass roughly south of that desert but also a bit of land south and east of the great desert of Arabia. Next (6), we have *Orientalia* or, as it is loosely called, "The Orient." This is southeast Asia, with a lot of sunken land farther to the southeast and multitudinous islands thereupon, down to this Wallace's Line. Its northern and eastern limits are very precise but puzzling to many, notably along the northern border of what is now Pakistan and India. The essential point to grasp [for our story] is that the Himalaya Mountains together with the great gutter of the upper Brahmaputra to their north lie wholly within this continent, while the southern edge of Eurasia [the last and final continent] begins along the great rampart of the Tibetan uplands.

Eurasia (7), is by far the largest continent. It starts on the west with Spitzbergen, Jan Mayen Island, Iceland, the Azores, and Canary Islands, and reaches to a line drawn up the northeast coast of Asia, from the mouth of the Yangtze River, west of the Aleutians and St. Lawrence Island, and through the Bering Strait. To the north, it includes all land and islands between these north to south lines, right up to the North Pole. Its southern limit forms a great inverted curve, with one northward-pointing kink in the middle. This curve runs, as may be seen clearly from this same Map XV, from southern Morocco across the Sahara to a point on the Red Sea about the border of the Sudan and Abyssinia. Thence, it crosses the Red Sea and southern Arabia to Cape Ras el Hadd; then across the Arabian Sea to the mouth of the Indus Valley. From there it proceeds almost due north to the Pamirs and then turns east and continues north of the vast plateau of Tibet to the Tsin-Ling Mountains of China, and finally to the mouth of the Yangtze about Shanghai. It is really a very compact block of land containing only two major flooded areas—those of the Mediterranean basin, with its ancient extensions to the Black, Caspian, and Aral seas, and the Red Sea.

These are the real continents and, although the first and the last have very much more in common than any others, they are fundamental units, each with its own character, life, and history. For this reason, whatever occurs on each has a significance that is doubly pertinent. If, for instance, something that *looks* identical is found on two of them, the odds are high that said two creatures [or plants] are not really alike except in appearance. Further, vegetational types which may be found on many, or more than one continent, may vary from one to another in their botanical constitution, but still "look" alike. We must always bear these facts in mind, and make due allowances for them.

The business of vegetation is really very simple. Our planet revolves around our star (the sun) on a fixed and flat plane. The axis around which our earth rotates is tilted to that plane by about 23 degrees. Thus, we get seasons which mirror each other both in time and in time-belts in the Northern and Southern Hemispheres. As a result, different amounts of sunlight bathe different *belts* around the earth, in different ways, and at different times of the year. Plants as a whole, feed on a combination of matter [dissolved in water] and energy [sunlight] and they have evolved in various manners to survive under various and differently changing conditions of sunlight during flexible periods of time. If you can imagine the earth without any seas or oceans and all at "sea level" but still having the same climate, weather, and such like atmospheric features that it does today [which, of course, is impossible, even theoretically], you would find that its vegetation would be arranged in a series of 20 major belts—or 24, with two vegetationless belts and circular blobs at top and bottom—half of which [in the Northern Hemisphere] exactly mirror the other half [in the Southern Hemisphere], and with a single double belt around the midriff. This formation is, as a matter of fact, exactly what our earth *does* have; but the belts are not all of the same width and don't all run neatly all around. To the contrary, while they invariably maintain a certain basic succession, they wave about from north to south as they go round the earth, and they constantly swell up or thin down, sometimes to the point of virtual disappearance.

These belts are, starting from the equator: the T-E-F or Tall Equatorial Forest; the H-D-F or High Deciduous Tropical Forest; the Orchard-bush; the Savannah; the Subtropical Scrublands; the Hot Desert; the Temperate Scrublands; the Prairies; the Parklands; the Temperate Deciduous Woodlands; the Boreal Coniferous Forests; the Tundra; the Barrenlands; and finally the Polar Icefields. Within these there are several prominent and many minor subdivisions but they need not concern us here, except to note that a rather important transition zone of mixed deciduous hardwoods and evergreen softwoods, or conifers, exists between the deciduous woodlands and the coniferous boreal belts; and that the T-E-F may be broken down into three very clearly recognizable sub-belts. All these major belts invariably lie in that order all over the land surfaces of this earth as you travel from the equator to either pole. However, their width and exact position (latitudinally) is, as I have already said, not the same when traveling down various meridians or longitudinal lines. What is the cause of this situation?

The question is a fascinating study in itself, but is not suitable for us here. It must suffice to sum it up with the simple statement that, despite all that may be said about climate, weather, winds, moisture, and other atmospheric factors, as well as geological and other geomorphological matters, there proves to be but *one* factor alone that causes these swings in the major vegetational belts. That is the major ocean currents. The incidence of the major ocean currents is displayed on Map XVI. The origin and conformity of these forms another subject in itself, that also cannot be pursued here, but which is basically brought about by the spin of the earth, which causes all blobs of liquid lying on its surface to revolve clockwise in the Northern Hemisphere and counter-clockwise in the Southern. The heating up is done in the equatorial belts; the cooling, as should be obvious, in the polar regions.

Another aspect of the distribution of vegetational types that is of first importance is that of the effects of altitude. The really amazing thing is that this does not have any effect on the basic arrangement, the succession, or even the regional variations in the major vegetational belts. Following any one, and coming to a mountain range, you will find that it just "ducks under" the uplands and appears again unscathed on the other side at the same altitude at which it met these uplands. Anything above sea level is, in fact, simply "dumped down" on this basic plan, as it were, and has no effect upon its general pattern. Only on the slopes of the mountains themselves do we see something else.

This also proves really to be a simple matter if you don't lose your nerve. Taking the most extreme possible case: if you start up a mountain that rises right on the equator, you will pass upward through all the major types (belts) of vegetation that you would pass through at sea level on your way to a pole, eventually arriving on a perpetual snowfield at about 17,000 feet. What is more, the farther toward either pole you go, the fewer belts you will pass through going upward from sea level, and the lower each one will be on mountains until, when you get inside the polar icefield region, everything will be covered all year round in snow and ice from its top down to sea level.

This is exactly what we find all over the world; invariably, and without exception. The montane floras [i.e. horizontal belts of different vegetational types] that you pass through as you climb are, however, called technically *zones*, simply in order to save wordage and to indicate that they are where they are due to *altitude* and not to *latitude*.

The over-all picture of the distribution of vegetational belts is displayed on Map XVI. This shows what conditions would be if all the land were at sea level. However, all land must be above sea level, and the moment you start to go up, things begin to change. Since the major belts are pretty large and wide, one actually has to go up some 600 feet before one may expect to pass from one noticeable zone to another. We may now compare this map with the pertinent available information of a purely geographical nature that we have about ABSMs as shown on Map XV. When we do so, moreover, we may well get quite a surprise.

We immediately see, and staring us in the face, a whole gamut of facts that have not previously been apparent. For instance, ABSMs, one and all, fall within the bounds of a rather limited number of narrow belts and more especially, even narrower zones within those belts. Past tradition of them—i.e. MLF—moreover, turns out to connect these special areas but never to "slop over" into surrounding belts or zones. Places where ABSMs have been reported, which are perpetually, or for long periods annually, covered in snow or ice, *all* fall within montane forest blocks. In fact, to sum up, ABSMs display in their distribution exactly the pattern expected of any group of animals [and notably of terrestrial mammals] and more particularly that of higher Primates. This is more than just merely significant; but, there is a further even more remarkable, and in some respects most convincing set of facts.

Possibly fossilized remains of primitive men, sub-men, sub-hominids, "super"-apes, and more lowly Primates will be found almost all over the land surface of the earth, but so far, we have merely scratched the surface of a few surface strata and in only a few places in our search for such relics. Yet, quite an amount of material *has* been unearthed, as we saw in Chapter 16. From what has been discovered, we see that there were once sundry pockets of higher primate evolution in various places. This may be no more than a surface appearance [or "emergence"] and due entirely to the fact that conditions suitable to the fossilization of the creatures concerned just happened to exist only at one time in those areas. However, even if all these Primate types were once universal, but are so far known in the fossil form only from one limited area, we can at least say that they *did* exist in that area.

Ignoring, therefore, what we don't know, and plotting what we do, we see that there were, at least at one time, various forms of ABSMs in various places, and that those places appear to be of great significance vis-à-vis the distribution of vegetational types. I should add that the distribution of both MLF and historical record also fall more than just neatly into the same pattern. To do another summing up, therefore, we may say that, just as current ABSMs conform perfectly to the rules of zoogeography and phytogeography, so also do they to the findings of paleoanthropology.

Armed with reliable facts such as these—facts basic, simple, and obvious—we may tackle the whole ABSM search in an entirely different manner. We no longer have to be dismayed by the seemingly heterogeneous plethora of details, apparent discrepancies, and outrageous suggestions that may have appeared to arise in the reports. These facts have a considerable cogency and a fine conformity. No longer are they just a mass of random jottings and silly statements. To the contrary, they speak enormously of the seriousness, honesty, and common sense of plain people; for, I cannot find a single case of anyone who alleges that he or she has information upon this subject who even suggests that his or her information came from any place *outside* any area in which its occurrence is logical, according to the above stated basic rules of Nature. In other words, all the reports come from places where such things are possible [or have been in the past], and all of them, as far as I can see, from places where, according to the best findings of the best scientific inquiry and effort, they are highly probable today. For instance, even the really extraordinary—and certainly at first sight, preposterous—reports from the bottomlands of the Mississippi drainage basin conform to these general principles, and, whatever one may think, they do not really, on proper analysis, outrage any valid zoological precepts. There are actually no exceptional cases. Take that of the puzzling suggestion that there are three distinct types of sub-hominid unknowns in the eastern Himalayan Region—the giant *Dzu-Teh*, the bestial *Meh-Teh*, and the little pigmy *Teh-lmas* of the lower valley-forests. One's first reaction is "don't be silly: why pile Pelion on Ossa? Isn't one bit of outrageous nonsense enough?"

When one comes to regard the distribution of montane vegetational types in the Himalayan area; and then map the discoveries of sub-men known from fossil evidence around that area; and finally adds to this the present distribution of other mammals in that area, one begins to see that there should be at least these three types thereabouts. So also with the little *Almas* and the large *Gin-Sungs* of the eastern Eurasian area. Nature "abhors a vacuum," and invariably fills all her niches; and there are slots in those mountain areas for just such a small and a large omnivorous type of primate mammal.

This brings us to another aspect of geography; and one over which there is as much if not more misunderstanding than there is over such simple matters as the disposition of continents and basic vegetational belts. A curious belief has grown up during the past half century to the effect that "the whole earth has been explored." This is not so. By far the greater part of it is entirely unexplored, very little of that part which has been, is mapped; a great deal of the earth which is mapped is never visited; while large parts of it are frankly unknown.

First, almost three quarters of the surface of this planet is covered by salt water and about 80 per cent of this goes to form the five great oceans. These are on an average about 2½ miles deep, and it is only now, since the last IGY, that we have even begun to obtain any over-all—let alone detailed— picture of the bottoms of the oceans. The seas are better charted and in some respects, we know more about their bottoms than we do of the land surfaces of our planet. Of the land, one ninth—or Antarctica at 5,700,000, plus Greenland at 840,000 square miles—out of a total of 58,000,000 square miles, is covered by great domes of solid ice. Another ninth is permanently frozen and supports endless coniferous forests that are not used—the immense *taiga* of Russia and Siberia, and the boreal forests of northern Canada. Much of these two largest forest areas are quite unknown and virtually impenetrable. Of the remainder, a third is desert [with surrounding sparse scrublands], and another third equatorial forest. There are still a few areas of considerable extent in the middle of the larger deserts that are not explored or mapped, and have only been passed through once or twice. The tropical forests are even less known. If you look at Map XVI, you will see the disposition of the tropical forests.

It is the closed-canopy forests that interest us most. Of these there are one major and two minor blocks in the Western Hemisphere; the same in Ethiopia; and, in a manner of speaking, the same again in Orientalia. Modern maps show all of these surrounded by place names, crossed by roads and even railways, and bespattered with names of rivers, towns, and mountains. In any standard atlas it looks as if the Amazon or the Congo Basins were as cluttered as the Mississippi Valley; while it looks as if this, in turn, was as fully occupied and as well-known as that of the Yangtze. Both concepts are not only misleading and misconceptions: they are downright rubbish.

If you will take a map of the Pacific in any standard school atlas, of say about the dimensions of *Life* Magazine; use a good magnifying glass and measure the dot on an "i" in the word "Pacific," and then calculate [or simply measure] its width on the scale given at the bottom, you will get a great surprise. I did this with a powerful magnifying device and some care, checking the actual distances on the printed map from other geodetic data, and I found that the dot actually covered 345 miles of territory [or water]. A place name, therefore, such as "The Tumuc-Humac Mountains" printed on a map of Surinam (previously Dutch Guiana), a country just about 200 by 200 miles in dimensions, can entirely fill up the hinterland (one third) of that country. To make matters in this instance worse, a certain Mt. Wilhelmina is usually marked in the middle of these Tumuc-Humac Mountains. It so happens, however, that the latter are actually a series of modest hills and uplands, while Mt. Wilhelmina is a complete myth; for, when planes flew over the place during World War II, where it was alleged to be, it was discovered that it was a large *depression*, in these "mountains," and almost at sea level! For this, if no other reason, I may just as well use this delightful little country for further examples of geographical don't-knows.

I spent a year there in 1938, collecting animals. Its coast is lined with a 30-mile-wide strip of impenetrable mangrove swamps. Behind, or south, of this lies a belt of coastal deposits with rich soil, on which are bauxite mines, the capital, and some small towns, farms, plantations, and a few roads. Behind this lies another 50-mile strip of continental plain. This is crossed from south to north by some enormous rivers at almost regular intervals. Strung along these, for about 100 miles inland, are isolated villages of the Djukas—free Africans who just walked away from slavery in the early days, and founded their own hegemony. These rivers are heavily forested for a few miles back from their banks, but in between them there are huge open areas of (short grass) savannah, as flat as tennis courts, with some clumpy copses of trees on them. In these live some Amerindian tribesmen (Arawak and Carib) very few of whom have even a single store-bought possession. Behind this belt, the land begins to rise into foothills, and there are mighty cataracts on the rivers. The whole country is clothed in a dense mat of "jungle" or T-E-F, often growing in four tiers one under the other, and constituting some of the tallest and most magnificent forest in the world. Here there are no Djukas, no Amerinds, *and* no paths. These foothills become increasingly steep, and the rivers run in narrowing gorges choked with another kind of tangled jungle; then they rise to these Tumuc-Humac [so-called] Mountains. They are one colossal jumble of low peaks, ridges, and deep gorges extending all the way from the Roraima range in British Guiana, through Surinam and French Guiana, and on into Brazilian Guiana. They are uninhabited [at least by humans], unpenetrated, and unused by anybody, and they have been crossed only once—in 1921 by a massive expedition led by a Dr. Stahel, which had to burn all its canoes on entering the gorges, to prevent its laborers from running home, and then build new ones, on the other side, to get out. On this other side, the whole business is reproduced in reverse, back down to navigable big rivers without cataracts, and with strips of "gallery-forest" bordering their banks.

When we were in that country, a very pale-skinned girl of great beauty was brought into the capital (Paramaribo) by some Djukas who had found her wandering about in the forest just above one of the cataracts. She was put under the care of this same Dr. and Mrs. Stahel, since he was the senior government biologist and his wife a trained ethnologist. In time, her story came out. She belonged to a tribe of people all pale-skinned like herself, who lived on the open savannahs *beyond,* or at the back of the Tumuc-Humacs on the Brazilian side. She implied that her people never even met the other Amerinds who dwelt along the rivers, but traded with them by leaving goods in cleared areas in the forest. She had run away and gotten lost.

Later that same year, the French, Netherlands, and Brazilian governments decided that they had better make a start in finding out where their borders really lay and how they joined. A large expedition was mounted and took 3 months to get back into these savannahs, going round the easy way. There they found this girl's people; and, sure enough, they were almost white, never went to the rivers, and had only one food plant which they stuck in the ground only when a large tree fell in the forest. Most extraordinary of all, they had never heard of white men, *or* black men, and did not even have a word for "sea" in their language. Yet, several thousand of them were living under 200 miles from the Atlantic Ocean.

From our back window, in a bush-house at a gold mine in the foothill forest of Surinam, we looked out in a direction that, on the best and largest maps, had not one single place name for over 2000 miles, all the way to the Matto Grosso. And all of *that* is covered by a great blanket of greenery like a vast bedspread—a little bumpy and sometimes raised into mild humps, as if a sleeper had one flexed knee, but otherwise absolutely homogeneous and quite impenetrable. Most of it is three layers thick, and on an average 150 to 250 feet tall.

This is the condition over the whole Amazon and Congo Basins. It pertains also to a great extent in the great peninsula of Indo-China, and to a lesser extent in Central America and the Colombia-Ecuador Northwest Pacific forest; it used to do so in the Tupi around Rio de Janeiro [but there it is now almost all cleared]; it pertains all the way from Senegambia to Guinea in west Africa back from the coast; again from Nigeria to the Nile and south to Angola; in Mozambique, and up the lowland east coast of Africa; in a strip down the east side of Madagascar; in bits of southern peninsular India and in Ceylon; all over inner Assam and over into the adjacent Chinese and Burmese territories; throughout a great part of Malaya, Sumatra, Borneo, the Celebes, parts of Java, and many of the smaller Indonesian islands; almost all over New Guinea; and in a fair-sized patch on Cape York in Queensland, Australia. None of this, apart from the borders of the waterways, is even mapped. Most of it has never been penetrated, not to say explored; and a very great part of it is just not visited at all or used in any way, even by what local people there may be around its peripheries.

Yet people talk about the human race having to harvest the sea to prevent mass starvation; having to colonize other planets; cut down their birth rate [a very good idea, but for other reasons]; or complain that there is "no land left to exploit." True, you have to be frightfully careful when you clear these wild places of their natural vegetative cover, because the soil may go with it; but our technology is quite capable of obviating this, *if* [and this is a very big IF] they'd only study vegetation per se in advance.

Almost the same applies to the limited one seventh of the land surface of our earth that we do inhabit and produce our food on. The greater part of this also lies fallow, and a very substantial part [the good Lord be praised] is also still forested. A lot of this forest is not used, seldom visited and, over wide areas, not mapped or even explored. You should take a drive around our own country some time. I did two years ago and it is amazing. Whole expeditions—and properly equipped at that—go into the mountains of Arizona and just don't come out. Helicopters go in to look for them and don't come out either. The forestry department has no decent maps of anything but the outside edges of some of our national forests; the Panamint Mountains in Nevada are unexplored and even game wardens admit that runaway camels of the Civil War period may still be living in them; there is the area described in Chapter 6 [the Klamath in northern California; 17,000 square miles in area] with only two roads through it, and which has never been properly surveyed. In Oregon, Washington and, of course, in all the provinces of Canada, there are enormous—nay, rather, absolutely vast—areas of forest that have never even been penetrated. There are no proper maps of the multi-thousand-square-mile bottomlands of the Ohio-Missouri-Mississippi-Yazoo river systems; just roads on road maps. Parties get lost in Maine, and no taxes have ever been collected in parts of the southern Smokies. There are people who have nothing to do with the United States not 30 miles from New York, on the Jersey Pine Barrens; and there are "hillbillies" 70 miles from New York, 90 per cent of whom have never even seen a *radio*—not a TV set, mind you. But let's turn the picture over and have a look at the other side.

The dredging up of a 5-foot, bright blue, Coelacanth fish—a creature of a group thought to have been extinct for 70 million years—with a vivid yellow eye, 4 inches across, was more than a mere surprise to zoologists; and that came out of a sea [not even an ocean, albeit]. Much worse has happened on land. Just 6 months before the time of writing (1960) a large herd of the Woodland Bison (*Bison bison athabasca*) were located in the Canadian Northwest Territories only about 100 miles from the reservation on which the [until then supposedly] last remnants of their breed had been maintained for over half a century south of Lake Athabasca. These enormous oxen are leftovers from the last ice-advance. The point I am trying to make here is, they were found right alongside a place where a mission station has been in operation for over a century, and not more than 50 miles from a new road, along which I drove in a standard model car while my partner typed beside me. The Woodland Bison is really an enormous ox,

MAP XVI. THE WORLD

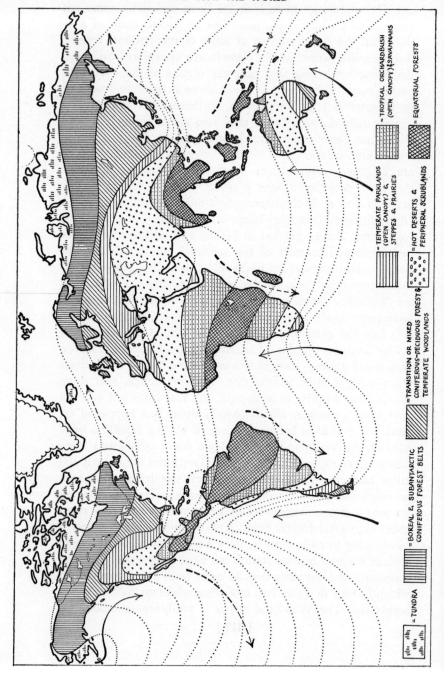

= TEMPERATE PARKLANDS
(OPEN CANOPY) &
STEPPES & PRAIRIES

= TRANSITION OR MIXED
CONIFEROUS-DECIDUOUS FOREST &
TEMPERATE WOODLANDS

= BOREAL & SUBANTARCTIC
CONIFEROUS FOREST BELTS

= TUNDRA

= TROPICAL ORCHARDBUSH
(OPEN CANOPY) SAVANNAHS

= HOT DESERTS &
PERIPHERAL SCRUBLANDS

= EQUATORIAL FORESTS'

but it is neither as big nor as fantastic as the creature that turned up in Indo-China in 1938, which I have already referred to several times.

This beast, the Kouprey (*Bos sauveli*), is the second biggest of the ox family—second only to the Gaur (*Bos gaurus*) of India. It now turns out to be quite common. They have large curving horns like Chillingham Cattle, but the males' horns are *tasseled,* starting about a foot from their tips. When this huge beast was first reported, "scientists" [our old friends, the orthodox, nontraveling, Anglo-Saxon zoologists] first called the whole thing a lie, and then said that, if they did exist, they must be "a cross between two other species." [Species of what, as usual unspecified, of course, in total disregard of their own contention that hybridization does not give rise to new forms.] Now that this incredible beast is properly known, it has even been suggested that it is a relative of the extinct west Eurasian Aurochs (*Bos primigenius*) from which our Western cattle are descended; and that it may even have been domesticated by the Khmers who built Angkor-Vat! Really, one sometimes becomes depressed!

The over-all point that I am trying to make is that, while we know nothing of a very large part of the land surface of

---

MAP XVI. THE WORLD

The most important feature of the land-surface of the earth to animals (and thus to men) is the type of vegetation that clothes it. There are seven major types—the equatorial closed-canopy forests; the open orchards and tropical savannahs; the scrublands and hot deserts; the steppes, prairies, and parklands of the temperate zones; the closed-canopy deciduous and coniferous forests of the higher latitudes; and the tundras and barrenlands of the polar circles. These girdle the earth, in that order, from equator to poles, in successive belts, but all of them waver to north and south and expand or contract, continuously, and in a variety of ways. These variations are due solely to the influence of the major ocean currents. Altitude has no effect on this belting; but on mountains the succession is repeated vertically, irrespective of latitude. ABSMs appear to occur only in mountainous regions and almost exclusively in those which lie in the forest belts. The one exception is eastern Eurasia.

our earth, we know even less of its inhabitants—vegetable, animal, and even, it sometimes appears, human. And yet people who have never set foot in so much [vegetative] as a wood, have the brazen effrontery to state that thus-thus-and-thus, or anything, or something, can't exist. Such statements are not pathetic; they are not just sad; they are downright dangerous.

19. Sundry Objectionable Facts

*Most of us dislike having to change our opinions.
So, while facts are facts, objectionable ones are
often deliberately misinterpreted.*

It will by now be obvious to anyone that facts such as those
already given, even if scattered piecemeal through thousands
of outlets and dozens of countries during more than a hundred years, could not fail to evoke some response. Nor have
they. Starting in 1920, they have produced violent reactions.

Though the story as a whole gained immediate popularity,
the reaction came primarily from the ranks of science, and
notably from zoologists. It was highly skeptical; and, in many
cases quite violent. Besides being dull, most professional skeptics are insufferably conceited, and in this affair have never
even bothered to collect the facts or examine them properly.
As is their wont they made positive statements, and before
we go any further, we must examine these because they form
a thick overlay of inaccuracy, illogicality, and illusion which,
if not exposed and analyzed, will distort any firm conclusions
you may wish to draw from the actual facts.

The pronouncements of most of the scientific skeptics and
"experts" are not caused exclusively by boneheadedness on
the part of those who make them; some are deliberately misleading and designed to promote further skepticism, without
any regard for truth. In the case of ABSMs, a whole gamut
of factors conveniently [for them] combine to promote skepticism in any case, and many of these are very fundamental.
Some are downright objectionable, but they cannot simply be
brushed aside on this account. They exist, and they lie at the
very core of the whole matter. Some of these factors may at

first appear to have little bearing on ABSMery but, assembled, they constitute a massive barrier to progress in the search and to any proper appreciation of its import by those who aren't in the know. I will put it this way.

Almost every time you open a newspaper you will find without much effort some crazy item that sounds not only odd but often illogical. Occasionally this is a lead story, but more often it is a filler. You read it and pass on; sometimes you go back and read it again and with an increasing sense of annoyance. Yet, we are for the most part pretty immune to such items, and have developed a habit of regarding them as just examples of irresponsible reporting. They amuse; and, if one does not take them too seriously, they are predominantly harmless. However, in some instances they are particularly aggravating because of their very persistence. These are the old chestnuts that, indeed, usually crop up in what is called "The Silly Season."

Among such are the matters of large unknown animals in lakes or the sea, commonly called "sea serpents" in the past, but now somewhat more properly sea monsters; children brought up by wolves; things called poltergeists that allegedly throw plates about; and the perennial UFOs or more popularly "Flying Saucers." Some years ago the magazine *Science Service* ran an article giving advice to science writers, in which was included a long list of "don'ts" and of subjects to avoid. This included the above items and a lot of other items that we all know; some truly silly and others just annoying. The editor of this article warned against dealing with these subjects on the grounds that they had all been *finally* and *utterly* discredited by "science."

This now famous list was compiled by Edwin E. Slosson, Director of *Science Service* and published on July 1, 1950. It states categorically that, among other items which are phony, and therefore taboo for all professional science writers, there are: "Seeds that grow after more than 300 years; especially that old chestnut about wheat in mummy cases." Wheat found in canopic jars [and placed therein in 1200 B.C.] has now been germinated on several occasions, but found, perhaps

[422]

rather disappointingly, to be nothing else than the lowest possible grade wheat, still grown by the fellahin in the Nile Valley. Further to this, seeds of the Flowering Lotus dug from a swamp near Pekin in North China, were germinated in the hothouse of Kew Gardens in London, and bloomed. This species of plant is not found today growing nearer Pekin than north Assam—a distance of 1500 miles away. The swamp from which these seeds were taken was originally believed to be some 500 years old. However, radio-carbon dating of other material taken from it, done in Japan, indicates that its lower layers, at least, are 24,000 years old. The seeds were in these layers and could be up to 40,000 years old! In this list of "don'ts" appears our poor, original ABSM.

Included also are some matters that either lie, or appear at present to lie, outside our logic. However, there are many, like ABSMs, that are quite logical and substantial, so that they ought to be open to physical examination and thus, ultimately, be obtainable. These, too, are nonetheless of a wide variety of likelihood in point of fact; by which I mean, some sound frankly very unlikely in the light of what we *do* know, while others seem perfectly feasible, if not probable, for the same reason. Unfortunately, science does not any longer make this distinction, but prefers to lump them all together, as in this article, and write them all off as *impossible*. This is not only silly, because, as time has shown it may prove to be wrong; it is *unscientific*.

Science is defined in the dictionaries as the pursuit of the unknown; yet science today is coming more and more to insist that it not be bothered with this, and it has reached a point where anything that is *not* already known is frowned upon. At the same time, there is a distinct tendency for science, per se, to become synonymous with technology, while the title *scientist* is becoming a class distinction founded on the occupation of the employer of the individual concerned. Thus, anybody who is not employed in or by very certain specific categories of organizations, whatever be his or her education, training, experience, and even published works, is referred to, and often scathingly so, as "an amateur."

Building radio sets or milking cows is not science; it is technology and technologists are manipulators. They are trained to do certain things and are conditioned to tackle only one thing at a time, preferring to be given a straight yes-or-no problem and be left to worry it out. If you suggest that they try to find but one correct answer out of many [or an infinity] of possibilities, their invariable answer is "If I had to investigate every crackpot theory that comes along, I'd never get any work done." Most regrettably, scientists today are tending to agree with the technologists in that they just don't want to be bothered with anything new that requires any novel effort or new thinking. There is also an extension of this attitude which denies even the possible existence of anything to which an answer has not already been given. Thus, all "unpleasantnesses," especially if brought up by the press or suggested by an "amateur" are often not just ignored but held up to ridicule as examples of dangerous practices.

One does not like to take "science" to task in this or any other way; and it is regrettable to have to treat it as if it were a sort of cult, but the state of affairs has become so irrational by the deliberate design of scientists themselves that this cannot be avoided. True scientists there are aplenty, but most of them appear to be so cowed by the system and its self-appointed hierarchy—which, I also regret to have to point out, is founded on a purely economic basis today—that they very seldom dare to speak out or give either their own or any truly scientific opinions. Then again, a not inconsiderable percentage of persons called or calling themselves scientists prove, on proper investigation, not to have any formal scientific training at all. Most regrettable of all, I have to state flatly that the percentage is vastly higher in this respect in the United States than in any other country. If anybody wishes to question this statement, let them go through any standard reference work in their public library, that will list the formal academic recognitions of each individual. From this, it will be discovered that a rather small percentage of those in directorial or other responsible positions in such institutions—such as museums and zoos—have any such [save

[424]

for "honorary degrees," which are not academic recognition]. At the same time, it will be found that there is no record of any practical scientific experience on their part either.

The result is that everybody other than this hierarchy is either overawed or beaten into submission by it, and is, as often as not, held up to ridicule to boot. The press meanwhile, trained to a high degree of skepticism for very good reasons, dares not make a decision on its own; runs to what it thinks is, or what it has been told is, *Science* for answers; and it publicizes these pronouncements quite unthinkingly. As a result, the public is increasingly less well informed on many vital matters and a most dangerous situation is being created generally—a situation that may very well lose us our position in the modern world.

The obvious and invariable question asked by the layman at this stage, is *what* scientists or "experts"? This is a difficult if not impossible question to answer but not for fear of giving offense. The most that can be said is that they were mostly if not all, professional [allegedly] zoologists, and most of them British for the simple reason that ABSMery was until quite recently a purely British affair. Then again, the press very seldom named those "experts" whom they quoted; again for the simple reason that said experts either refused permission for them to do so or were acting as spokesmen for corporate institutions or official organizations and were not therefore *permitted* to do so. Most of the criticism of these pronouncements [which were themselves all highly critical] was directed at the British Museum (Natural History), as being the official mouthpiece of zoology in Britain; but, although that institution always "officially" denied the possibility of there being ABSMs, this did not perhaps reflect the real opinions of all its professional staff. Then there is another aspect to this.

Actually, almost no professionally trained or professionally employed scientists have published on the subject—even critically. This may not be so much evidence of caution as an indication that they wholly and almost universally considered the matter so impossible as not to be worthy of mention in print. Apart from Elwes' brief notice in the Proceedings of the Zoological Society of London, I don't really know of any mention in British journals. The Hollanders in the Indies did so publish; and, of course, the Russians, Mongolians, and Chinese have issued quite a body of material. There have been others in some countries who have made passing references in their books; works that may be regarded as textbooks; but these are almost, one and all, mere references to the existence of the "problem."

Despite the above, and apart from those scientists whose assistance I acknowledged at the beginning of this book, there have always been some—and this number is increasing every year—who have taken an unbiased view of the matter and who have not denied it in toto. At the same time, not even these people would say so to the press, so that the latter fraternity has always had to fall back on the accepted but misleading term "experts say . . ." The greatest trouble arising from this situation has been that any "Tom, Dick, or Harry," trained or working scientist or not, has been able to say almost anything he likes, off the record; even speaking for an institution, provided what he said was in accord with the agreed or expressed policy of that institution. Thus, it is the over-all attitude of the sciences [and notably zoology] that is to blame and at fault. Science may criticize, but if it positively denies anything, it should at least state a case *and* give valid [scientific] reasons for this. This it has not done; and, worse, it has never really reviewed or investigated the facts, the alleged facts, or even the reports.

As an example, witness the real shock sustained by the lay public on the announcement of the Sputnik I launching. This was only the beginning; Lunik I, only a reminder; the Coelacanth fish, a mere hint. Much worse is to come; and we may expect it from all angles and at an increasing tempo. Another example, exactly comparable to the Coelacanth, but of much greater human potentiality, is that of ABSMs. This is, moreover, a rather special case.

A landing by a UFO, piloted by some super-intelligent entity from some other celestial body, would seriously jolt our whole world; but to get one of our own ancestors, thought to have been extinct for thousands of years, for study, would have almost as profound effects. If, moreover, that creature turned out to be so intermediate in character and characteristics between man and beast as to be ineligible for either class, it would have even profounder effects—at least at first —than the arrival from space of thinking creatures of an entirely different origin and culture, because it would touch what is perhaps the rawest spot in our consciousness. This is our religious belief.

Let us not forget that not so many years ago a thinking man was persecuted and physically attacked for having so much as taught the Theory of Evolution and by inference our kinship with the apes, other primates, and lower animals. *Man,* more than one Holy Writ states categorically, is made in the image of God, whereas "the animals" are not. What therefore are you going to do with a living creature that not even a scientist can say is either one or the other, specifically and definitely? Quite apart from the legal, ethical, and social problems involved—such as whether he [or it] has a vote and other citizens' rights, whether shooting it is murder or hunting, and whether you lock it up in a cage or invite it in for afternoon tea—you have to face the much more basic question as to whether it is in the image of God or not; and, if it is, *what* image of *what* god, and what the latter's qualities may be. Further, if it stands half—or anyway between us and the other mammals, are we to believe Holy Writ or Charles Darwin?*

The matter of ABSMs, while utterly intriguing is, also perhaps more than any other enigma, open to another very human sentiment; namely, that an unsolved mystery is much preferable to a solved one. As I have threaded my sometimes weary way through the maze of facts about this business for over a quarter of a century, I have often wondered, and sometimes with a real jolt, whether some of its greatest protagonists *really* want the matter solved or one of the creatures caught. It seems that people will go just so far, but the

* See Vercors' splendid book, *You Shall Know Them.*

moment they see any real possibility of a solution, they find some subconscious excuse to draw back.

All these factors combine to create a very widespread and united front for the skeptics in this business. But they are not the only ones. There is still pure ignorance; and by this I do not mean of the scientific variety, but of a perfectly legitimate nature. In order to avoid giving offense, I should stress that just because most of us can't build a television set, it is no indication that we are uneducated. It is nevertheless an example of "ignorance" on our part. Nobody expects anybody to know everything, and nobody *can* know everything, but this puts upon all of us a limit as to what we are qualified to talk about. What is more, even specialists can hardly any longer be expected to know the whole of their own field. In addition, there is a very great deal more to be learned about everything than is at present known, and this applies to our earth as much as to anything else. In fact, we really know astonishingly little about the latter (see Chapter 18). Finally, there is, today, still another bugbear; namely, the compartmentalization of knowledge.

To put it crudely, very few specialists in one field know even the rudiments of any other, although, among "experts," there is a sad tendency to act as if they do. The rift is not just between, say, the physical and the biological sciences. My own speciality happens to bridge botany and zoology, and I am constantly and consistently unable to discuss it with either zoologists or botanists, simply because a zoologist who takes any botanical matters under consideration is an extreme rarity, while the average botanist finds no use for animals except as a minor ecological factor. The amount of plain "ignorance" even among the most learned is quite terrifying, though the truly learned are always the first to admit this. We have no quarrel with the learned nor with the true scientist: our clash is with these so-called "experts."

This is itself a thoroughly loose term, and the whole concept of "expertism" is based on false premises. You can be a specialist in several things, but you cannot really be an expert in anything; and when someone says he is—or he is said to be, in some matter, and particularly a matter the person concerned denies exists—it is manifest that something is very wrong. The press is principally to blame for the widespread use of this cliché and it is a very dangerous procedure to which the public should be alerted. The very word should be suspect and, if used at all, should be fully qualified. Thus, if somebody says he is an "expert" in the matter of ABSMs he is a liar; if others say that he is an expert on anything, it is incumbent upon them to state just why they say so; and this entails stating in just what fields he may have been trained and have specialized. In the ABSM case, such specialists as mountaineers and hunters do not qualify to pronounce upon the matter, apart from making straightforward factual reports which is everybody's [and anybody's] prerogative. Further, even if a zoologist has specialized in the known animal life of, say, the eastern Himalayas, it does not qualify him to make statements about the extinct sub-hominid Pithecanthropines of the other parts of the Orient; though, as a zoologist, he ought to display sounder judgment on such matters than, for instance, a botanist.

We come then to still another hurdle. This is a very odd one indeed. In some respects it seems illogical but it is nonetheless a fact, as the press, above all others, can attest. Explanation of it is often attempted on psychological grounds or by a general appeal to the fallibility of human nature. Personally, I have always felt that this is avoiding the main issue and is nothing more than an intellectual "out." I refer to the extraordinary manner in which strange, odd, and especially inexplicable happenings gang up, both in time and space. To try and explain what I am talking about let me attempt a purely hypothetical case.

Let us suppose that some county newspaper reports the incidence of innumerable apparently spontaneous fires breaking out in some isolated farmhouse. If the reports persist, some larger newspaper may send a reporter. Arriving in the area, he may stumble upon some grotesque local political situation. Then, something like a murder may take place, involving still another story of, say, alleged witchcraft. Then the funsters

[429]

get in the act and pull some incredible stunt that starts the local constabulary off on a wild-goose chase. But, if the outsider persists, he will, it seems, almost invariably turn up something else again, quite unrelated to any of the previous shenanigans, and probably quite unknown to or even suspected by the local inhabitants. Ask any reporter. For the life of me I cannot name one single news-story that I have ever been on that, starting with an oddity, did not bring to light half a dozen other enigmas and a similar quota of hoaxes, accidents, and other red herrings.

When we come, therefore, to examine our principal matter on hand, and its reception by public, press, technology, and science, we must bear in mind that it offers an extraordinary range of possibilities for intelligent skepticism, and that there is much legitimate reason for plain, honest people to be skeptical. However, the scientific skeptics have gone too far and too fast, so that in the end they have become frankly asinine and brought into existence a powerful counterforce. This is a solid skepticism of the scientific skeptics themselves, now so widespread and potent that anybody criticizing any aspect of the business immediately becomes suspect himself. This is a healthy but also a dangerous development.

As I tried to make clear in my brief introductory history of ABSMery, it was the coining of the phrase *"abominable snowman"* that first brought this matter out of the seclusion of what was till then regarded as native folklore and made it front-page news. There had been hundreds of thousands, if not millions, of strange and bizarre tales told before that time by travelers of every ilk ever since prehistoric days but few had ever made "news"; while news dissemination itself had previously been slow, ponderous, and was in no way as effective then as it is today. The scientific world had really never been called upon to face a public outcry of these dimensions and urgency, while they had always before had not only the press, but the public on their side in coping with unwanted and awkward items. Never before 1920 had the public clamored so insistently for an explanation of what was still then only a "newspaper story"; and never before had

[430]

the originator of the story held a position quite like that of Colonel Howard-Bury who, to boot, was on a mission of deadly seriousness to his country. The attack on Mt. Everest and its conquest was a prestige matter to the British; it had official blessing; and, it was of enormous popular appeal. Persons in charge of it simply could not be ignored.

The result was that those zoologists to whom the press applied for guidance and an answer to the new riddle responded more or less out of habit and to form. They were frankly unaware of the potential of this business and seemed to have thought that it could be brushed aside like the other objectionable matters that were brought to it from time to time. They simply denied it. This is to say that, without any consideration at all, they glibly announced that it was a lie.

This was a staggering thing to do and one that could only have been possible in a hierarchical realm, conditioned to issue categoric pronouncements that would not, or were deemed should not, be questioned. But they had forgotten several new factors already in existence, such as the Everest business and the new-found power of the press. They also seem not to have realized the growth of real skepticism among the public. Even the press was a bit staggered by these denouncements of a popular hero.

Accused, in turn, of being liars themselves, those who had pontificated immediately either retracted or elaborated their previous statements, they came up with two alternatives. Either, they said, the whole thing was a case of mistaken identity by somebody, however worthy in other fields, but who was not a trained zoologist and who therefore could not be expected to know what he saw or interpret correctly anything of a zoological nature; or, they said, it was a hoax. There was not very much anybody could do about the first suggestion because it cannot be denied; while there was much evidence that untrained personnel did in fact often make mistakes in identifying things in other specialized fields. It was not till later that it was pointed out that an experienced mountaineer is more likely to interpret correctly what he sees in mountains—where seeing is notoriously unreliable—than a zoologist who has never been on a mountain; just as a sea captain is more likely to know what he sees in the sea than any landlubber. [Moreover, it was then observed, that the seaman had better be able to do so, or ocean travel should be abandoned until he did, for the safety of ships depends wholly upon their captains' ability in this respect: and, if they are going to start mistaking bits of seaweed, floating logs, or deflated Navy blimps for sea monsters, they are not going to be able to pick up marker buoys, or even to make port.] There are lots of people and categories of people, much better qualified to identify animals in the field than zoologists, and especially that breed of the latter who made these particular pronouncements, most all of whom had spent their lives in museums! The suggestion of a hoax presented quite other possibilities.

The idea that this whole thing might be of such a nature is naturally both intriguing and satisfying to the average person, because it explains something unpleasant and "explains away" many things that are highly objectionable. The only aggravating feature to it is that the confounded things themselves persist in cropping up again and again and from ever wider sources when, one would have supposed, the whole matter had been settled once and for all. The trouble is that very few have examined the premises of the hoax theory carefully. Let us analyze this idea.

First, if the whole thing is a hoax, it must be the oldest one in history because exactly the same things have been reported by Western Europeans since at least the mid-15th century (see Chapter 14). Second, they, or it, must be the product or products of a thoroughly international organization, because they have appeared on five continents throughout the ages. Third, the organizers must have had a positively enormous amount of money at their disposal at all times, because all the concrete evidence has turned up not only in the most out-of-the-way places, but almost precisely where the toughest explorers on the one hand and the most expensive oil companies on the other have either not previously been, or where they have found it almost impossible to go. Fourth, their or-

ganization must surpass in security techniques anything ever
devised by any other organization, private or public, because
not one iota of suspicion, let alone evidence of, their existence
has ever turned up or even been claimed. Fifth, their oper-
atives must, throughout the centuries, have been chosen and
trained with a skill that is really quite unbelievable, for they
have managed to get into the most impossible places and
have done things there, persistently and without ever being
seen even by the locals, with such devilish cleverness as defies our imagination. How, for instance, did they manage to
lay out a set of bipedal foot-tracks in fresh snow just before
Messrs. Howard-Bury, Shipton, *et alii,* together with the best
local Sherpa mountaineers happened, by mistake, to turn aside
up a certain pass, and to select a stretch of territory so rugged
and difficult that even the Sherpas gasped in admiration at
their mountaineering skill. This is all odd enough, but I have
a further parenthetical question.

Why?

To manufacture a lot of fossils to fool a professorial ped-
ant is one thing, but to pull a hoax, world-wide, for centuries, to fool nobody in particular is not only senseless and
illogical, it is just plain fantasy. The hoax theory is, in fact,
so stupid that it hardly warrants mention, and when it is presented, as by Messrs. Peissel and Thioller (see *Argosy*) as being a deliberate conspiracy to promote "tourist" trade on the
part of a scattered bunch of Buddhist monks in Nepal, it becomes far more laughable than the real story itself.

When the public insisted that the thing was *not* a lie, and
the simplest logic showed that it was not a hoax, the zoologists resorted to what was to them "Holy Writ," and they
trundled out their biggest intellectual guns to try and prove
it. Since the tracks existed, as they now had to admit, they
stated that the whole thing was perfectly normal, and in no
way odd, being no more than cases of mistaken identity. They
then proceeded solemnly to recount the list of animals known
to exist in and around the upper Himalaya and [to be on
the safe side] chose a goodly selection of them as being the
makers of the tracks. These included the Giant Panda [which

is not found within a thousand miles], outsized Gray Wolves, certain larger species of Langur Monkeys, the Snub-nosed Monkey (*Rhinopithecus*), the Snow Leopard, and even some "large bird" [type unspecified] but, above all, a creature they called the Isabelline Bear. This is a very mysterious creature itself, as was explained in Chapter 2.

Much of the world "bought" the bear theory and there is an endless literature on the subject, most of it absolutely not worth reading, since none of the writers seem ever to have taken an impression of any bear's tracks nor even have bothered to look in any of half a hundred books [some published up to 50 years ago] that give drawings, measurements, and even excellent photographs of such things. ABSM tracks are *not* bear tracks. Nor are they those of any of the other animals listed above and suggested by the skeptics. I need not go further into this irksome business; and you may refer for further information to Appendix B. Since I am not among those who affirm that the average intelligence of the citizen of this country is that of a 12-year-old, and since I have the profoundest respect for the intelligence and more so the perspicacity of 12-year-olds in any case, wherever they were born; and, furthermore, since I believe that any child of much less age can tell the track of an eagle or a monkey from that of a man, I refuse to further insult anyone's intelligence. But, believe it or not, this is just exactly what the skeptics did. It is one of the most deplorable bits of chicanery that I know of ever perpetrated on the public in the name of science.

After Eric Shipton published his photographs of ABSM tracks in 1951, none other than the British Museum (Natural History) mounted a large exhibit in the main hall of their building in London, and advertised the fact that they had done so. This exhibit showed pictures and casts of Mr. Shipton's discovery alongside casts and photographs of bear tracks and those made by Langur Monkeys and then had the brazen effrontery to state that these proved that the ABSM tracks were those of bears—"or monkeys," as they neatly put it! This was outright deception, and not just a foolish prank or due to a lack of knowledge. The whole exhibit was designed specifically to debunk Mr. Shipton's findings and the whole ABSM business, but, so completely out of touch with reality were those who perpetrated this hoax that they apparently really thought that their word was enough to fool the public. One can hardly credit such stupidity, let alone the duplicity of anybody who would place the casts of a bear, a man, a langur, and an ABSM [especially Shipton's enormous and most unusual find] alongside each other, and then try and tell the public that these *proved* that the last was a bear— let alone "either a bear *or* a monkey." What in the name of anything do such people think—or don't they? Personally, I cannot believe that it was just stupidity and simple lack of education. The British Museum is administered as a part of the British Civil Service and for all that may be said of that organization, it is certainly not stupid. Mistakes it may make, but to attempt a *hoax* of this nature—and that is exactly what it was—is not in their book. The Empire would have collapsed long before that time if it had been; and, besides, the British public is as sane as any other.

This most disgraceful of all hoaxes backfired. It was virtually the end of the skeptics, for even the press overcame its inverted scruples, and howled. It marked a turning point in the history of ABSMery, just as important as Colonel Howard-Bury's telegram in 1920 and the decision of the Soviet Government to investigate the matter in 1958. In fact, this really ended the regime of the skeptics and opened the field to intelligent appraisal by honest men. From then on, people demanded the facts and were no longer interested in the mouthings of "experts" or the tricks of officialdom. So, I now turn to a final appraisal of those facts.

20. Certain Abominable Conclusions

Just saying that something does not exist neither disproves that it does, nor does it make the thing go away. Explaining something away is not the same as explaining it.

We have now done with fantasy and come, as I stated at the end of the last chapter, to *Fact*. This is a trilogy composed of reports, evidence, and objects. One might suppose that from this point on it will be plain sailing. Alas, this is very far from the case for there are even more pitfalls along the road through this field than there are in the bewildering world of make-believe, ignorance, and prejudice that we have just waded through. And these traps are much more deadly because they at first appear to be quite logical.

Saying that something does not exist proves nothing. Showing what something is *not*, does not prove what it is. Even proving what a thing is does not exclude the possibility of the existence of other things. Then, there are the old saws about having not gone to China and thereby proving statistically that China does not exist; and the corollary, that Tibet is China because you can prove that it is not any other country. A full understanding of such matters, and of paralogic in all its forms, is essential to a proper understanding of our problem because it rules not only in that negative world of skepticism which we have just been through but is also very prevalent in the positive world into which we are now going to plunge. This is the realm of newsmen, policemen, and lawyers —and ABSMs at this time are still a police job.

Policemen are true experts in the processes of paralogic and in the classification, behavior, and ecology of its exponents.

They are also specialists in the study of another terrifying breed—that of witnesses. It is logical to suppose that the only way to solve a crime is to catch the culprit, but this is not the final answer to such a problem. Still less is it the *only* solution that is possible of any crime. In the case of ABSMs it is often said that the only way to prove their existence is to catch one. This is a valid statement, but not a true one; and on several counts. I warn you, we are now heading into a real jungle.

ABSMs are not yet objects for scientific research. In science there is supposed to be a high standard of ethics but no place for sentiment in its wider and proper sense. By this, I mean that truly scientific research is supposed not only to be completely honest, but utterly devoid of all prejudice. It is concerned solely with facts and in it there is supposed to be no place for fancies, which is to say, credos. Scientific evidence is supposed to be in the form only of proven fact, and the basis of its methodology is the criterion that all such facts must be reproducible on demand by anybody, anywhere, and at any time. Only then may beliefs or opinions be expressed—which is to say, hypotheses put forward and theories erected. These, in turn, then have to be proved.

This puts a tremendous strain on *evidence* in science, and it is in appraising evidence that scientific methodology most often breaks down. Here the line between fact and fancy is sometimes hazy, and that is where scientists, being only human, display their greatest tendency to prejudice. Nothing is more aggravating than coming across convincing evidence that your pet theory is fancy rather than fact; and proof that this is so, sometimes sends even the truest scientist off his rocker. It is his faith that is shaken and the world has until very recently run on faith, not on facts. It may be clearer now just why I stepped aside in the last chapter to discuss a number of matters which may not have appeared to have had much direct bearing upon our main theme.

Anybody can tell a story; but, true or false, once told, that story itself becomes a "fact." If this is presented as fiction, there is no further trouble, but if it is put forward as *fact*, we run into complications. Immediately we want to know who said so, and on what grounds. Is there any evidence, and, if so, can it be proved; which means can it be reproduced or, if it is a physical object, produced? And this raises the question: how do we go about appraising evidence once we have been informed of its existence?

The ABSM reports, as given above in Chapters 1–14, are claimed by all those who told them to be fact, but there are many who have stated that they are fancy, if not pure fantasy. Both opinions are permissible but it is incumbent upon both parties to prove their case, and here we run into some pretty legalities and a realm quite outside that of scientific research, because evidence, proof, and even physical objects turn out to be just as difficult and elusive as stories, and research through these channels cannot as yet be directed at the ABSMs themselves. We are still in the realm of *search;* and although this may be prosecuted in what is called a scientific manner, it is, as I say, primarily a police job and one that should be run on crime-detection lines. Further, analysis of evidence and appraisal of witnesses must be conducted by methods laid down by the legal profession, otherwise it will become bogged down in a morass of paralogic.

Thus, there are two ways in which both the asserters and the deniers of ABSM stories can go about proving their contentions. Both can bring acceptable evidence to prove that they are right or that the other party is wrong. This may, in the latter case, sound like proof by default, but this is a very peculiar case and at present in a really unusual stage of prosecution. There is chicanery afoot on both sides, and in both ways. Certain reports of ABSMs have been outright fakes, others pure mistakes; but so have certain of the attempted disproofs of them. We are dealing here not only with human credibility and fallibility but with outright "crime" in the intellectual sense, in that paralogic has been deliberately employed by one party at least: namely, the skeptics or deniers of the facts.

The zoologists who took it upon themselves to act as spokesmen for established science as a whole, introduced this at the outset. Having stated flatly that the whole thing was a lie, they put forward what they expected us to believe was evidence in the form of the "China does not exist" bit. Their method was to collect as universal a poll as they could of zoologists and others who would affirm that such a thing as any ABSM was *impossible,* and they then proceeded to say that, since this opinion was unanimous and [in their opinion] nobody else was qualified to express any other opinion on the matter, it *was* impossible, and therefore a lie. Balked in this when the other party furnished better evidence *for* ABSMs, they then applied the "Tibet is China" proposition by amassing evidence that visible signs of ABSMs—and in particular the tracks in snow in the Himalayas—were *not* anything except those of bears; whereby, *ergo,* they *were* bears. But there they ran into the impasse described at the end of the last chapter. The proof of their paralogical "evidence" not only simply would not stand up, it turned out to be damning to their whole contention and opinion. Had the zoologists carefully followed scientific methodology instead of prejudice, and produced a valid case supported by acceptable and provable evidence, the business would have remained legitimately debatable. As it is, they did not do so, and, what is more, they threw all their eggs in one basket, for they would not admit any possibility of any ABSM existing.

Actually, this aided the search in many ways, most notably by virtually closing the matter to further debate. From then on, the entire onus of proof devolved upon them, the scientific skeptics, and today they are faced not only with disproving the existence of ABSMs, and the validity of *all* the stories about them, plus the credibility of the persons who told them, and the evidence they produced, but also, at the same time, they must *prove* their own position. All of this is more significant than the general public realizes, for it brings a lot else besides mere ABSMs into question.

If the proponents of any discipline clearly demonstrate that they are unreliable, or worse, dishonest in their own specialty, their opinions and pronouncements in all others become suspect. If, moreover, their line is a closed bailiwick, is specialized, and is therefore incomprehensible to others, it becomes very highly suspect. How are we to know just what is going on? If zoologists can be so viciously obtuse about this subject, just how right are they about things that they *do* claim to know? It is not a pleasant thought; and it casts aspersions upon other disciplines.

Anthropologists very sensibly kept mum about the whole ABSM business till a very late date. Whether ABSMs might be undiscovered human tribes, Tibetan outcasts, hermits, or some kind of animal, or whether they might be living remnants of otherwise extinct sub-men, or sub-hominids, they refrained from stating. They let the zoologists bounce about out on their limb. But once the "Now-you-can-see-for-yourself-it's-only-a-bear-or-a-monkey" story broke, they went to work quietly and without fanfare. This is not to say that either all anthropologists, or the science as a whole [which is something quite different], immediately stood up and cheered for the "pro" side. Quite the contrary; most of them made noises every bit as grouchy as the zoologists, and some of them became just as puerile, for it hit them too on a raw spot, and right in their own compound. Luckily for them, however, they had never said that such a thing was *impossible;* though this was not for any lack of thinking so. The idea was so completely unholy to them that they had never even considered the necessity for saying anything at all. Meanwhile there were, luckily, those among their ranks who took a completely different and truly scientific view of the whole business, and it is primarily to these specialists that we turn for guidance in appraising the evidence that is produced by the honest searchers for the truth.

The first thing we have to do in assessing the whole question of ABSMs is to make a clear distinction between "reports" and "objects." Both are in their way facts, but the two are otherwise in altogether different categories. A report, whether written down or not, cannot really be proved; an object does not have to be proved, though it may need explanation. A great part of ABSMery is regrettably reportorial in nature. Its most outstanding defect is our lack of ability to assess the validity of these reports, let alone their contents. There are lots of reports floating around, some even in very solid print, the mere origin of which cannot be discovered. Nor is this all. There are some most definite statements that have been copied over and over in recent times, the origin of which apparently cannot be traced. In this respect I should mention the widely known report attributed to "a famous British explorer named Hugh Knight" which is almost a linchpin in the whole Himalayan ABSM story. This is a classic, and has been repeated in almost every article on the subject for years; yet, I have to state flatly that, despite long probing, I have been completely unable to find the original statement; or, what is more, have any of us who are sincerely interested in this matter even been able to ascertain whether anybody named Hugh Knight ever really existed. In other words, we have not only to question, probe, and assess the reporters, but also those who report on reporters. And there is often no real way to assess either of them.

The most we can hope for is some assurance that a person was *considered* to be reliable. But this term has a very wide connotation. The most upright people tell outright lies on occasion, sometimes deliberately and with the best intentions, as in intelligence work; while anybody can go mad or even *be* mad, though the fact is never known. Then again, any reporter can make a mistake, while there are all sorts of influences at work that may cause anybody to fabricate stories or to convert, divert, or twist stories that they, in turn, have heard. The whole business is, in fact, more than a psychological jungle.

However, there is one thing that can be done with "reports." This is to subject them—when you have enough of them—to various statistical analyses. Statistics are at least impersonal [if not always reliable], and they do not really need to take any account of the reporter, his reputation, or his veracity. If you get enough reports on anything, from diverse enough locations in time and space, and [on correlating them] find one or more agreements, above a certain number, you will know that you have identifiable factors in operation. Coincidence is a strange thing, but it eventually runs out, statistically, and simply by the law of averages. Thus, if foottracks with five very unique characteristics are reported from ten different countries for over 200 years, you may fairly safely, and scientifically, say that there is a cause other than sheer mendacity on the part of those who reported them. Footprints with one odd feature, turning up in 5 countries over a 5-year period, might be explicable by coincidence or be the outcome of an initial story read by persons who happened to be in each of those 5 countries. Foot-tracks with 5 oddities spread over 50 countries in over 500 years is another matter altogether. The assessment of the reports on ABSMs is not of itself so significant; rather it is the remarkable similarities, in certain circumstances or in certain areas [vegetational provinces, for instance], and throughout time, that are so.

Somewhat similarly our approach to folk-tales has changed considerably during the past century. From being regarded almost as historical record, their value first dropped not only to nil but beyond, into the realm of the misleading. Then it started to mount again to a position of esteem, and today, there is a tendency to take folklore under very serious consideration, for a great deal of it has proved on proper analysis, and in the light of new methods of interpretation, to be valid history, simply expressed in another format, or upon a logic other than our [Western] currently accepted one. The ancients did not, and living primitive peoples *do* not, subscribe to our ways of thinking. They simply have not developed them, but they nevertheless attempted and still attempt to record facts. For instance, the migration of swallows was once "explained" in northern lands by asserting that they all went down to the bottoms of ponds and slept in the mud during the winter. Today we know this to be nonsense but the fact that they all went away in the fall and that they later return remains true, and has been proved. Thus, if the Chinese long ago stated that there were men-bears or bearmen in Szechwan and eastern Tibet, it did not mean that they said there were crosses between men and bears to be found there but, simply, that there was a kind of creature thereabouts that could best be described as being halfway between a man and a bear in appearance. It was—and still is —called a *Gin-Sung;* and by all accounts looks very much like a large, broad-shouldered man wearing the skin of a bear, otherwise known as the *Dzu-Teh.*

This brings us to another category of recorded evidence; namely, the truly historical. This is the field of bibliographical research (or search) and constitutes another wilderness of bewildering confusion. Again it has been Bernard Heuvelmans who has led the way into this further jungle. Those who have followed have been most assiduous and the outcome has been startling. The Bibliography appended to this book does not really give any indication of the volume of reference to ABSMery in its widest sense because I have been forced to omit the details of whole categories, and lump them under a single item. This published material varies enormously and is spread over a really immense period of time and throughout a very wide variety of literatures. The greatest volume of reference is in the category of the travelogue, but most of this is casual, passing, and usually brief. Quite a number of the authors did not even realize the significance of what they were recording. The second largest category is that of ethnological, ethnographic, and socio-anthropological works, some of which are positively crammed with reports and comments on the basic question of ancient and extinct humanoids, varying all the way from alleged lower animals with human characteristics to very definite humans with characteristics of lower animals. Most of these are presented as MLF but, when viewed in another light, are manifestly straightforward accounts of the previous existence of creatures in the area of ABSM type.

Purely biographical evidence, of course, merges with what I call the secondhand account; namely, one derived from the statements of others. Much of the information recorded in modern travelogues is of this nature; the author stating that he was told by so-and-so that, at such-and-such a place, in the year this-that-or-the-other, a person or persons said they saw something. The assessment of such statements is really quite impossible, because anybody can say that he was told almost anything, without running the risk of being called a liar. What is more, he can start out by saying that the person who told him was obviously making it all up in the first place. Nonetheless, the statistical method of analysis may again be employed here; and by doing so, some very strange things come to light.

Firsthand reports, especially when published over the signatures of a "big name," whose activities are well-known, and can be traced, are quite a different matter. These, even when not supported by any form of pictorial or concrete evidence, have been the principal stimulants to the whole ABSM business. In this category must be placed many travelers in what are called modern times, who have left published records in which they make definite statements that they either saw an ABSM, its tracks, droppings, or other parts, or who said they heard it or inspected such corollary evidence as the moving of cairns on mountaintops. The most astonishing aspect of the roster of such reporters is not so much their actual number, but the proportion that they form of *all* travelers who visited the countries concerned. Equally surprising is the almost universally high standing and reputation for probity of these reporters. This is particularly noticeable among those who have written of the Himalayan region. Almost everyone who has been there has reported something concrete and definite about ABSMs, be they geographers like Ronald Kaulbach, mountaineers like Eric Shipton and Sir Edmund Hillary,* doctors, anthropologists, political and forestry officers, and all manner of other specialists such as have already been mentioned. At this juncture, I should point out that few of the scientific "skeptics" have ever been within sight of the Himalayas and indeed most of them have never been out of Europe or America.

Northern California is a very forceful case in point. There the matter goes to extremes because even local people who

* Hillary's early reports, that is.

have been born in the forests concerned, seem never to have ventured more than a few hundred feet into those forests, yet they may solemnly state that anything alleged to have been seen therein by those who *have* penetrated them is either a lie or the product of a hoax. The situation was frankly preposterous when I visited that area in 1959. Intelligent people who had lived all their lives not 30 miles from where the tracks of the *Oh-Mahs* were turning up night after night in the mud on a new road, not only refused to go and look at them, but were quite violent in their denunciation of the road builders who were moving their families out because of them, calling them fakers, liars, and other much less pleasant things. People in the nearby town of Eureka were at the same time in an uproar because their local newspaper had printed straight accounts of what these road builders had said. The citizens denounced the editor, and even the local police issued deprecatory statements about him.

The truth is that many people do not want such reports, and, more precisely, they do not want to have to read any as fact. Given as fantasy, they are quite prepared to accept them. Yet, there are several things that almost all firsthand reporters seem to have in common. These are integrity, a reputation for honesty, and above all, provable firsthand experience of the country concerned, to say nothing of the matters reported. The skeptics, on the other hand, are almost without exception—if not entirely so—persons who have never been near the scene of events, while quite a number of them prove to have a reputation for a prejudicial outlook, hidebound ideas, an ax to grind, or a desire for self-publicity. Unfortunately neither party is, except in a very few and exceptional cases, scientifically trained, or especially experienced in those matters and disciplines most needed for a proper interpretation of the facts observed. This does not, of course, apply to the Russians, Mongolians, and Chinese because the only people we have heard from on the subject in print from those quarters *have* been scientists, and they seem in most cases to have been deliberately seeking scientific evidence of this matter.

In assessing firsthand accounts, therefore, I personally tend

[445]

to give the benefit of any doubts to the reporters, and more especially when this is bolstered by either direct concrete evidence or the statistical method of analyzing details of their stories. In fact, I just refuse to call such people as Ronald Kaulbach, Gerald Russell, and Professor Porshnev, liars; and I just as forcibly refuse to question the details of their observations. How dare anybody do so, who does not have their training and experience and who, above all, has never been to the area where they made their observations? Most of the skeptics are actually crackpots, yakking away in a vacuum of make-believe. They do not have the facts; they often don't even read or examine them; they are not trained to interpret them; and they have preconceived notions, often on everything. Moreover, these are usually quite erroneous, even deliberately so.

This ends my reportorial contribution to the subject of ABSMery, but I find that I have a few pages left over. I shall therefore employ these for some comment and even some opinions. I am constantly—and quite legitimately—asked what I personally make of all this. Frankly, I welcome an opportunity to reply and perhaps to sound off a little. Straight reporting is, to me, the only really satisfactory occupation that there is; but there are times, I must admit, when one gets the itch to not just comment but to pontificate. After so many years in this morass, the business looks this way to me:

First, it is my humble opinion that ABSMery is not only a valid but a concrete subject for investigation. Unlike such wholly unsubstantial things as, say, poltergeists or even such unapproachable ones as UFOs, they have always seemed to me to be not only quite possible but extremely probable. In fact, the longer I live, the more I read, and particularly the farther I travel, the more convinced I become that they do exist. However, I have a strong personal feeling that they [as a whole, or as an item of existence] have been not only grossly misunderstood but misinterpreted.

My central belief is in a way just like that of the skeptics—to wit: that there is really nothing odd about the whole business. In this, however, my reasons for such an attitude are almost diametrically opposite to those of said skeptics. They say that there is "no problem" because all the tracks are made by bears or other known animals; I, on the other hand, would affirm that there is no problem because we have ample evidence of all manner of sub-men and sub-hominids in the past; have living examples of many Primitive humans still in existence; and still know very little of a major part of the surface of our planet. For these reasons—and because of the discovery of all manner of huge forms of life right up till the time of writing and even on our own continent (*vide:* the new herd of Woodland Bison)—I cannot see any possible valid argument *against* the continued existence of ABSMs. This attitude has naturally been enhanced by my good fortune in having been able to wander all over the earth since childhood and actually to see for myself the real conditions pertaining in many lands. I *know* that most of the lands in the world are still more than half empty of humanity, and are simply unexplored, in any real sense.

This notion, of course, conflicts absolutely with general world opinion, ranging upward to the topmost echelons of the United Nations. I'm sorry, but, with all due deference to world organizations and to all sincere persons in every field, I have to give it as my considered opinion that it is rubbish. Indeed, we humans—i.e. Modern Man or *Homo sapiens* as we have chosen to call ourselves—are rapidly approaching the Malthusian limit but this is not for lack of space. Nor is it primarily because we are a disease-ridden bunch of semi-educated breeding machines, lacking sufficient know-how and mass technical skills. To the contrary, it is almost solely due to the fact that we are basically a *gregarious* species of primate mammal. There have been famines in Russia when you could hardly walk across the street for the droves of fat ducks. A "famine" can have sundry meanings, some of them having nothing to do with *famine.* To the Russian peasants of bygone years it meant simply a breakdown in the supply of bread.

But, you may say, countries such as China and India are

[447]

different. Surely they have famines there so ghastly that men eat mud. For all their monstrous population and poverty, there are in both lands still lush and enormous areas that are not agriculturally used. Certainly the report that "wild people" had been found in the southern Chinese upland *Massif* should have been sufficient to demonstrate this. Nor are our highly industrialized Western countries any different. I need not reiterate the examples that I have already given of the extent and number of true wildernesses on our own continent.

Thus, there is actually more than enough room for all manner of as yet uncaught and unidentified creatures—even of very large size—to be running around completely unknown to us, and sometimes right on our doorsteps. Proof of this contention need not be sought beyond the case histories, aforementioned, of the Okapi, the Lado Enclave "White Rhino" or Cotton's Ceratothere, the Kouprey, and the Woodland Bison. Therefore, the *possibility* of even a dozen kinds of ABSMs being around, and in not inconsiderable quantities, is not impossible: it is quite probable. Personally, I think that it is a certainty, and from my half a lifetime of studies of Nature in operation, and especially of the distribution of her lifeforms, I believe that it is almost necessary—in order to fill all her niches; something, it seems, she must always do.

By the same token, and at the same time harking back to a previous statement, I feel that the whole business of ABSMery has been misinterpreted even by zoologists and anthropologists, in that both continue to subscribe to some unwritten and invalid set of rules that grew up sometime in the last century about what *can* and *cannot* be, plus what is and what isn't. Actually, if you come to review what *is* known and accepted about the rarer, odder, most obscure, and unknown races of people that *do* still exist today, you will find that they are really legion, and that we already have pretty fair candidates for not a few of the smaller ABSMs. Where we are to draw a line between these Primitives and outright relic races of sub-men and sub-hominids, I have not the slightest idea. Personally I cannot draw any such line, and I don't know upon what criteria to try to do so.

My notion is that, if only we could all clear our minds of our many preconceived ideas about what is possible and what isn't, and then take a hard look at what we know *is*, we would find that there is really no "problem" here. We are dealing with Hominids, ranging all the way from Modern Men who don't wash much to, maybe, creatures so primitive that they have never known speech, fire, or even tools. Nobody any longer denies that such creatures once existed, and nobody denies that Gorillas, Chimps, and Mias still exist, though we class them as altogether more primitive than Hominids. If the latter have survived, why not the former?—more especially when those former undoubtedly had at least the glimmerings of what we call co-ordinated "intelligence" as well as purely animal wits, or instincts.

Thus, my answer—and I do not mind how far out on however slim a limb I go in saying this—is that I think there are at least three main types of ultra-primitive men, and/or sub-men, and/or sub-hominids, still alive today. These I would say are, first, sundry pigmy types of very near-human or completely human composition; second, some remaining Neanderthaler types in eastern Eurasia; and, third, some very primitive and large creatures almost absolutely without any "culture" in any sense of that term, in northwestern North and Central America, perhaps in South America, the eastern Sino-Tibetan uplands, and in Indo-China. Then, I am even more sure that there still remains something else.

This is the great, bestial, *Meh-Teh;* the unwitting originator of the whole business; the original "Abominable Snowman"; and the most mysterious, though best-known, of all. As I have said repeatedly, I don't know any more than anybody else what this might be, but I'll bet not just the proverbial dollars, but any gold bars I might acquire to stale doughnuts that it exists, and all over a very wide area. From what has been reported about it, and even more from an analysis of its tracks and footprints, it is my conviction that it is the remnant of a most ancient side-branch of both our own and the apes' family tree and more likely from the twig of the apes than from our lot.

I have not by any means said all that I could say, and I have really reported only a small part of what I might on this matter, while my files keep growing even as I write, but I shall say no more. My personal opinions probably will not and certainly should not influence those of others. I have tried to give all the facts possible within the compass of a book, and all I ask now is that you draw your own conclusions.

Appendices

Appendix A
ABSMal Connotations

One of the greatest headaches to laborers in ABSMery has been, almost since its inception, its concomitant philology. The names for ABSMs that have been recorded are seemingly without end, while very few of the recorders of these names have been professionally trained or even amateur philologists. Most of them did not understand the language in which the name was given to them. Also, it was not until a quarter of a century ago that an international agreement was reached upon the transliteration of both the written and the spoken word in all languages. There is now a sort of universal alphabet—known as the P.C.G.N.* System—by which anybody can transliterate almost any noises made by men; even to the series of incomprehensible glottal clicks used by the Bushmen. However, hardly anybody uses it; and true linguists, philologists, and etymologists frown upon it a little.

The names used today for the sundry ABSMs in various parts of the world have been discussed in the body of my story, and in some cases their origins and meanings, or supposed origins, were also touched upon. The North American names are not as yet properly recorded—despite the herculean labors of Mr. J. W. Burns—so that any further attempted interpretation of their philology or etymology is at present worthless. Those from Central and South America are probably beyond the ken of any living scholars; and the same may be said about the few names such as *Sehité, Muhalu,* and *Agogwe* of Africa. I have been told, but cannot assert, that the first and last mean simply "little wild men." In the southeast Asian

* Permanent Committee on Geographical Place Names.

area, apart from the *Kra-Dhan* which seems to mean "great monkey," we have only *Tok, Kung-Lu,* and *Sedapa.* I have asked around about these but, although being the recipient of the usual plethora of fascinating material that one always obtains upon applying to any philologist for anything, I have not received clear answers as to what the words may mean. *Tok* does, however, seem to mean "big mouth" in one Kachin dialect.

This leaves us with *Eurasia,* and it is in this area—as it always has been with regard to the names applied to ABSMs—that all searchers have always been most interested, and seemingly most confused. The inhabitants of the great "Gutter" of the upper Brahmaputra are of Tibetan origin. Across the Himalayas themselves, moreover, there has been a blending of central Asiatic and Indic tongues. Thus, in respect to philology, the Himalayas and eastern Eurasia are connected, though they are in different "continents"; and they must be taken together.

Even I, without any training in or even understanding of linguistics, have for years been able to appreciate that there is a monumental muddle and misunderstanding of both the languages of, and of the names used in these areas for, almost everything alive. Some fifty different words (spellings only in some cases) for ABSMs have got into our literature through the writings of those who have visited the Himalayas. When we come to Tibet, Mongolia, and Siberia, the whole business gets completely out of hand. Yet, so important are the "proper" names of these creatures, and so valuable is the information that may be gained from an analysis of them, that every serious worker in the field has for years been appealing for some proper exposition of them. This has been attempted several times but those "scholars" who undertook the studies appear to have been thoroughly incompetent because anybody, even without any knowledge of the subject, can readily see that said "scholars," in many cases, obviously did not have any understanding of the languages concerned.

As a result, I prepared a list of all the names for ABSMs

[454]

from these two regions [eastern Eurasia and the Himalayas]
that I could find in the published literature of all languages,
including the Russian, and passed this to the Rabbi Yonah N.
ibn Aharon (see Chapter 17) who is one of the few persons
conversant with the principal dialects underlying the languages that are spread all over this vast area—i.e. from the
east Russian border to China, and from the south Siberian
border to the Himalayas; and the great arc bowing northward from Ethiopia in the west, via Arabia, Persia, and
southeastern Russia, to central Eurasia in the east. He has
been kind enough to prepare the following statement on the
subject which will, I trust, not only settle a number of ran-
kling questions, but also the general confusion. At the same
time it will help scholars in the field of ABSMery. He writes
as follows:

A CONTRIBUTION TO THE PHILOLOGY OF ABSMery

By Yonah N. ibn Aharon

The study of words relating to the ABSM differs from other aspects of
the problem in that the researcher has the advantage of a tangible starting point from whence to launch his investigation. When we remember,
however, that the great majority of these words have been reported to
us by persons to whom even the fundamentals of phonetic transcription
constitute a mystery almost as profound as that of the ABSM itself, the
difficulties of the situation are evident.

In the first place, most of those who have reported these words, do not
realize the scholarly apparatus at the disposal of the specialist, and so
neglect even to make sure from what language the word comes. This
vicious combination of improper transcription and uncertain origin does
much to impede a reliable definition of many of these words, because the
homophonic properties of about 80 percent of them render them interchangeable among several languages and even language groups, with
a corresponding divergence of meaning which is, of itself, a comedy of
errors. One of the most ambiguous of them for example, can mean anything from silver spoon in one language to sour turnips, in another! [See
Teh-lma below.]

The vast reaches of Central Asia,* of course, constitute the prime source

* By this term should be understood central and eastern Eurasia, and
the Himalayan ranges of the continent of Orientalia (see Map XV).

of Snowman literature, and of words relating to ABSM studies. It is principally with this area, then, that we shall be concerned in what is to follow. Although geographically Central Asia is one of the most complex areas of the world, its philology is far simpler than most people realize. There are only three important languages with which we must trouble ourselves in the pursuit of satisfactory definitions of words originating in this area—Tibetan, Mongolian, and, to a minor extent, Nepali. There are two reasons for this relative paucity of language groups. The first is the extreme antiquity of the Tibetan and Mongolian socio-cultural groupings. The second is the fact that a single literary tradition has held sway over most of these peoples almost since the time they settled down in their present homes.

Once proper allowance has been made for local habits of pronunciation, neither the Tibetan nor the Mongolian dialects present any problems of philological verification, through the use of dictionaries, chrestomathies, and written source material. In both areas, no attempt has been made to preserve these peculiarities of local speech in the written language (although this is amended every few centuries in order to conform more closely to the changes in the speech habits of educated persons since the previous revision), but the dialects are in no case beyond the authority of the written speech. Indeed, differences of pronunciation are rarely the occasion of anything more problematical than good-natured remarks about the strange accent of the people in the next valley. Grammar has changed only slightly since the most remote times, and accents tend to change more slowly in rural areas than in the cities.

Each dialect is, of course, characterized by certain kinds of sound. Unfortunately, the ear of the average Westerner is rarely able to resolve these sounds, even to the extent of telling us whether they were or were not aspirated, or what quantity we should give to the vowels. Fortunately for our purpose, Tibetan has only one vowel quantity (its vowels correspond to those of the short vowel in Italian), but the untrained ear of most reporters has led them to make even more mistakes with regard to the vowel sounds of the ABSMal vocabulary, than with regard to the consonants. This makes it necessary for the researcher to consider each consonantal combination in the light of all possible joining vowel combinations. The fact that the consonants may have even more values than the vowel is of relative unimportance, because dictionaries exist for both Tibetan and Mongolian, with the words classified according to their final consonant, as well as their initial letter. The value of these special dictionaries has been largely ignored by Western scholars, but provides the determining criteria for some of the definitions provided in this discussion.

The important principle to be observed in evaluating these words is that which is related to the direction of language movement. Specifically, even the sketchiest knowledge of Nepalese history will reveal the intellectual dependence of the Nepali people on Lamaism, a Tibetan religion of Buddhistic origin. This fact has, however, been completely ignored by a number of rather self-confident Indian and Western writers who have, in perfect good faith, been seeking the word-meanings of Nepali names for the ABSM either in Turner's excellent *Nepali Dictionary,* or else in the dictionaries of other Indic languages which they think to be related to Nepali.

The fact is that Nepali is a rather cosmopolitan speech, considering the isolation of the country. Nepali has absorbed hundreds of Mongol words (via Urdu) and almost as many English words as has Hindi. Nepali is an idiomatic, colloquial speech, well suited to the needs of the people. Its grammar is not too difficult, and it is written in the very same Deva-nagari script that serves the rest of the languages descended from Sanskrit. A great deal of the talk about the rare dialects of Nepal generated by men like Prince Peter of Greece, is simply not true. W. R. J. Morland-Hughes, in his convenient little *Grammar of the Nepali Language* notes that Nepali is also known as "Gurkhali" (language of the Gurkhas), "Khaskura" (language of the Khas), and "Parbatiya" (mountain language). He might have added many more to the list of mystifying names for this pleasantly uncomplicated vernacular.

The proper source for the ABSMal words of Nepal is in the *Tibetan Lexicon* of Jaschke (London, 1889) or any of the many excellent Tibetan-Sanskrit dictionaries that have appeared in the last thousand years. The Tibetan-Mongolian dictionary printed for the Finno-Ugrian Institute of Helsinki is also to be recommended when available. It will be found that many of the Mongol names are merely translations of the Tibetan names originating in the monasteries of Sikkim, Ladakh, and Tibet itself.

When, how, and why Buddhism came to Tibet, it is not necessary to explain, except to say that with Buddhism came a hybrid form of Sanskrit known as Pali. The Tibetan scholars, however, were not content merely with the loan of several thousand Pali and Sanskrit technical terms, but they set about making Tibetan compounds and formulae to serve in their place. The translation of the Buddhist canon into Tibetan took place at an early date, and the number of Sanskrit words to be found in these books is smaller than in some of the corresponding European translations. But this absence of loan-words was not true of translations made from the Tibetan into Sanskrit, Lepcha, Ladakhi, and the other languages of Buddhist North India. In these, Tibetan words were to be found in plenty,

[457]

and what is more, the people to the south were not always sure of the meaning of these words, a problem that has been solved for them only during the last century. Hindu scholars * found themselves in this position more frequently than Buddhist scholars, who had access to a better class of dictionaries. The effect of this literary exchange on Nepal and its dependencies was formidable. To this day, Nepal continues to borrow words from the Tibetans, to the great distress of the Indian telegraph offices, who are called on to handle such messages. And whenever a Nepali is at a loss for words, he is more likely than not to throw in a Tibetan phrase, much as the English are addicted to bad French in times of stress.

Another factor which must be considered before we pass on to our glossary is that of the Comparative Philology of the Indic and Tibetan pronouns. The resemblance of the third personal forms of Nepali and Tibetan are remarkable, and of great importance to our subject. We must also advert in more detail to the significance of Lamaist Buddhism for the philology of ABSMal words. In southern Tibet and Nepal, there subsists a great religious tradition which has for its focal point the mystery of the Sangbai-dagpo, or "Concealed Lords." This religion certainly antedates Lamaism, and is obsessed with the transmigration of the human soul into the bodies of the lower anthropoids. The ABSMs are revered by the adherents of this sect, and the heads, hands, and feet of deceased specimens find their way into their ritual. The effect of this animistic doctrine on Tibetan Buddhism should not be under-estimated. Its effect on the ABSM mystery has, moreover, been felt in two ways: firstly, it motivates the local people to protect these creatures from the quest of the European, and to mislead Westerners wherever possible by passing off the remains of other animals for those of the ABSM. Second, it has resulted in their unwillingness to speak the true names of the ABSM, in much the same way that a Jew is not allowed to mouth the name of the God of Israel. Thus, the names that find their way into the literature almost all fall into the classification of indefinite pronouns or else generic terms that describe other species as well as the ABSM. This is also the case with the Mongolian words.

* One of the most notorious examples of such presumption is one "Sri Swami" Pranavananda (See *J. Bombay Nat. Hist. Soc.*) whose anxiety to discredit the existence of the ABSM led him to betray a complete ignorance of the U-chan, or written Tibetan, as in his contention that *teh* (3rd per. sing. rel. pron.) means "bear"!

PRINCIPAL PUBLISHED NAMES FOR ABSMs IN THE EAST EURASIAN AND HIMALAYAN AREAS

(1) *ADAM-JA-PAIYSY*
Said to originate from the Kunlun Mountain region.
A Mongolian word, in transcription to Tibetan. (No relation with the Hebrew *adam*.)
Meaning unknown. Its origin could be highly various.

(2) *ALBAST* (also, *Alboosty*)
From the Mongol *alub* (traveler), and *usud* (water).
Thus, "One who moves over (lives in) wet places."

(3) *ALMAS* (also, *Almasty*)
Mongol *ala* (to kill), and *mal* (cattle).
Thus, "One who can kill stock (cattle) animals."

(4) *CHUTEY* (and by various spellings)
Tibetan *Tssu* (a diminutive), *teh* (it thing).
Thus, "The little (living) thing."

(5) *DZU-TEH*
Tibetan *Dzhu* (big, or hulking), *teh* (it thing).
Thus: "A hulking (living) thing."

(6) *GOLBO* (also *Golub-yavan; Guli-biavan; Gul-biyavan; Kul-bii-aban; Uli-bieban; Yavan-adam.*)
Mongolian terms composed of *gul* (inf. auxiliary), and *bayi* (to stand) or *yabu* (to walk), as in *yabugul* (to send). *Dam*, probably "dharma" (but not the Semitic "*adam*")—i.e. "An Entity or Manifest Being." *Note:* this word does not always occur in combination with the *golbo* stem.
Thus; "A living entity that (also) stands upright and walks."

(7) *HUN-GURESSU* (also *Khun-goroos; Kumchin-gorgosu*)
Mongolian *khun* (man), and *kur* (to reach), *orgen* (long). The verb *osu* (to multiply) should also be mentioned in connection with this and associated terms.
Thus, "The Man-like One with extra long arms."

(8) *JELMOGUZ-JEZ-TYRMAK* (pronounced *dzhel-moghul-dzh-tura-muk*)
Mongolian *dzhel* (big), *moghul* (great), *dzh* (a conjunctive), *tura* (hills —a Tibetan loan-word), and *muk* (a generic suffix).
Thus, "Great big living things not found in all but in most hilly places."

[459]

(9) KANGMI (also [?] *Chumis; Meh-Teh; Mige; Migo; Migu; Mih-Teh; Rimi,* etc.)

Tibetan *gyang* (scrambling, pulling), and *mieh* (3rd pers. sing. n., for "Man"). Tibetan *kang* (foot) should also be noted in this connection, but, assuming the transcription to be colloquial in origin, *gyang* is more probable.

Thus, "The Man-Creature that scrambles." (And, note that this is exactly as described by the Dzungarians anent these, as described in Chapter 13, who said that their ABSMs climbed like spiders.)

(10) KAPTAR

In part as in (8) above. Mongolian *qupa* (climbing or clutching), *tura* (hills—a Tibetan loan-word). Urdu uses the word *qupatur* for "illegitimate son" and *qaputra* for "degenerate or unworthy ones." In our transliteration, "khoop-turr."

Thus, "A primitive climbing creature of the hills—One who 'hooks on to' rocks, etc. But definitely not an animal."

(11) KSY-GIIK (*Ksy-Gyik* and possibly *Kish-giik*)

Mongolian *kusegchin* (pejorative; in the sense of a "curse"). (The Urga lexicon does not agree with Poppe's rendering), and *ngui* (blackness).

Thus, "The black-colored Abominable Ones."

(12) MEH-TEH (also Me-Teh)

For the *Me,* see also *Kangmi, Mige, Migo, Migu, Mihteh, Samjda.* For *teh,* see also *Dzu-Teh, Mih-Teh, Yeti.*

Tibetan *mieh* (3rd. pers. sing. n., for "man"; corresponding to Nepali *mi,* meaning "it"—something more closely related to *m//la* [a being] than to *po*). Tibetan *teh* (living thing) corresponds to Nepali *ti* (a 3rd pers. rel. pronoun, implying "remote distance," as in Spanish *aquel.* This word, in fact, shows up in the literature in much the same context as the English word "vermin" or American "varmint."

Thus, "A manlike living thing, that is *not* a human being."

(13) METOH-KANGMI (i.e. Col. Howard-Bury's original signal)

This could be either: (a) a pure mistake for *Meh-Teh, Me-Teh, Miehteh,* etc., or (b) *MIEH-TÙH,* meaning "One who can carry off a Man."

[460]

(14) *MI-GE*

For *mi*, see (12) above. The Tibetan *mieh*, or the Nepali *mi*, meaning a "man-thing" as opposed to either a Man or a Woman and *ge* (an object).

Thus, "A (neuter) Man-Thing or creature; but definitely a corporate, living one."

(15) *MI-GO*

Apart from the *mi* (Tibetan; see above), the *go* is probably *gyo* meaning "fast" or "quick."

Thus, "The fast-moving Manlike Creature."

(16) *MI-GU*

Again *mi* as above—perhaps the Tibetan *mi* plus again *gyo* but equally probably *gyu* meaning "cave."

Thus, "The cave-dwelling manlike creature."

(17) *MI-TEH* (see also Me-Teh)

This is actually as here written a Tibetan loan-word, a version of *mieh* plus *teh* (as in Me-Teh). However, it is probably a mere mistransliteration by the "westerner" who first recorded it.

(18) *NYALMO*

The Tibetan transcription of the Pali *niyalam//la*, a Buddhist theological term describing a class of manlike beings. This is thus as near a "proper name" for any Asiatic ABSM that we have.

(19) *RACKSHI-BOMPO*

Tibetan *rag* (to claw), *dzh* (a conjunctive), and *bompo* (to handle) note; adjective, handlike, in ETib.

Thus, "The handy one (or with prominent hands) who claws at things."

(20) *RI-MI*

(For *mi*, see (12) above), and *ri* (mountain).

Thus, "A Man-Creature of the Mountains."

(21) *SOGPA*

Tibetan, transcribed into Khmer syllabary as *dzhogbo*, implying "a mystery." (*Note:* This name is used only on the south slopes of the Himalayan ranges where the ABSMs are not indigenous.)

Thus, "The Mysterious Ones."

[461]

(22) *TEH-LMA*

From *teh* meaning in Tibetan literally (as in Cockney English) "That-there thing" (i.e. a 3rd. pers. rel. pronoun implying remote distance) and *Lma*, a corruption arising from an incorrect reading for the ligature *m//la* which, in Pali, means a class of "manlike beings."

(23) *ULI-BIEBAN* (see also (6) above).

Mongolian *ulus* (strange people or tribe), and *bieban* (adjectival for "that which walks upright" with the implication of "nonetheless"). Thus, "Strange or Not-quite-right people who nevertheless walk upright."

(24) *YAVAN-ADAM* (see also (6) above).

Mongolian *yaban* (walking), and *dam* (being).
Thus, "A living entity who walks upright."

(25) *YETI* (see, in part, (12) above).

This is rapidly becoming the key name used by foreigners for ABSM in the Himalayan region. The Nepali pronoun *ye*, and the Tibetan *teh*, produce the provincial *yite*, *yihda*, and *yehda*. Herein, there is no connection with the Mongolian *yati* (cold), the ye being a dual form of *yi* (these) in literary Newari (see Turner's dictionary). Nepalese *ti* may not be current among the hill people—see letter read before the Benares Congress in Observance of the 2500th Anniversary of the Buddha by one Anag. dzhi-blo-Langyarup. From *teh* (see above) meaning "living thing," we get a concept that again can be expressed only in Cockney English. This is simply "That-there Thing."

(26) *ZERLEG-KHOON*

See possibly in part also (7) above; but, further information would be appreciated since *zerleg* has a most profane meaning in Karaturki, and the recorder of this expression probably was not conversant with this tongue.

NOTE: Bears in Tibetan are *"dara-unjeh."*

[462]

Appendix B
The Importance of Feet

The study of footprints and foot-tracks—the difference between which was discussed earlier in this book—form the subject of a very precise science called *Ichnology*. This discipline is employed in quite a number of fields, notably in police work and in palaeontology. The identification of the tracks of living animals in snow and mud has, of course also been an art in hunting since time immemorial, and it is of great interest to the field naturalist. Its more psychological aspects were also discussed when we first introduced the matter of ABSM tracks. We should now consider the details of this discipline.

Tracks [the word I shall use from now on, unless dealing specifically with a single print] are caused by gravity. The first requirement is that the object on top that presses down be composed of a denser material than that upon which it is pressed, but this does not mean that tracks will invariably result. Thus, a body made of steel if rolled across a sheet of lead need not leave tracks. There appears here another factor—that of weight—which results in the beginning of the erection of a complicated formula. Above a certain weight the upper body will leave tracks, but the point at which it starts to do so is also dependent upon the compressibility of the under body, or surface. Then again, tracks can be either pressed or punched into a substance: in other words energy, in addition to mere gravity, may be exerted. In this case the point at which an impression begins varies according to quite a number of factors which fortunately need not concern us since they lie in realms of engineering that do not apply to purely biological matters. Nevertheless, one must bear in mind that there is a considerable difference in appearance between a print made by pressing and one made by punching an object downward into a surface. The former will be found to be surrounded by little cracks all running inward to the "gutter" of the print, while the latter will be surrounded by a sort of levee or ridge.

One of the easiest ways to spot an artificially made print is to find such an *impact ridge* around a print where there is no cause for it under the natural conditions pertaining on the terrain. By this is meant, where the creature [and this does not apply to machines] had no cause to *jump*. Jumping results in the application of "weight" to the normal gravity and so is equivalent to "punching." Thus a creature running, rather than walking, will leave differently formed individual prints, and when going downhill, they will be quite other than when it is going uphill. Already the matter becomes, as you may readily agree, complicated. But there are several further complexities.

Perhaps the most notable is the area of the object which makes any print and, in the case of animals, the number of such objects [i.e. hands, feet, tail, or other appendages] employed in so walking, running, hopping, jumping, or otherwise progressing. The other most important factor is naturally the nature of the material or surface into which the tracks are impressed. This is itself an enormously complicated subject.

Tracks may be left in gravels or sands, all of which are of course much denser than any animal foot that passes over them. This is due to their particulation or "looseness." They are dry, and such substances range widely in consistency from what are called screes, of sometimes enormous boulders piled against the sides of mountains, to dry plaster of Paris. Much finer substances have now been artificially developed but in Nature we need not concern ourselves with anything much finer than what we call a fine dried silt, which is a little coarser than dry plaster.

There are two other types of solids in which imprints may be left. These are, first, materials, such as lead, that are themselves malleable or what is commonly called "soft" and, secondly "solutions" in the widest sense of that word. This means, popularly speaking, *wet*. We need not overly concern ourselves with the first since such substances may be regarded as nonexistent in Nature—that is to say in the nonhuman world. The second is of course the most important type of material in which the tracks of animals appear—and this goes for snow, which in many respects holds an intermediate position, since, dependent upon temperature, it may be either a particulated material, or a mere wet mixture. Prints can thus be left in three kinds of substances—(1) dry, as sands, (2) wet, as in muds, or (3) snow, which has to be separately classified.

There are those, such as the technologists of police laboratories and road-construction engineers, who know so much about the factors just named and the results of making impressions in various substances that it would really startle you. On another hand, certain palaeontologists have made profound studies of this subject, and most notably in connection with the interpretation of fossil footprints of early amphibians, reptiles, mammals, and birds. The findings of all three groups of specialists of course coincide for, as I say, Ichnology is a very precise science. However, most unfortunately, totally insufficient application has been made of their findings in ABSMery. Also, there has been an extraordinary lack of appreciation of the basic theorem of Ichnology which is, simply stated, that a print can be left only by an object that exactly "mirrors" it—a term that should be self-explanatory. This brings us to the questions arising from the conformation of vertebrate animals' feet.

For some reason—and there may well be *no* real reason— it just so happens that the first creatures with backbones to crawl out of water on to land dispensed with all but five of the rays (digits) on the four appendages they retained. This gave all of us land vertebrates a basic pentadactyl pattern— i.e. four "legs" each with five "fingers" or toes. True, there are some animals like whales that have somehow again reduplicated the number of phalanges in their digital extremities. This is, actually, one of the most extraordinary things in zoology as it flies in the face of one of the basic precepts of genetics; namely, that a characteristic once lost cannot be resuscitated from the same source. [The additional phalanges of the Cetaceans are phylogenetically developed by reduplication.] This basic five-fingered and five-toed pattern speaks much for the unitary origin of all land vertebrates—the amphibians, reptiles, birds, and mammals.

Many forms of all these major groups have at one time or another lost one or more phalanges on various digits, whole digits, or even whole limbs. The snakes, for instance, have lost everything. The results may be most readily seen in the [mirrored] prints left by the hands and feet of the various types; and from such prints a great deal about their makers may, by inference from known types, be reconstructed. This brings us to the classification of feet and specifically of mammalian extremities. Happily we do not have to go into this vast question and may concentrate upon those of the Primates—i.e. the *Order* to which we belong, together with the apes, monkeys, lemuroids, and a few more abstruse types like the Feathertails.

The Primate foot [as opposed to their hands, which we may also from now on ignore] is pentadactyl—i.e. five-toed—but is otherwise of a variety of forms. These forms may be classified in various ways but two sets of factors are of particular interest to us. The first is whether they are what is called wholly *plantigrade* or not; the second, whether the big, great, or first toe is opposed to or lies alongside and points in the same direction as [and/or is bound to] the other toes. A plantigrade foot means that its owner stands and walks with the whole under surface, from the tips of its toes to its heel—which is to say the last bone of the ankle—on the ground. Some Primates, such as the tiny Tarsiers, do not do this, the ankle bones being greatly elongated. Men (Hominids), Apes (Pongids), and most Monkeys (Simioids) are plantigrade. Further, none of these three groups of higher Primates have long enough nails to produce clawlike excrescences which touch the ground and leave noticeable marks though those of some Cynocephaloids (i.e. Baboons and Macaques) may do so when they are running or galloping. There are few—if any, as a matter of fact—other mammals that do *not* leave claw marks. Then again, claws and nails although having a similar origin anatomically, are not identical structures: but this also need not, fortunately, concern us further.

In dealing with Primate footprints we are therefore primarily concerned with the nature and position of the big or great toe. Among the photographs is one showing a wide variety of Primate feet. From this it will be seen that the only one that does not have an opposed big toe is the Hominid group. Nonetheless, this digit in Hominids also varies considerably in the degree to which it is set off from the other four toes. From both the prints found in the cave at Toirano in Italy and from the skeleton of a whole foot found in the Crimea, we now know that the big toe of the Neanderthalers was rather widely separated. There are people living today who have feet not unlike those of Neanderthalers —notably certain Amerinds from the extreme southern end of South America (see a photograph in an article by Dr. Carleton S. Coon, in *Natural History,* January, 1961) and some Australoids. However, there is no indication, even among those which we know, of any evidence whatsoever of a truly "opposed" big toe, among any Hominid.

One known fact about abnormalities among human feet is nonetheless of some significance to our story. This is that shown in the two photographs of a strange type in which the second toe is longer than the first, sometimes more massive, and also widely separated from the remaining three toes. This is the more odd in that the second and third toes of normal Modern Man are partially webbed. If a foot, normal or abnormal, of this nature developed [or even merely occurred] one would have supposed that the second and third toes [together] would have become widely separated from, on the one hand the big toe, and on the other, the remaining two. When this abnormality occurs, both the big and the second toes tend to curl downward and inward not unlike those of the *Meh-Teh.* But they are still not opposed.

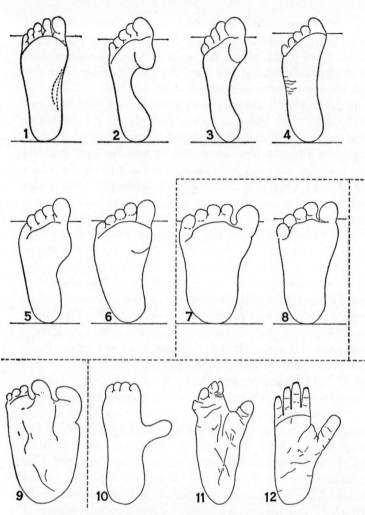

(1) HUMAN Adult (West Caucasoid). Imprint in clay mold. (2) HUMAN Adolescent, 14½ years (West Caucasoid). Wet imprint of left foot on hard surface. (3) HUMAN CHILD, 10½ years (West Caucasoid). Wet imprint of left foot on hard surface. (4) HUMAN Infant, 2½ years (West Caucasoid). Wet imprint of left foot on hard surface.

(5) HUMAN Adult (Cromagnon Man). From clay floor of cave in France. (6) HUMAN Adult (Southern Amerind). From mud of river bank Chishue, Patagonia. (7) SUBHUMAN (Neanderthaler). From moist clay floor of cave, Toirano, Italy. (8) ABSM (Guli-yavan type). From a sketch of track in mud, Kirghiz S.S.R.

(9) ABSM (Meh-Teh). From photo of cast made from photo of print in snow, by Eric Shipton. (10) APE (unknown form). Sketch made by Charles Cordier in the Congo (over-all, 30 centimeters). (11) APE (Lowland Gorilla). From photo of cast of foot. (12) APE (Chimpanzee). Outline of extended foot from plaster cast.

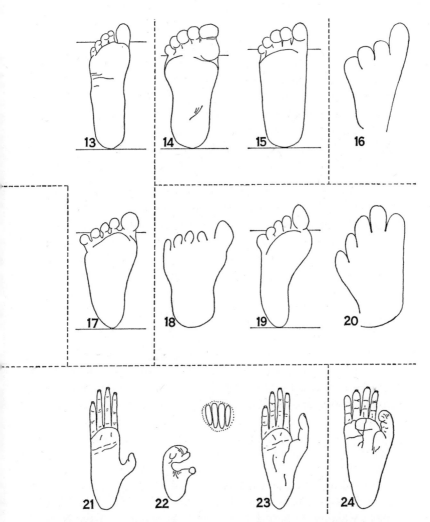

(13) HUMAN Adult (Bushman). From a sketch made from a cast. (14) ABSM (Sasquatch). From tracing of deep print in firm wet clay. (15) ABSM (Dzu-Teh). Drawn by author under supervision of Gerald Russell. (16) ABSM (?) (Kakundakari). Sketch made by Charles Cordier in the Congo. (Over-all, 10 centimeters)

(17) HUMAN (Malayan Negrito). From photo of cast. (18) ABSM (Sedapa). Traced from print found in Danau Bento Swamp, Sumatra, by Dr. Jacobson. (19) ABSM (Teh-Ima). From sketch made by the author, under supervision of Gerald Russell. (20) ABSM (Apamandi). Sketch made by Charles Cordier in the Congo. (Over-all, 20 centimeters)

(21) APE (Orang-utan). Outline of extended foot from plaster cast. (22) APE (Orang-utan). Imprints of knuckles and foot, ex Huevelmans. (23) APE (Silvery Gibbon). Outline of extended foot from plaster cast. (24) MONKEY (Old Male Rhesus). Drawn from live specimen, ex Osman Hill.

Let us now analyze or try to analyze the prints left by the four types of ABSMs. These, we will arrange in the following order: (1) the Proto-Pigmies, (2) the *Almas*, (3) the Neo-Giants, and (4) the very different *Meh-Teh* type. It will be seen from the sketches of the outstanding types of known human feet [accompanying this Appendix] that those of the living pigmy human types such as the Negrillos and Negritos show a distinct tendency toward a very short broad foot with rather large [in proportion] and widely splayed toes *and a very constricted* or narrow heel. It is hardly even a "step" from them to the *Sedapas, Teh-lmas,* and other alleged pigmy ABSMs. Such a tiny, human-type, plantigrade, flat foot does not however begin to approach the form of the various bears, and not even the Malayan Sun Bear (*Helarctos*).

Coming to the prints allegedly left by what I have called the *Alma* type, one finds that they are hardly in any respect different from those of the Neanderthalers left in the cave in Italy. This, in turn, perfectly accords with the now expressed belief of the Russian scientists that the former are but living representatives of the latter. Despite their relatively low plantar index [i.e. the number of times their width goes into their length] they are hardly at all nonhuman: in fact, they can be matched by the prints left by not a few Modern Men living today and most notably by persons who have never worn shoes or other footgear. We hardly have to discuss these tracks any further, except to mention that such have been rumored from many places other than eastern Eurasia—such as, I may say, northwest North America, South America, and Africa. But how are we to tell whether such prints—if they ever really existed—were left by some wild thing or by local men, walking without footgear and happening to have Neanderthaloid-type feet?

Our real problems begin with the Neo-Giants. Here, I want first to try and wipe away a lot of dross. It has been said, and repeatedly, that such tracks and prints have been found all over northern Indo-China, and on northward through the arc formed by the uplands and mountains of Szechwan via the Tsin Lings to Manchuria. The same—up to 20-inch-long humanoid-type prints—are also said to have been found in the Matto Grosso and in Patagonia in South America. This may be so but after, lo, these many years, I have been completely unable to obtain any photograph or even sketch of one that is stated to have been made on the spot. I have several "sketches" made by members of expeditions to these places but sketches made in retrospect after questioning and mostly in my apartment in New York, but nothing "original." The only areas from which I have been able to obtain such first-hand photos or sketches, *and* plaster casts has been from British Columbia and northern California.

These prints at first look wholly human and, I may say, a bit ridiculous. However, on further analysis they display other qualities. As may be seen from the accompanying sketch, they show one or two extremely odd characteristics that are definitely not typically human. As photographs of whole tracks of these monsters have shown, they walk with their feet pointing straight forward: not pigeon-toed, like bears, or in any way out-turned like most men. Thus, they must flex about [or across] a line that forms an arc at right angles to that line of travel behind the toes. Then this means that, although the whole foot is enormous and at first looks very long, it is really a very *short* and broad one [with an index of only 1.61].

A friend of mine, Mr. Fred Laue, long in the shoe business, has worked out the trade equivalents of these proportions. Working from one of our plaster casts with an over-all length of 15¾" and a width of 7", a No. 21 shoe would be needed but no less than 13 sizes in width greater than the widest shoe made—i.e., a EEE. Second, it means that the curious double inner pad under the ball of the foot has something to do with the basal joint of the big toe and not with the end or outer end of the first metacarpal. If this is so, and the foot bends or flexes along the arc that runs between the two pads all the other toes are not just extraordinarily but so exceptionally long that they cannot be of the typically human form. But, if they are that long, why don't they splay or, alternatively,

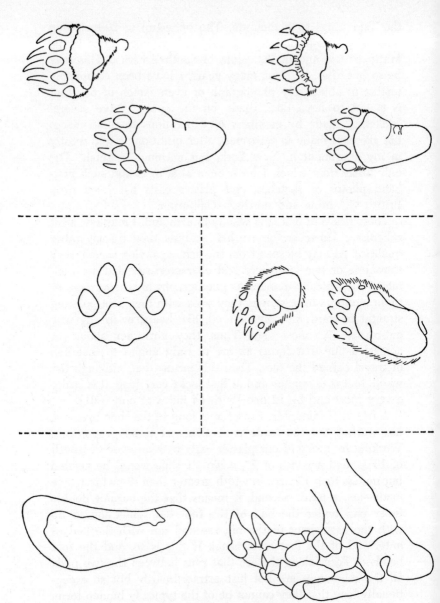

See facing page for legend

[472]

why does not the mud in which the tracks were left well or squeeze up between them as it does with even a normal human foot?

In all the tens of thousands of prints that have now been found and examined of these creatures this has never been observed to have occurred. The only thing that I know of that could prevent this—and at the same time produce such a sharp left-to-right ridge under the flexure of the toes—is an almost complete *web between all of them*. Thus, although the Neo-Giant prints at first look almost completely human they are not. They have a double first subdigital pad; they are extremely short and broad for their size; and, the second to fifth toes seem to be conjoined. The significance of these points should by now be appreciated.

This leaves us with the most abominable problem of all—the *Meh-Teh*-type prints—and abominable this is, indeed. These simply do not fit into any pattern. They are definitely not Pongid in that the big toe though enormous is not truly apposed. Similarly, they are no more Hominid in that said big toe *is* "set off." However the second toe looks for all the world like a semi-apposed digit. Such a condition is not known in any mammal. As we already pointed out, the development of such a condition is somewhat more likely to be able to be undertaken by a Pongid [with an already offset and apposed big toe] than by a Hominid without one. These prints are *not*

(*Top row, left*) ASIATIC BROWN BEAR (*Ursus arctos*). Left hind and right fore foot imprints. (*Top row, right*) NORTH AMERICAN BLACK BEAR (*Euarctos americanus*). Left hind and right fore foot imprints.

(*Center row, left*) PUMA (mountain lion) (*Profelis concolor*). Right hind foot. (Fore and hind prints of the large cats are very similar; so also are all prints of the Great Cats.) (*Center row, right*) NORTH AMERICAN POR-CUPINE (*Erethizon dorsatum*). Left hind and right fore foot imprints.

(*Bottom row, left*) THE FAMOUS FOOTPRINTS OF CARSON CITY JAIL. The outlined portion was "restored" by Dr. Harkness. This was a left foot print in sandstone. The tracks were not found in the jail yard but some miles distant. (*Bottom row, right*) EXTINCT GROUND SLOTH (*Mylodon*). Skeleton of a left foot seen from the side.

those of a string of foxes all jumping into the same hole in the snow as we by now know all too well; also, and this has perhaps not yet been sufficiently stressed, the prints show very clear indication of a rather complex musculature which, although so far unknown in any other animal, accords perfectly with what would be expected if one developed such a foot and was bipedal.

Most but by no means all of the *Meh-Teh* tracks have been found in snow. The others have been in mud and sometimes first in one and then in the other in a continuous line. This absolutely and positively disposes of another proposition that has constantly been put forward; namely, that the prints were made by a comparatively small creature and were then subsequently enlarged by melting of the snow with or without regelation. Considerable work has now been done on this phenomenon [John Napier writes me that he carried out experiments in the Juras in the winter of 1960–61. Unfortunately his findings will not be available in time for inclusion here.] It is true that small prints may become large ones in this way, and it is an extraordinary fact that the tracks seem or appear in some uncanny way to grow to fit them. By this I mean that, optical illusion or whatever, the stride seems to grow. This of course is physically impossible whereas exactly the reverse should occur because, as the prints get bigger and bigger, their peripheries must get closer together. I observed this most closely at my farm by excluding all from going near a set of tracks made by my small wife wearing close-fitting boots in fresh firm snow from the house to the trash-disposal affair beyond our lawn. In a few days her prints had grown to enormous proportions and looked incredibly sinister and as seen from an upper floor window appearing for all the world like those of a positive giant—and with a giant stride. Actual measurement however showed that the stride had of course remained as originally laid down. Nevertheless, none of these things, actual or illusory can occur in *mud*.

The *Meh-Teh* tracks and prints are, in fact, by far the most puzzling of all, and especially since such persons as Shipton,

Bordet, and others obtained clear photographs of them taken from directly above. Here is obviously a bipedal creature of considerable size and weight that inhabits the Himalayas and the ranges north of the Tibetan Plateau. It was the original "Abominable Snowman" and it comes out as the last "abominable enigma."

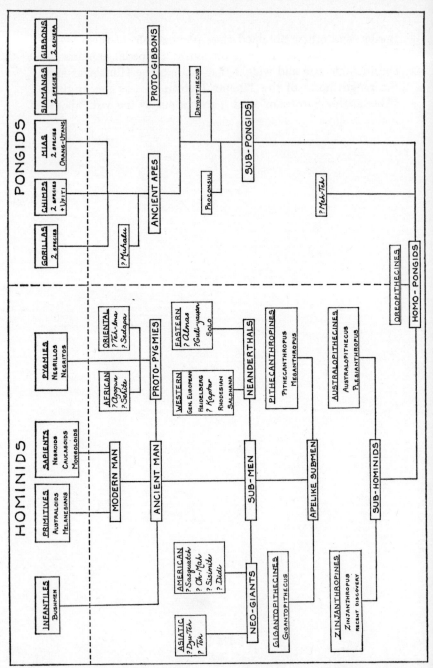

THE FAMILY TREE OF MAN AND THE PONGIDS, WITH THE ABSMs INSERTED

Appendix C
Where We Come In

ANTHROPOIDS
- **Hominids**
  - Modern Men
  - Primitive Men
  - Neanderthalers (Almas?)
  - Proto-Pigmies (Sedapas, etc.?)
  - Neo-Giants (Oh-Mahs?, etc.)
- **Pongids**
  - Meh-Tehs?
  - Gorillas
  - Mias ("Orang-utans")
  - Chimpanzees
  - Siamangs
  - Gibbons

SIMIOIDS
- **Cynopithecoids**
  - Baboons and Drills
  - Macaques and Rhesus
  - Mangabeys (of Africa)
- **Coloboids**
  - Guerezas (of Africa)
  - Langurs and Leaf-Monkeys
  - Nosey-Monkeys
- **Cercopithecoids**
  - Military Monkeys
  - Guenons
  - Swamp Guenons

PARASIMIOIDS
- **Ceboids**
  - Howler Monkeys
  - Spider-Monkeys
  - Woolly Spider-Monkeys
  - Woolly Monkeys
  - Capuchins
- **Pithecoids**
  - Squirrel-Monkeys
  - Bearded Sakis
  - Sakiwinkis
  - Uacaris
  - Douroucoulis
- **Hapaloids**
  - Titis
  - Callimico
  - Lion Marmosets
  - Pinchés
  - Bald Pinchés
  - Marikinas
  - Moustached Tamarins
  - Black Tamarins
  - Oustitis
  - Plumed Marmosets
  - Pigmy Marmosets

PROTOSIMIOIDS
- **Tarsioids**
  - Tarsiers

LEMURS
- **Lemuroids**
  - Greater Lemurs
  - Lesser Lemurs
  - Dwarf Lemurs
  - Mouse-Lemurs
  - Phaners
  - Weasel-Lemurs
  - Monkey-Lemurs
  - Dog-Lemurs (Avahi)
  - Aye-aye
- **Lorisoids**
  - Lorises
  - Pottos
  - Angwantibos
  - Galagos
  - Needle-clawed Lemurs
  - Pigmy Galagos

INSECTIVORES
- **Tupaioids**
  - Tree-Shrews
  - Feathertails
- **Macroscelids**
  - Elephant-Shrews

Appendix D
Others Involved

The known mammalian fauna of the Tibetan Plateau, the Pamirs, and the Himalayan ranges is very extensive. In the last area we find a combination of typical Eurasian and Oriental forms, together with several unique types not found elsewhere. We are concerned, in our discussion, only with mammals, in that other animals do not have hair; and it has been hair above all else [other than footprints and tracks] that have been brought forward in exposition of ABSMs; and especially in these areas. Moreover, for some reason—but one which does not seem quite logical to me—it has been the larger mammals that have been suggested as the possible origins of these hairs which have been said to be of *yetis*.

This actually need not be final for there are many medium-sized and small mammals the hairs of which are just as long, and sometimes almost as sturdy as those alleged to be from ABSMs. Some alleged ABSM remains—i.e. the scalps displayed by Sir Edmund Hillary (see Appendix E)—have been proved to have been made from the goatlike animal called the Serow; while some of the Californian hairs have been identified, on the one hand as those of a Moose [not found in the area, incidentally], and on the other as "those of a large cat" [i.e. the belly-hairs from a Puma, *Profelis concolor*]. Therefore it is of first importance to list those larger mammals known from the Tibetan and Himalayan areas that might be used to make "yeti artifacts." The most outstanding of these are:

Primates

Hoolock Gibbon (*Hylobates hooloch*)
Langurs (*Semnopithecus schistaceus, entellus,* etc.)
Snub-nosed Monkey (*Rhinopithecus roxellanae*)

Carnivora

Snow-Leopard (*Panthera uncia*)
Giant Panda (*Ailuropoda melanoleucus*)
Lesser Panda (*Ailurus fulgens*)
Wolf (*Canis lupus, subsp.*)—Eurasian form (*not* the Indian
   Wolf *Canis pallipes*)
Red Bear (*Ursus arctos, subsp.*)
Blue Bear (*Ursus arctos pruinosus*)
Snow Bear (*Ursus arctos isabellinus*)
Himalayan Black Bear (*Selenarctos thibetanus*)
Sloth-Bear (*Melursus ursinus*)

Artiodactyla

(1) Camelines (*Camelidae*)
    Bactrian Camel (*Camelus bactrianus*)
(2) Cervines (*Cervidae*)
    Red Deer—Hangul (*Cervus kashmirianus*)
    Red Deer—Shou (*Cervus wallichi affinis*)
    Thorold's Deer (*Cervus albirostris*)
(3) Bovines (*Bovinae*)
    Yak (*Bos grunniens*)
    Mithan (*Bos frontalis*)
(4) Antelopines (*Gazellinae*)
    Zeren (*Procapra gutterosa*)
    Goa (*Procapra picticaudata*)
(5) Goat-Antelopes (*Saiginae*)
    Saiga (*Saiga tatarica*)
    Chiru (*Panthalops hodgsoni*)
(6) Rock-Goats (*Rubicaprinae*)
    Goral (*Naemorhedus cinereus*)
    Serow (*Capricornis sumatrensis thar*)

(7) Sheep-Oxen (*Ovibovinae*)
   Takin (*Budorcas taxicolor, B. tibetanus,* and *B. bed-fordi*)
(8) Goats (*Caprinae*)
   Ibex (*Capra ibex, subsp.*)
(9) Sheep (*Ovinae*)
   Bharal (*Pseudovis nahura*)
   Argali (*Ovis ammon*)

Appendix E

Sir Edmund Hillary's Scalp—
A News Story from Nepal

This is, as of the time of writing, the most recent and the most outstanding story in ABSMery. Also, it has received a great play in the world press, resulting in all manner of strange and curious asides. It has also turned out to be quite a saga on several counts, with both overtones of misunderstanding and undertones of an almost farcical nature. The whole thing really proved nothing.

Sir Edmund Hillary, "Conqueror of Everest," left in 1960 for Nepal and the Chinese border, on a somewhat grandiose expedition with novel equipment and, eventually, assembled a large retinue in the field. The expressed objectives of this expedition were: first, to study human physiology under conditions of altitude, and, secondly, to gather information on what Sir Edmund now customarily calls "The Yeti." This expedition was sponsored and backed by the World Book Encyclopedia Company of the Field Enterprises Corporation of Chicago, *and* among its personnel were Mr. Marlin Perkins, Director of the Lincoln Park Zoo in Chicago (widely known to Americans for his long-lasting television series entitled "Zoo Parade" over NBC network for 10 years) and a Mr. Lawrence Swan of San Francisco State College, described by the press as a "Himalaya fauna expert."

In December, 1960, the press announced that Sir Edmund and these two gentlemen, together with the British reporter, Desmond Doig (who speaks a Sherpa dialect), and the headman of a village named Khumh-Dzhungh (called Khumjung earlier) named Kunyo Chumbi, were arriving in the United

States by air. This they did, landing in Chicago in the early part of that month. It was announced that they had brought with them a scalp (see Chapter 15), which they said its owners (the Khumh-Dzhungh-ers) stated to be that of a *yeti*. Further, the daily newspapers said that this scalp was going to be put on exhibit in Chicago for examination by a panel of scientists a few days after their arrival. The names of some of these scientists were given.

It was further announced that Sir Edmund and his associates felt they had evidence as to the origin of this "cap" and wished to submit this to scientists for consideration, and that hairs pulled from the scalp would be given to all those in the invited group who wished them and had proper laboratory facilities for examining them and comparing them with hairs of other animals. Unfortunately, there was a heavy snowstorm over much of the country which made it virtually impossible for those invited to travel. However, the object was examined by Dr. Philip Hershkovitz of the Chicago Natural History Museum, and he obtained several sets of hairs from the scalp and mailed them out along with others marked "Animal X" to those interested. Sir Edmund meanwhile appeared on television with the scalp on a revolving table, and the charming Kunyo Chumbi who did a little dance.

Asked on one of these shows (Dave Garroway's morning news show, *"Today,"* taped on the 13th, and at which I was present at the New York end) whether he thought ABSMs existed, Sir Edmund replied that he thought they had once done so, but that they had for some time been extinct. (Later, as I understand it from the press accounts, when asked how come said extinct beasts might leave physical evidence in the form of fresh footprints and so forth today, this group of investigators stated that all had been caused by or were the residue of known animals.) A further showing of the scalp was announced for the morning of the 16th, and the same scientists again invited, now that the snow was over and travel easier. However, on the morning of the 14th, the party decided to fly to Europe with the scalp, and so no American savants apart from Dr. Hershkovitz obtained sight of it. However, it was alleged that no less than 8 square inches of it were cut off here and sent to various laboratories for them to make serological (blood) comparisons! As seems to be the invariable rule in ABSMery, no results appear to have been published.

The "scalp" was then exhibited in London and Paris, where a showing was arranged at the Musée d'Histoire Naturelle, under the auspices of Prof. Millot of Coelacanth fame. Sundry specialists and "experts" examined it. Sir Edmund showed it on a television program with Bernard Heuvelmans and was very gracious in pulling some hairs from it and shaving off some pieces of skin with a penknife. In the meantime, I had received some hairs reliably reported to be from this scalp (though none of the Animal X). These I compared with a number of other hairs sent to me, and alleged to be from these Nepalese "caps." They were not all the same—not even those said to come from the Khumh-Dzhungh "scalp." However, Dr. Heuvelmans obtained hairs from various Himalayan and Tibetan animals, and immediately came to the conclusion that those he had been given from the scalp were from a goat-like animal. He further tracked down the exact kind of caprine concerned, a Serow (*Capricornis sumatrensis thar*). I was constrained to agree with him, and, to vindicate his diagnosis, *Life* Magazine came out with an article a week later in which it was said that this was the animal from which the cap had been made according to Sir Edmund Hillary. (It had been rumored that this fact was not going to be released until the 17th of February, presumably to give the scientists time to make their comparisons.)

But, during all this, something else came to light, which was that Sir Edmund was carrying with him in a suitcase a *second* cap. This he told the authorities in Paris he had had made for him earlier that year from a circle of skin from the back of a Serow he had shot. It had been molded on a wooden block just like that used by milliners, and just as we had made a similar conical cap from the scrotum of a Hereford bull in America a couple of years before. No previous mention had been made of this second (new) "scalp," and when it was, some doubts naturally arose.

If Sir Edmund was carrying such an object, that he knew

[485]

definitely to have been made from the pelt of a Serow, then presumably those from "Animal X" could be from this animal. The identification of single hairs is not easy, but was duly accomplished. However, what everybody wanted to know was whether the old cap that had been exhibited was what the Khumh-Dzhungh-ers said it was, and from which hairs had been taken previously by others, or was it, itself, an imitation, for it also could be fresh and smoked to look old. Did hairs taken from it match those taken from the original relic which he was reported to have borrowed?

Investigation into these matters brought to light these rather astonishing facts: (1) that the cap the village of Khumh-Dzhungh has always owned is said to be a "fake" in that it was made "in imitation" of one owned by their rivals—the villagers of Dhyangh-Bodzhei (Thyanboche, as it is commonly written), and (2) that *both* are said by their owners not to be scalps "*of* Yetis" but to have been made "*in imitation of* Yetis," while (3) neither were held in much regard, were not "religious relics," and were not considered particularly valuable. Yet, it was reported, Sir Edmund Hillary's expedition had been obliged to leave a sizable deposit in Nepal against the return of this object, and there was said to have been a time limit put upon this which Sir Edmund failed to make, due to the weather preventing a helicopter landing. It had been said that this relic was so valuable that the estimable Kunyo Chumbi had to travel with and keep an eye on it at all times. Yet—and this is the part that seems inexplicable—hairs were pulled from it and bits cut off, to the extent of 8 inches in the U.S., it was alleged, and handed to laboratories not only here, but also in Paris and apparently in London. This seems to me to be a strange way to handle a "priceless religious relic." Also, I would like to know, where are the results of this effort?

The whole news story was of course a fine piece of publicity, with the "cap" itself and Mr. Kunyo Chumbi being admirable background. It was also billed and primarily designed as a real chance for scientists to get a look at one of the alleged ABSM relics, but apart from identification of the hair, which coincided with the expedition's findings, what happened to the rest

[486]

of the findings? Perhaps the results of the serological comparisons are still to be published, but so far nothing has been said. And, even if such results show that the skin is also from a Serow, what will it prove about ABSMery generally? This expedition had gone off with great fanfare and in a blaze of publicity which included an article by Sir Edmund Hillary in the New York Sunday *Times* in which it was stated that, although it was a secondary objective of the enterprise, the famed explorer was going to get an answer . . . or else. However, despite the "scalp," this objective unfortunately was not attained.

The reasons were, *first,* that the group had hardly gone, before it was back; *second,* it went right through the real habitat of ABSMs—namely, the montane forests; and *third,* it is stated to have consisted of a small army that might well have scared even human beings into moving over into the next valley. Failing to obtain any truly concrete evidence that any ABSM exists, it was apparently decided that evidence of their nonexistence might be demonstrated by the old method of debunking one of its aspects.

Then, there is another odd thing. The hairs handed out by this expedition have been compared with some taken from the (allegedly) same "scalp" in Khumh-Dzhungh by Prof. Teizo Ogawa. No results have as yet been published, but it is notable that Prof. Ogawa's microphotos of the Khumh-Dzhungh cap do not agree with those of hairs taken from other caps, owned by other Nepalese villages.

To reiterate, therefore, the cap that Hillary displayed appears definitely to have been made from a Serow, but (1) was it the "real" Khumh-Dzhungh scalp seen by others or (2) a copy of it, and (3) if the only and original one, is that "genuine" or only made in imitation of other such caps? We simply do not know the answers, and probably even Mr. Doig, who speaks the language fluently, does not.

Sir Edmund later made some really astonishing statements, notably that the fresh foot tracks (which he had himself several times previously reported) were all made by a line of foxes "stringing"—i.e., all following a leader, and all jumping with all four feet exactly into the same spot as that leader, after which the snow melted, so that all the holes so made ended up with exactly the same shape and size, stretching for miles. This, of course, is a ridiculous suggestion, since the prints as photographed by Shipton clearly show the musculature of both left and right feet. It is also completely negated when said tracks are found in mud, which of course cannot melt. Finally, when pressed on these and other observations, Sir Edmund admitted that he still could not explain the weird noises said to have been heard by many people and alleged to have been made by ABSMs. (And, parenthetically, I might add that I cannot get an answer out of anybody as to just what kind of fox inhabits these upper fastnesses of the Tibetan Rim.)

Thus, this whole affair has achieved nothing, even by way of disproving the existence of the *Meh-Teh,* or the true origin of the so-called "scalps" held or alleged to be held in Nepalese monasteries and villages. It has done nothing to prove the nonexistence of ABSMs. But it has led a great body of the public to believe that the whole concept of such creatures existing has finally been debunked. At the same time, it inadvertently provided an opening for something much more vicious. This was a story issued by *Tass,* the official Soviet news agency, and reads as follows:

Wednesday, Jan. 11, 1961. END OF A MYTH. No Snow Man Had Ever Existed. The Scalp and Footprints Were a Fox's. Exposures of the Newspaper *Nazione.*

The hirsute scalp, originally alleged to belong to the Snow Man (Yeti) and brought with great pomp from Nepal to the United States and Europe, has started on its return trip to the Himalayas after being studied by experts in Chicago, Paris and London.

When placed in the experts' hands by its custodian, an elder in the village of Khumdjo Chumbi, and Edmund Hillary, the head of the Himalayan Scientific Expedition, the scalp was found to be not the main argument in proof of the existence of the Snow Man, but rather the Coffin of that myth.

Before sending the scalp back, Hillary announced that the experts' report on the subject would be published this month in Chicago. Hillary said that in general he felt that a reasonable and logical answer had at last been found to the problem of the Snow Man.

What is the answer? It was expounded to representatives of the Western press by the well-known biologist Lawrence Stone* who participated in Hillary's Himalayan Expedition. Stone said that the investigation had established that the scalp brought from the Khungdjung Monastery was a fox or goat's scalp and had never belonged to any "abominable Snow Man." Stone also spoke of the results of the study made of the Snow Man's footprints. He said that he himself had been confident that they were really the marks of the Snow Man until he was presented with clear proof that they had been formed by four foxes' footprints which had merged in the melting snow. Stone concluded by saying that the legend of the Snow Man had made an amusing story and that he regretted to have to kill it. But, he said, the time had come for the true facts to be made known and they were that no Snow Man had ever existed.

The story might end at this point—were it not for the despatch from Katmandu sent in by the Special Correspondent of the Floerentine † newspaper *Nazione*, Corrado Piccinelli and published in that paper the other day. In it he says that while Hillary was supposed to be looking for the Snow Man, he was actually spying on China. The Italian journalist makes fun of those who believe or pretend to believe that the famous New Zealand mountain climber headed that large expedition because he was concerned to find proof of the existence of the Snow Man. The Italian journalist writes:

"It isn't that at all. What is really the case is that his scientific expedition of 600 men [!] is there chiefly to draw up exact maps of that inaccessible region, due to the absence of such maps until now . . . and finally to establish the truth of the rumour that communist China has fired rockets, missiles and sputniks."

It would appear, then, that the footprints on the untrampled snows of the Himalayan peaks belongs to creatures even craftier than the fox. (*Izvestia.* In full.)

* Lawrence Swan.
† i.e. Florentine.

Bibliography

Aelianus, C. [III Century] (1832). *Aeliani de natura animalium libri XVII.* Verba ad fidem librorum manuscriptorum constituit (et annotationibus illustravit F. Jacobs. Intextae sunt curae secundae postumae J. G. Schneideri Saxonis). Adjecti sunt indices, rerum et interpretatio Latina Gesneri a Gronovio emendata. (Cum addendis et conjecturis ineditis J. J. Reiskii.) Jenae: imprensis Friderici Frommanni, 2V.

Agassiz-Harrison The Advance Newspaper (1960). "Sasquatch Watches Berry Picker Near Nelson," Oct. 20, B.C., Canada.

Ali, S., & Santapau, H. (1954). "The Abominable Snowman." *J. Bombay nat. Hist. Soc.* Bombay. 52, 594–8.

Almazov, A. (1957). "Does the snow man exist?" *Evening Moscow,* Apr. 24, p. 3.

American Karakoram Expedition, 1st, 1938 (1939). *Five Miles High.* New York (Dodd, Mead & Co.)

American Naturalist (1882). "The Recent Discoveries of Fossil Footprints in Carson, Nevada." v. 16, pp. 921–23. Phila.

Barnier, H. (1958). "Sur les traces de l'homme des neiges." *L'humanité Dimanche,* Feb. 9.

Bartels, M. (1881). "Über Menschenschwanze." *Arch. Anthrop.* vol. 13, pp. 1–41. Braunschweig.

Bartels, M. (1884). "Die Geschwanzten Menschen." *Arch. Anthrop.* vol. 15, pp. 45–132. Braunschweig.

Ibn Batuta (1829). *The Travels of Ibn Batuta.* Translated by Rev. Samuel Lee. pp. 187–8. London.

Beauman, E. B. (1937). Letter to the Editor: "The Abominable Snowman." *The Times,* July 17. London.

Bolinder, G. (1959). *We Dared the Andes.* London. (Abelard-Schuman).

Bolm, D. (1957). *Das Schnitthaarbuck.* München. (Mayer Verlag).

Bontius, J. (1658). *Historiae Naturalis et Medicae Indiae Orientalis.* Amsterdam.

Bordet, P. (1955). "Traces de yeti dans l'Himalaya." *Bulletin Museum nat. histoire natur.* vol. 27, no. 6, pp. 433–439. Paris.

Bordet, P. (1956). "Traces de yeti dans l'Himalaya." *Montagne et alpinisme*, no. 7, pp. 206–209.

Boule et Vallois (1952). *Les Hommes Fossiles*. 4th ed. Paris. (Masson).

Bourdillon, J. (1956). *Visit to the Sherpas*. London.

Bowler, L. P. (1911). *Gold Coast Palaver*. London.

Brandt, K., & Eisenhardt, H. (1939). *Fahrtenund Spurenkunde*. 6th ed. Berlin. (P. Parey)

Brasser, J. C. (1938). *Mes Chasses dans la Jungle de Sumatra*. Paris.

Britton, S. W. (1955). "Man Walks Upright Because of Snow." *Science News Letter*, Apr. 23, Washington.

Broom, R. (1950). *Finding the Missing Link*. London.

Broom, R., & Robinson, J. T. (1950a). "Man Contemporaneous with the Swartkrans Ape-Man." *Amer. J. phys. Anthrop*. vol. 8, pp. 151–5, Philadelphia.

Broom, R., & Robinson, J. T. (1950b). "Ape or Man?" *Nature*, vol. 166, pp. 843–4. London.

Broom, R. (1952). "Swartkrans Ape-man Paranthropus crassidens." *Transv. Mus*., Mem. No. 6.

Bruce, Gen. C. G. (1924). *L'Assant du Mont Everest, 1922*. Macon. (Protat Freres).

Bruller, J. (1953). *You Shall Know Them*. Translation of Les Animaux Denatures. Boston. (Little, Brown). (Pseud. "Vercors").

Bucks, I. (1959). Letter to the Editor: "The Snowman." *Man*, 337. London.

Buddhacarita (1934). (Pali Theological Treatise, a well-known Buddhist Text. Available in English translation by Rhys David.) Benares.

Burns, J. W. (1929). "Introducing B.C.'s Hairy Giants" *Maclean's Magazine*, Apr. 1, Toronto, Ontario, Canada.

Burns, J. W., & Tench, C. V. (1940). "The Hairy Giants of British Columbia." *Wide World Mag*. Jan., pp. 296–307, London.

Burton, M. (1956). "The Snowman." *The Illustrated London News*, Nov. 3. London.

Busson, B., & Leroy, G. (1955). *Les derniers secrets de la terre*. Paris.

Byrne, P. (1958a). "Frogs Lure Abominable Snowman." *New York Journal-American*, June 5.

Byrne, P. (1958b). "Searching for the Abominable Snowman." *New York Journal-American*, June 15.

Cepilli, T. (1954). "Does the man of the snows exist?" *In Defense of the World*, no. 4, pp. 92–94.

Chaigneau, A. (1954). *Empreintes et voies des animaux gibier et des Musibles*. Paris.

Chard, J. R. S. (1936). *British Animal Tracks*. London. (C. A. Pearson).

Chenu, J. C. (1850–8). *Encyclopedie d'Histoire Naturelle*. Paris.

Chevalley, G. (1953). *Avant-premières à l'Everest*. Paris.

Clark, L. (1953). *The Marching Wind*. London. (Hutchinson)

Clifford, Sir H. (1897). *In Court and Kampong*. London.

Cobb, E. H. (1952). "The Abominable Snowman." *Country Life*, III, 163. London.

Combe, G. A. (1926). *A Tibetan on Tibet*. London. (T. Fisher Unwin Ltd.).

Coon, C. S. (1954). *The Story of Man*. New York. (Alfred A. Knopf).

Cope, E. D. (1883a). "The Nevada Biped Tracks." *American Naturalist*, v. 17, pp. 69–71. Philadelphia.

Cope, E. D. (1883b). "The Carson Footprints." *American Naturalist*, v. 17, p. 1153. Philadelphia.

Ctesias (1612). *"Indica Opera"* in Photius: Myriobiblon. Geneva.

Dammerman, K. W. (1924). "De Orang Pandak van Sumatra." *Trop. Natuur*, 13, 12, 177–82. Weltevreden.

Dammerman, K. W. (1930). "The Orang Pendek or Ape Man of Sumatra." *Proc. 4th Pacific Sci. Congr.* vol. III, Biological Papers, Batavia-Bantoeng.

Dammerman, K. W. (1932). "De Nieuw-ontdekte Orang Pendek." *Trop. Natuur*, 21, 8, 123–31. Weltevreden.

Davis, H. (1940). *Land of the Eye*. New York. (Henry Holt & Co.)

Delaquys, G. (1952). *L'homme de neiges*. Paris.

Delavaille, J. H. (1862). "Les Hommes a queue." *Hist. nat.* 1st series, p. 243. Paris.

Denman, E. (1954). *Alone to Everest*. New York.

Diringer, D., D. Litt. (Univ. of Florence). *The Alphabet, A Key to the History of Mankind* with a Foreword by Sir Ellis Minns, Emeritus Professor of Archaeology, Cambridge, Second Edition, Revised. New York (Philosophical Library, n.d.)

Dollman, G. (1937). Letter to the Editor: "The Abominable Snowman." *The Times*, July 19. London.

Downing, R. (1954). "The Snowman: The Strangest Story Yet." *Daily Mail*, Jan. 7. London.

Drennan, Prof. M. R. (1953). "A new race of prehistoric men: The Salhanha skull, found with stone and bone tools, and the fossil remains of extinct South African animals." *The Illustrated London News*, Sept. 26. London.

Dubois, E. (1894). Pithecanthropus erectus: eine menschenahnliche Ubergangsform aus Java. Batavia.

Dutt, N., M.A., B.L., Ph.D., D.Litt. (1934). *The Pancavimsatisahasrika Prajnaparamita*. No. 28 in the Calcutta Oriental Series; Accepted as thesis (pt. ii) for the Degree of Doctor of Literature in the University of London. London. (Luzac).

Dyhrenfurth, G. O. (1952). *Zum dritten Pol.* München (Nymphenburger Verlagshandlung).

Dyhrenfurth, Prof. G. O. (1955). *To the Third Pole.* London. (W. Laurie)

Eickstedt, E. Freiherr von (1942). *Rassenkunde und Rassengeschichte der Menschheit.* Stuttgart. (Ferdinand Enke Verlag)

Elwes, H. J. (1915). "On the Possible Existence of a Large Ape, Unknown to Science, in Sikkim." *Proc. zool. Soc. Lond.* p. 294, London.

Enjoy, P. D. (1896). "L'Appendice Caudal dans les Tribus Moi." *Anthropologie,* 7, 531–5. Paris.

Estoppey, G. L. (1959). "Qui est l'abominable homme des neiges?" *Les Alpes,* 35, Lausanne.

Evans, C. (1956). *Kangchenjunga.* London. (Hodder & Stoughton)

Finot, L. (1928). "A propos des Moi a queue." *Bull. Amis Vieux Hue,* no. 4, pp. 217–21, Hanoi, Haiphong.

Fitter, R. (1960). "Mask provides new clue to the Snowman." *The Observer,* Oct. 23, London.

Frank, L. (1802). *Memoire sur le Commerce des Negres.* Paris.

Franklin, S. (1959). "The Sasquatch." *Weekend Magazine,* Apr. 4, v. 9, no. 14. Canada.

Frechkop, S. (1937). "Le Pied de l'Homme." *Mem. Mus. R. Hist. nat. Belg.* 13, 19, Brussels.

Frechkop, S. (1938). *Exploration du Parc National Albert, Mission G. F. de Witte (1933–35).* Brussels. (Institut des Parcs Nationaux du Congo Belge)

Friedenthal, H. (1911). *Tierhaaratlas.* Jena. (Fischer)

Fürer-Haimendorff, G. (1954). "Scalp of an Abominable Snowman?" *The Illustrated London News,* vol. 224, no. 5997, p. 477, London.

Gant, D. (1956). *Ascent to Everest.* Translated from English. Moscow.

Gatti, A. (1936). *Great Mother Forest.* London. (Hodder & Stoughton Ltd.)

Gemelli-Careri, G. F. (1745). *A Voyage Round the World* in J. Churchill: *A Collection of Voyages and Travels.* vol. 4, p. 410. London.

Gromier, E. (1948). *La Vie des Animaux sauvages de l'Afrique.* Paris. (Payot)

Gupta, M. (1958). "The Snowman: Does he exist?" *The Illustrated Weekly of India,* vol. 79, Feb. 2, pp. 10–11.

Haeckel, E. (1865). *Generalle Morphologie der Organismen.* Berlin.

Hagen, V. W. von (1940). *Jungle in the Clouds.* New York. (Duell, Sloan and Pearce)

Hardie, N. (1957). *In highest Nepal, our life among the Sherpas.* London. (G. Allen & Unwin)

Hardy, M. E. (1920). *The Geography of Plants.* Oxford. (The Clarendon Press)

Herwaarden, V. (1924). "Een Ontmoeting met een Aapmensch." *Trop. Natuur,* 13, 103–6, Weltevreden.

Heuvelmans, B. (1952a). "L'Homme des Cavernes a-t-il connu des Géants mesurant 2 a 4 Métres?" *Science et Avenir,* no. 61, Mai, Paris.

Heuvelmans, B. (1952b). "L'Homme des Cavernes a-t-il connu des Géants mesurant 3 a 4 Métres?" *Science et Avenir,* no. 63, Mai, p. 207, Paris.

Heuvelmans, B. (1955). Sur la piste des bêtes ignorées. t. 1, ch. 6, Paris.

Heuvelmans, B. (1958). "Oui, l'homme-des-neiges existe." *Science et Avenir,* no. 134, Avril, p. 174, Paris.

Heuvelmans, B. (1958). *On the Track of Unknown Animals.* London. (Rupert Hart-Davis)

Hill, W. C. Osman (1945). "Nittaewo, an Unsolved Problem of Ceylon." *Loris,* 4, 251–62, Colombo.

Hillaby, I. (1957). "The Kwangsi Jaw." *Amer. J. phys. Anthrop.* 15, 281–5, Philadelphia.

Hillary, E. (1955). *High Adventure.* London. (Hodder & Stoughton)

Hillary, Sir. E. (1960). "Abominable—and Improbable?" *New York Times Magazine.*

Hooijer, D. A. (1949). "Some Notes on the Gigantopithecus question." *Amer. J. phys. Anthrop.* 7, 513–18, Philadelphia.

Hooijer, D. A. (1951). "The Geological Age of Pithecanthropus, Meganthropus and Gigantopithecus." *Amer. J. phys. Anthrop.* 9, 265–2.

Hooker, J. D. (1854). *Himalayan Journals.* v. II, pp. 14–15, London. 2V. (J. Murray)

Hoosier Folklore (1946). "Strange Beast Stories." v. 5, p. 19, March. Indianapolis. (Published for the Hoosier Folklore Society by the Indiana Historical Bureau.)

Hooton, E. A. (1937). *Apes, Men, and Morons.* New York. (G. P. Putnam's Sons)

Howard-Bury, Col. C. K. (1922). *Mount Everest, the Reconnaissance.* London.

Hunt, A. (1957). "The West Coast's Abominable Treemen." *The Globe Magazine,* July 6. (Toronto Globe and Mail)

Hunt, J. (1954). "Siege and assault." *The National Geographical Magazine,* v. 106, no. 1, pp. 1–44.

Illustrated Times (1956). "Travelling Snowman; Abominable Snowman tracks." Jan. 23, London.

Izzard, R. (1954). *The Innocent on Everest.* Mount Everest expedition, 1953. New York. (E. P. Dutton & Co. Inc.)

Izzard, R. (1955). *The Abominable Snowman Adventure.* Account of the Daily Mail Himalayan expedition, 1954. London.

Jackson, J. A. (1955). *More Than Mountains.* London. (George G. Harrap & Co. Ltd.)

Jacobson, E. (1917). "Rimboeleven in Sumatra." *Trop. Natuur,* 6, 69. Weltevreden.

Jacobson, E. (1918). "Nog eens de Orang pandak." *Trop. Natuur,* 7, 173. Weltevreden.

Jaschke, H. A. (1889). *Tibetan-English Dictionary.* London.

Jaschke, H. A. (1954). *Tibetan Grammar.* New York. (Frederick Ungar)

Jeimin Jibao (1959). "The mystery of the snow man will be uncovered." An interview by the correspondent of the newspaper Jeimin Jibao with Prof. B. Porshnev (in Chinese). Apr. 18.

Kalmpffert, W. (1956). "Abominable Snowman is a bear?" *Science Digest,* no. 4.

Kaulback, R. (1937a). "20 Months in Tibet." *The Times,* July 12, London.

Kaulback, R. (1937b). "The Mountain Men." *The Times,* July 13, London.

Keane, A. H. (1920). *Man Past and Present.* Cambridge. (The University Press)

Keel, J. A. (1958). *Jadoo.* London. (W. H. Allen)

Keith, Sir A. (1929). *The Antiquity of Man.* vols. I & II. London. (Williams and Norgate, Ltd.)

Kisliakow, N. (1959). "Snow man? No, a legend!" *Literature and Life,* Aug. 23.

Kleinenberg, E., & Porshnev, B. (1959). "Studies in the Soviet Union in connection with the snow man. *Huanmin Jibao,*" Feb. 11 (in Chinese).

Koenigswald, G. H. R. von (1935). "Eine fossile Saugetierfauna mit Simia aus Sudchina." *Proc. Acad. Sci. Amst.* 38, (8), pp. 872–9. Amsterdam.

Koenigswald, G. H. R. von (1947). "Search for Early Man." *Nat. Hist. N.Y.* 56, I. New York.

Koenigswald, G. H. R. von (1956). *Meeting Prehistoric Man.* New York. (Harper & Brothers)

Komsomol Truth (1959). "Facts and assumptions." An interview with A. Pronin and D. Evgeniev. Jan. 19.

Korolev, A. (1957). *Beyond the Border of Asia.* pp. 70–71. Moscow.

Korovnikov, V. (1959). "The Mystery Snow Man." *Soviet Russia,* Feb. 25.

Lal, R. N. (1947). *The Student's Practical Dictionary.* Containing Hindustani Words with English Meanings in Persian Character. Eleventh Edition, Revised. Allahabad, India.

Lambert, R. S. *Exploring the Supernatural:* The Weird in Canadian Folklore. Ch. 10, pp. 167–180. London. (Arthur Barker, Ltd., n.d.)

Lane, F. W. (1955). *Nature Parade.* 4th ed. revised. London.

Le Bret (1885). "Notice sur l'Existence d'une Race d'Hommes a Queue." *Arch. gen. Med.* February, Paris.

Le Conte, J. (1883). "Carson Footprints." *Nature,* vol. 28, London.

Le Double, A. F., & Houssay, F. (1912). *Les Velus.* Paris.

Le Gros Clark, W. E., F.R.S. (1950). *History of The Primates.* London. (British Museum [Natural History])

Leroi-Gourhan, Prof. A. (1952). "Among the ten most ancient human remains yet discovered: early Mousterian jawbones found at Arcy-sur-Cure on a unique site continuously occupied for 140,000 years." *The Illustrated London News,* Nov. 29. London.

Letavet, A. (1958). "Was it a Snowman?" *Moscow News,* Jan. 22.

Lewis, F. (1919). "Notes on Animal and Plant Life in the Vedda Country." *Spolia zeylan.* 10, 119–65, Colombo.

Ley, W. (1955). *Salamanders and Other Wonders.* Ch. 4: The Abominable Snowmen. New York.

Lochte, T. (1938). *Atlas der menschlichen und tierischen Haare.* Göttingen, Leipzig. (Schoeps)

Lonnberg, E. (1932). "Ett avalojat swindelforsok meddenfelan de lanken." *Fauna och Flora,* 249–54, Uppsala.

Lyell, D. D. (1929). *The Hunting and Spoor of Central African Game.* London. (Seeley, Service & Co.)

McGovern, W. M. (1924). *To Lhasa in Disguise.* London. (Thornton Butterworth)

McGovern, W. M. (1924). *To Lhasa in Disguise.* New York. (The Century Co.)

MacIntyre, Maj.-Gen. D. (1889). *Hindu-Koh.* Edinburgh. (W. Blackwood & Sons)

MacIntyre, N. (1936). *Attack on Everest.* London. (Methuen & Co.)

Maier, R. (1923). "De Orang pandak of Orang pendek." *Trop. Natuur,* 12, 154. Weltevreden.

Maitre, H. (1912). *Les Jungles Moi, Exploration et Histoire des Hinterlands Moi du Cambodge, de la Cochin-chine, de l'Annam et du Bas-Laos.* Paris.

Mariagin, G. (1958a). "Almas—the snow man?" An interview with Prof. Rinchen. *Literature and Life,* Nov. 12.

Mariagin, G. (1958b). "Almas—the snow man?" Reply from B. Porshnev. *Literature and Life,* Nov. 16.

Marques-Riviere, J. (1937). *L'Inde Secrete et sa Magie.* Paris.

Marsh, O. C. (1883). "On the supposed human foot-prints recently found in Nevada." *American Journal of Science,* Ser. 3, vol. 26, New Haven, Connecticut.

Mason, G. F. (1943). *Animal Tracks.* New York. (William Morrow & Co.)

Masters, J. (1959). "The Abominable Snowman." Harper's Magazine, January.

Mayet, Dr. B. (1942). *Recherche de l'espece animale par l'examen des poils des fourrures.* Lyon. (Bosc Freres, M. et L. Riou)

Megasthenes (1846). *Indica, fragmenta* in E. A. Schwanbeck: Megasthenis Indica. Greek text. Bonn.

Mendis, W. (1945). "Where the Pygmies of Ceylon Lived." *Loris,* December, p. 262, Colombo.

Merejinsky, U. (1959). "From the ancient manuscripts." *Evening Kiev,* Oct. 23 (in Ukrainian).

Merrick, H. (1954). "Abominable Snowman." *New Statesman and Nation,* v. 27, no. 6, p. 152. London.

Moore, G., M.D. (1957). "I Met the Abominable Snowman." *Sports Afield,* May, New York.

Morin, M. (1953). *Everest du premier assaut à la victoire.* Grenoble.

Morland-Hughes, W. R. J. (1947). *Grammar of the Nepali Languages in the Roman and Nagri Scripts.* London. (Luzac)

Moscow Komsomolets (1958). "Snow man—Who is he?" An interview with the science editor Prof. of Anthropology, A. Shmakov. Jan. 26.

Murie, O. J. (1954). *A Field Guide to Animal Tracks.* Boston. (Houghton Mifflin)

Murray, W. (1953). *The Story of Everest.* New York.

Murray, W. (1953). "Is there an Abominable Snowman?" *Scots Magazine,* May, pp. 91–97.

Mus. J. Phila. (1915). "Guatemala Myths." vol. 6, no. 3, pp. 103–115.

Nebesky-Wojkowitz, R. von (1956). *Where the Gods Are Mountains.* London. (Weidenfield & Nicolson)

Neff, E. D. (1948). "Nature's Model-T Man." *Nature,* December. London.

Nevada, a guide to the Silver State (1940). Writers program. American guide series. p. 16.

Nevill, H. (1886). "The Nittaewo of Ceylon." *Taprobanian,* February, pp. 66–8, Bombay.

Newman, H. (1937). *Indian Peepshow.* London. (G. Bell & Sons)

Noel, Capt. J. (1927). *The Story of Everest.* Boston. (Little, Brown & Co.)

Northern Neighbors (1959). "Abominable Snowman again." November, Ontario, Canada.

Noyce, W. (1954). "The Ascent of Everest." *Asian Rev.,* April, p. 128, London.

Oakley, Dr. K. P., & Hoskins, Dr. C. R. (1950). "New Evidence on the Antiquity of Piltdown Man." *Nature,* Mar. 11, pp. 379–382. London.

Obruchev, S. (1957). "The footprints of the snow man in the Himalayas." *News of the Geographical Society,* v. 87, edition 1, pp. 71–73.

Obruchev, S. (1957b). "New materials about the snow man yeti." *News of the Geographical Society,* v. 89, edition 4, pp. 339–342.

Obruchev, S. (1959). "Present day evaluation of the snow man." *Nature,* no. 10, London.

Ostman, A. (1957). "I Was Kidnapped by a Sasquatch." *Agassiz-Harrison The Advance,* Aug. 22, British Columbia, Canada.

Pai-HSin (1958). "Information of a Chinese moving picture director Pai-HSin, how he had seen the snow man." *Beiden Jibao,* Jan. 29 (in Chinese).

Palladius, Bishop (1665). *Palladii de Gentibus Indiae et Bragmanibus.* Latin & Greek texts. London.

Parker, W. (1960). "Europe: How Far?" *The Geographical Journal,* vol. cxxvi, part 3, London.

Passarelle, C. (1958). "Alerte au yeti (vers la certitude)." *Science et vie,* April, no. 487, pp. 86–92.

Peissel, M. (1960). "The Abominable Snow Job." *Argosy Magazine,* December.

Pei Ven-Jun, U Ju-Kand, Gow Min-Gen (1958). "The riddle of the snow man." *Huanmin Jibao,* Feb. 27 (in Chinese).

Pei, Wen-Chung (1956). "New Material on Man's Origins." *China Reconstructs,* August, Peking.

Pei, Wen-Chung (1957). "Find Man-Like Ape." *Science News Letter,* Oct. 5, Washington.

Petzsch, H. (1958a). "Der Schneemensch ist kein Hirngespinst." *Liberal Demokratische Zeitung,* Jan. 23.

Petzsch, H. (1958b). "Nochmals der problematische Schneemensch." *Liberal Demokratische Zeitung,* Feb. 15.

Pianikov, P. (1959). "It is not a legend." *Komsomol Truth,* Feb. 1.

Pijl, L. van der (1938). "De Orang Pendek als Sneeuwman." *Tropische Nat.* 27, 53, Weltevreden.

Pliny the Elder (1601). *The Historie of the World.* Translated by Philemon Holland. vol. I, p. 157, London.

Poliakov, A. (1957). "Only footprints and legends." *Physical Culture and Sport,* no. 4, pp. 20–22.

Polo, M. (1818). *The Travels of Marco Polo, a Venetian, in the Thirteenth Century, being a Description by that Early Traveller of Remarkable Places and Things in the Eastern Parts of the World.* Translated with notes by William Marsden. London.

Poppe, N. (1954). *Grammar of Written Mongolian.* Studies on Asia, Far Eastern and Russian Institute. University of Washington at Seattle/

Porta Linguarum Orientalium Wiesbaden (Otto Harrasowitz [WGB GmbH])

Pora, E. (1959). "Poate exista yéti—omul zapexilor?" *Natura*, v. 2, no. 2.

Porshnev, B. (1958a). "Does the snow man exist?" *Evening Moscow*, Nov. 13.

Porshnev, B. (1958b). "Does the snow man exist?" *Evening Moscow*, Nov. 29.

Porshnev, B. (1958c). "Legends? But they may be the real thing." *Komsomol Truth*, June 11.

Porshnev, B. (1959a). "Search for a snow man." *Komsomol Truth*, Feb. 15.

Porshnev, B. (1959b). "From the pages of ancient manuscripts." *Komsomol Truth*, June 7.

Porshnev, B. (1959c). "The riddle of the snow man." *Contemporary East*, no. 8, pp. 56–57.

Porshnev, B. (1959d). "The riddle of the snow man." *Contemporary East*, no. 9, pp. 52–55.

Pranavananda, S. (1950). *Exploration in Tibet*. Calcutta.

Pranavananda, S. (1955). "Abominable Snowman." *The Indian Geographical Journal*, 30, 3, 99–104.

Pranavananda, S. (1955a). "The Abominable Snowman." *J. Bombay nat. Hist. Soc.* 54, 358–64, Bombay.

Pranavananda, S. (1955b). "Footprints of Snowman." *J. Bombay nat. Hist. Soc.* pp. 448–50, Bombay.

Prjevalsky, N. (1876). *Mongolia, the Tangut country, and the solitudes of Northern Tibet*. London. (S. Loro)

Pronin, A. (1958). "An encounter with a snow man." *Komsomol Truth*, Jan. 15.

Publications (4 booklets) (1958–9). The Special Commission formed to study the problem of the Snowman by the U.S.S.R. Academy of Sciences. Compiled under the supervision of Prof. B. F. Porshnev and Dr. A. A. Shmakov. Moscow (in Russian).

Quatrefages de Breau, A. de (1887). *Les Pygmees*. Paris.

Rahul, R. (1960). "Stories Increased Search for Yeti." *The Christian Science Monitor*, Dec. 17.

Ramstedt, G. J. (1902). *Bergtscheremissische Sprachstudien*. Helsinki. (Societe Finno-Ougrienne)

Ratsek, V. (1959). "According to me . . . no more of him." *Physical Culture and Sport*, no. 4, pp. 20–23.

Rendall, E. B. (circa 1957). *Healing Waters*.

Reuben, D. E. (1955). "The Abominable Snowman." *J. Bombay nat. Hist. Soc.* 54, 3, Bombay.

Rinchen, Prof. (1958). "Almas—Mongolian relative of the snow man." *Contemporary Mongolia*, no. 5, pp. 34–38.

Rockhill, W. W. (1891). *The Land of the Lamas.* London. (Longmans, Green Co.)

Roerich, N. *Altai Himalaya,* a travel diary. London. (Jarrolds, n.d.)

Roosevelt, T. (1893). *The Wilderness Hunter.* New York. (G. P. Putnam's Son)

Rosenfeld, A. (1959). "About some dated (out dated) ancient beliefs of the people of Pamir, in connection with the legend of the snow man." *Soviet Ethnography,* no. 4.

Rosenfeld, M. (1931). *Across Mongolia in an automobile.* Moscow.

Russel, Dr. S. (1946). *Mountain Prospect.* London. (Chatto & Windus)

Ruttledge, H. (1934). *Everest 1933.* London. (Hodder & Stoughton)

Ruttledge, H. (1937). *Everest: The Unfinished Adventure.* London. (Hodder & Stoughton Ltd.)

Sanderson, I. T. (1950). "The Abominable Snowman." *True Magazine,* May, New York.

Sanderson, I. T. (1955). *How to Know the American Mammals.* New York. (Hanover House)

Sanderson, I. T. (1957). *The Monkey Kingdom.* New York. (Hanover House)

Sanderson, I. T. (1959). "The Strange Story of America's Abominable Snowman." *True Magazine,* December, New York.

Sanderson, I. T. (1960). "A New Look at America's Mystery Giant." *True Magazine,* March, New York.

Science (1956). "Abominable Snowman." 123, 1024.

Science (1956). "Again the Abominable Snowman." 126, 3210, 22, July 6, Washington.

Science News Letter (1956). "Abominable Snowman." 69, 26, 405, June 30, Toronto.

Schaefer, E. (1951). *Grandes Chasses sur le Toit du Monde.* Paris.

Schiltberger, J. (1879). *The Bondage and Travels of J. S. in Europe, Asia, Africa, 1396–1427.* London. (Hakluyt Society Works, v. 58)

Schlegel, H., & Muller, S. (1839–44). Bijdragen tot de Natuurlijke Historie van den Orang-oetan (Simia satyrus) (in Verhandelingen over de Natuurlijke Geschiedenis der Nederlandsche Oederzeesche Bezittingen, door de Leden der Natuurkundige Commissie in Indie en Andere Schrijvers). p. 12, Leiden.

Seton, E. T. (1958). *Animal Tracks and Hunter Signs.* New York. (Doubleday)

Shalimov, A. (1957). "The snow man in Pamir?" *Around the World,* no. 7, pp. 29–32.

Shipton, E. (1938). *Blank on the map.* pp. 202–3, (Ch. by H. V. Tilman) London. (Hodder & Stoughton)

Shipton, E. (1951). "A Mystery of Everest: Footprints of the Abominable Snowman." *The Times*, Dec. 6, London.

Shipton, E. (1952). *The Mount Everest Reconnaissance Expedition, 1951.* London. (Hodder & Stoughton)

Shipton, E. (1952). "Everest: 1951 Reconnaissance of the Southern Route." *The Geographical Journal*, June, London.

Shipton, E. (1953). "The expedition to Cho Oyn." *The Geographical Journal*, 119, 2, 126–133, London.

Shipton, E. (1955–56). *Men against Everest.* Englewood Cliffs, N.J. (Prentice Hall)

Shipton, E. (1956). "Fact or fantasy." *The Geographical Journal*, 122, 3, 370–72, London.

Simonov, E. (1957). "Snow man—myth or reality?" Literaturnaya Gazeta (Literary Newspaper). Notes by S. Obruchev, A. Alexandrov, A. Litovet and N. Sirotkin.

Simonov, E. K. (1959). "To the highest elevation on the globe." *Geography in the school*, no. 1, pp. 29–38.

Simonov, J. (1959). "Po stopach snĕžnéo muže." *Veda a technicka mladeze*, no. 10, v. 2, pp. 12–15.

Simpkins, F. J. (1952). "Himalayan footprints." *The Field*, July 12, London.

Skeat, W. W., & Blagden, C. D. (1906). *Pagan Races of the Malay Peninsula.* vol. II, pp. 281–3, London.

Smythe, F. S. (1930). *The Kangchenjunga Adventure.* London.

Smythe, F. S. (1937). "Abominable Snowman." *The Illustrated London News,* Nov. 13, p. 848, London.

Smythe, F. S. (1938). *The Valley of Flowers.* London.

Snaith, S. (1937). *At Grips with Everest.* London. (The Percy Press)

Solecki, Prof. R. S. (1960). "Neanderthal Men in the Heart-Land of Human Evolution: Discoveries at Shanidar Cave, in Northern Iraq." *The Illustrated London News,* May 7, p. 772, London.

Soule, G. (1952). "The World's Most Mysterious Footprints." *Popular Science,* December, p. 133.

Speakman, F. J. (1954). *Tracks, trails and signs.* London.

Spittel, R. L. (1933). *Far-off Things.* pp. 61–74, Colombo.

Spittel, R. L. (1936). "Leanama, Land of the Nittaewo." *Loris,* 1, 37–46, Colombo.

Stansfeld, Lt.-Col. H. S. (1952). "Himalayan footprints." *The Field,* Aug. 2, p. 194, London.

Stanukowitch, K. (1957). "Golub-Jawan." *News of the Geographical Society,* v. 89, pp. 343–345, edition 4.

Staudinger, P. (1932). Einige Angaben über den Orang Pendek oder Orang Letjo. S. B. Ges. naturf. Fr. Berlin.

Stonor, C. (1955). *The Sherpa and the Snowman*. London. (Hollis & Carter)

Stonor, C. (1958). *Sherpas and the Snow Man*. Moscow.

Straus, W. (1956). "Abominable Snowman." *Science*, 123, 3206, 1024–1025, June 8, Washington.

Struys, J. (1791). *Voyage en Tartarie, Perse, Inde*. vol. 1, pp. 88, 100, Rouen.

Stuart, J. (1959). "Canada's Abominable Snowman." *Argosy Magazine*, December.

Technika for Youth (1959a). "Who is he?" no. 4, pp. 31–34.

Technika for Youth (1959b). "Who is he?" no. 5, pp. 37–39.

Teilhard de Chardin, P. (1947). "Sur une Manibule de Meganthropus." *C. R. Soc. geol. Fr.* 15, 309–10, Paris.

Tenzing, N. (with Ullman, J. R.) (1955). *Tiger of the Snows*. New York. (G. P. Putnam)

Teuwsen, E. (1936). *Fährten und Spuren*. Neudamm. (J. Neumann)

Tilman, H. W. (1937). *Snow on the Equator*. London. (G. Bell & Sons Ltd.)

Tilman, H. W. (1938). "The Mount Everest Expedition of 1938." *The Geographical Journal*, v. 92, pp. 481–498, London.

Tilman, H. W. (1946). *When Men and Mountains Meet*. London.

Tilman, H. W. (1948). *Mount Everest, 1938*. Cambridge. (University Press)

Tilman, H. W. (1949). *Two Mountains and A River*. Cambridge. (University Press)

Tilman, H. (1952). *Nepal Himalaya*. Cambridge. (Univ. Press)

Tombazi, A. (1925). *Account of a photographic expedition to the southern glaciers of Kangchenjunga in the Sikkim Himalaya*. Bombay.

Topley, Mrs. M. (1960). Letter, *Man*, 60, no. 33, February, London.

Tschernezky, W. (1954). "Nature of the Abominable Snowman." *Manchester Guardian*, Feb. 20.

Tschernezky, W. (1958). "On the nature of the Snow Man." Added to the book of R. Izzard, *Following the Footprints of the Snow Man*. Moscow.

Tschernezky, W. (1960). "A reconstruction of the foot of the Abominable Snowman." *Nature*, 186, 4723, 496–7, May 7, London.

Tulpius, N. (1641). *Observationum Medicarum, Libri Tres*. Amsterdam.

Turner, S. (1800). *An Account of an Embassy to the Court of the Teshoo Lama, in Tibet*. London.

Tyson, E. (1699). *Orang-utang, sive Homo Sylvestris, or the Anatomy of a Pygmie Compared with that of a Monkey, an Ape, and a Man*. London.

Udin, T. (1959). "Five expeditions to the Himalayas." *Evening Moscow*, Feb. 7.

Ullman, J. R. (1947). *Kingdom of Adventure: Everest*. New York. (W. Sloane Assoc., Inc.)

[503]

Urbain, A., & Dekeyser, P. L. (1953). "Le Gorille." *Naturalia,* no. 3, December, Paris.

Urysson, M. (1958). *Das Geheimnis des Yeti.* Freie Welt, Heft 17, Apr. 24, s. 14–17.

Usinger, A. (1954). *Fahrten, Spuren und Gelaufe.* München.

Vandivert, W. & R. (1957). *Common Wild Animals and their Young.* New York. (Dell)

Vercors (see Bruller, J.)

Verma, B. K. (1959). "Man's Nearest Living Cousin: The Yeti." *The Illustrated Weekly of India,* Sept. 27.

Virchow, R. (1885a). "De la Queue chez l'Homme." *Zbl. Chir.* p. 539, Leipzig.

Virchow, R. (1885b). "L'Appendice Caudal chez l'Homme." *Gaz. hebd. Med. Chir.* p. 29, Paris.

Vlcek, E., M.D. (1958). "Co vime o sněžnem muži?" Ziva Roenik VI (XLIV), Brezen, no. 2.

Vlcek, E., M.D. (1959). "Old literary evidence for the existence of the 'snow man' in Tibet and Mongolia." *Man,* v. 50, August, London.

Waddell, Maj. L. A. (1899). *Among the Himalayas.* London. (Archibald Constable)

Waddell, Maj. L. A. (1905). *Lhasa and Its Mysteries.* p. 482, New York.

Wals, H. (1938). *La Chasse a Java.* Paris.

Weidenreich, F. (1945). "Giant Early Man from Java and South China." vol. 40, part 1, New York. (Anthropological papers of the American Museum of Natural History)

Weidenreich, F. (1946). *Apes, Giants and Man.* Chicago.

Weiner, Dr. J. S. (1955). *The Piltdown Forgery.* Oxford. (The University Press)

Weinert, H. (1946). "Über die Neuen Vorund Fruhmenschenfunde aus Afrika, Java, China und Frankreich." *Z. Morph. Anthr.* Stuttgart, xlii.

Weinert, H. (1949). "Menschenaffen oder Affenmenschen, das Unlosbare Problem der Neuen Africa-Funde." *Orion,* 4, pp. 169–74, Munich.

Weinert, H. (1951). "Zur Frage des Gigantenproblems der Summoprimaten." *Anat. Anz.,* 98, pp. 279–87, Leipzig.

Weinman, A. (1945). "Nittaewo." *Loris,* 4, no. 2, pp. 337, Colombo.

Wendt, H. (1960). *Out of Noah's Ark.* Boston. (Houghton Mifflin)

Westenek, L. C. (1918). "Orang Pandak (Boschmenschen) of Soematra." *Trop. Natuur.,* 7, p. 108, Weltevreden.

Wood, W. W. (1948). "Snowman or Monkey." *Country Life,* 103, p. 1234, London.

Wormington, H. M. (1957). *Ancient Man in North America.* Denver, Colo. (The Denver Museum of Natural History)

Wynter-Blyth, M. A. (1952). "A Naturalist in the Northwest Himalaya." *J. Bombay nat. Hist. Soc.*, 50, pp. 559–72, Bombay.

Wyss-Dunant, E. (1953). "The first swiss expedition to Mount Everest." *The Geographical Journal*, September, London.

Zaborowski, S. (1897). "Les Hommes a Queue." *Bull. Soc. Anthrop.* series 4, 8, pp. 28–32, Paris.

Zerchaninov, U. (1959). "Following the footprints of Kaptar." *Komsomol Youth*, Aug. 29.

Zorza, V. (1960). "Farewell, Soviet Snowman." *Manchester Guardian Weekly*, Feb. 11.

Index

Aaron, 381
Abakan Mts., 314
Abidjan (Guinea), 187, 190
Abyssinia, 389, 406
Acheulian Man, 89
Achin (Sumatra), 214
Adam-ja-paiysy, 459
Adiopodoumé, Inst. of Ed.
    (Guinea), 187
ADVANCE, THE (Agassiz, B.C.),
    50, 75, 79
Afghanistan, 237, 298, 299, 300
Afghans, The, 301
Africa, 85, 182, 187, 209, 211,
    240, 253, 271, 297, 321, 357,
    368, 375, 386, 388, 416, 453,
    470
Africans, 32, 74, 85
Africanthropus, 353
Afropavo, 182
Agassiz, B.C., 40, 49, 66, 72, 75
Agogwe, 191, 194, 349, 357, 360,
    368, 399, 453, 476
Aharon, Yonah N. ibn, 287, 378,
    455
Ailuropoda melanoleucus, 480
Ailurus fulgens, 480
Akalché, 152
Akunakunas, The, 193
Alabama, 92, 98
Ala-Dagh Mts., 286, 296
Alai-Tagh Mts., 300, 311
Ala-Shan Desert, 23, 325
Alaska, 23, 30, 40, 41, 43, 100,
    325, 327, 339, 372, 392
Ala-Tau Mts., 23
Albany (Ore.), 140
Albast (Alboosty), 459
Albasty, 291
Aleutian Islands, 405, 406

Algeria, 197, 386
Allen, Mrs. Elizabeth (Betty),
    104, 131, 134, 135, 136
Alligator Lake (B.C.), 40
Alligators, 40
Almas, 74, 291, 308, 317, 318, 319,
    320, 326, 327, 341, 349, 357,
    360, 369, 412, 459, 476, 477
Almasty, 291, 459
Alouatta, 160
Altai (The Grand) Mts., 23, 312
Alto Planos (of the Andes), 169
Altyn Tagh Mts., 286, 311
Amazon Basin, 23, 155, 156, 167,
    169, 182, 413, 416
Ameghino, Prof. Florentino, 173
Ameranthropoides loysi, 171
American Geographical Society,
    151
Am. Mus. Nat. Hist. (N.Y.), 370
Americans, The, 306
Amerinds, 30, 32, 34, 36, 38, 42,
    44, 46, 65, 69, 70, 74, 75, 79,
    84, 85, 90, 99, 103, 113, 116,
    119, 124, 127, 136, 137, 143,
    153, 164, 173, 174, 177, 178,
    179, 232, 242, 251, 289, 303,
    327, 328, 339, 341, 343, 372,
    387, 388, 414, 415, 467, 468
Anadyr Mts., 325
Andaman Islands, 253, 377, 395
Anderson, Alexander C., 29
Andes, The, 154, 156, 165, 169,
    182, 359, 388
Angkor-Vat, 419
Angola, 182, 201, 416
Annam Mts., 238
Anteater, Giant, 114
Anthropoids, 141, 142, 206, 268,
    287, 297, 341, 477

[507]

Bear—(*Continued*)
Sun, 349, 470
Beauman, Wing-Com., E. B., 261
Béć-Bóć (*Bekk-Bok*), 244
Becci, Prof. Nello, 180
Beds (made by ABSMs), 60
Beemis, Mrs. Jess, 128
Bei-shung, 274
Belize (B.H.), 165
Bell, Sir Charles, 5
Bengal, Bay of, 253
Bengal Province, 4
Benkoelen (Sumatra), 215
Berbice River (B.G.), 180
Bering Straits, 23, 41, 90, 177, 327, 369, 372, 406
Bhang-Bodzhei (Pangboche), 334, 335
Bharal, 481
Bhutan, 237, 334
Bhyundar Valley (India), 261
Biafua Glacier, 261
Bible, The, 378, 382, 383, 391
Biet's Monkey, 239
Bigfeet, 21, 69, 120, 127, 130, 131, 142
Big Muddy River (Ill.), 95
Binturong, 242
Bireh Ganga Glacier, 260
Bishop's Cove, B.C., 35
Bison bison athabasca, 417
Bison, European, 289
Woodland, 355, 417, 447, 448
Biswas, Dr. A., 15
Black Ape, 215, 272
Blackfellows (Australoids), 336, 395
Black Sea, 286, 288, 407
Blanco, Cape, 113
Blood (of ABSMs), 224, 336
Bluff Creek (Cal.), 104, 123, 125, 132
Bodaudi (Pamirs), 313
Bodele Basin, 185
Boekit Mts. (Sumatra), 214, 218, 220
Bondande, The, 206
Bongo Antelope, 206
Bonnie (Ill.), 95
Bordet, The Abbé Pierre, 263, 271, 475

Borneo, 213, 214, 253, 396, 416
Bos frontalis, 480
Bos grunniens, 480
Bos primigenius, 239, 419
Bos sauveli, 239, 419
Bosphorus Strait, 286
Bottomlands, 83, 91, 95, 97, 98, 411
Bowler, Louis, 204
Brachytanites klossi, 213
Brahmaputra Gutter, 257, 267, 276, 301, 406, 454
Bramapithecus, 353
Brazil, 9, 154, 166, 174, 414, 415
Breazele, Bob, 131
Brelich's Monkey, 239
British Columbia, 9, 21, 23, 27, 30, 33, 40, 41, 45, 46, 49, 50, 65, 69, 76, 78, 98, 103, 113, 172, 281, 327, 341, 342, 344, 392, 471
British Guiana, 414
British Honduras, 149, 157, 163
British Museum (Nat. Hist.), 363, 434, 435
Brooks, Dr. George K., 245
Brown, Charles Barrington, 179
Brown, Kenneth C. (Cal), 157, 158, 160
Browne, G. M., 229
Budorcas, 481
Buinaksk (Dagestan ASSR), 295
Bumthang Gompa (Nepal), 261
Burgoyne, Cuthbert, 191
Burma, 237, 242, 254, 394, 416
Burns, J. W., 29, 36, 38, 46, 70, 72, 79, 453
Bushbabies, 209
Bushbuck, 189
Bushmen, 193, 194, 197, 213, 366, 367, 368, 377, 387, 397, 453, 469, 476
Byrne, Peter, 18, 263, 265, 269

Caātinga (Brazil), The, 154, 156, 166, 169, 173
California, 18, 20, 21, 30, 34, 46, 69, 100, 103, 104, 113, 115, 118, 121, 122, 124, 129, 146, 149, 333, 337, 338, 340, 342,

344, 345, 403, 417, **444**, 471, 478

California Acad. Sci., 114

Calls (of ABSMs), 55, 59, 69, 72, 78, 79, 94, 95, 107, 140, 144, 145, 157, 158, 160, 161, 174, 175, 179, 262, 265, 292, 295, 314, 315, 317, 320, 332, 342, 357, 384, 388, 444, 488

Cambodia, 237

Camels, 290, 320, 480

Camel, Bactrian, 480

Camelus bactrianus, 480

Cameroons, 193, 30**3**

Camerun, 183

Campeche (Mex.), 157

Canada, 7, 23, 34, 45, 46, 91, 99, 111, 121, 232, 339, 355, 380, 412, 417

Canadians, The, 40, 288

Canary Islands, 389, 406

Canis lupus, 480

Capilano, The S.S., 35

Capra ibex, 481

Capricornis, 279, 335

Capricornis sumatrensis thar, 335, 480, 484

Capridae, 336

Caribbean, The, 151, 154

Caribs, The Amerind, 178, 414

Caribs, The Black, 164

Carnivorous ABSMs, 161, 171, 228, 242

Carson City (Nev.), 113, 115, 473

Carvings (of ABSMs)—*see De-pictions*

Cary, Oliver, 195

Caspian Sea, 286, 288, 304, 407

Cascades, The, 25, 100, 113, 117, 338

Cassowaries, 211

Casuarius, 211

Caucasoid Race, The, 193, 284, 285, 291, 324, 367, 368, 369, 377, 384, 389, 395, 396, 468, 476

Caucasus, 19, 23, 286, 287, 288, 290, 291, 296, 298, 307, 326, 386, 392, 402

Cavally River (Guinea), 189

Cax Vinic, 157

Cebidae, 171

Ceboid Monkeys, 477

Celebes Island, 215, 253, 272, 416

Central America, 89, 151, 153, 154, 156, 157, 253, 357, 358, 388, 398, 416, 449, 453

Ceratotherium cottoni, 195, 448

Cercocebus, 249

Cercopithecoid Monkeys, 477

Cervus albirostris, 273, 480

Cervus kashmirianus, 480

Cervus wallichi affinis, 480

Ceylon, 209, 212, 237, 368, 392, 393, 416

Chaldeans, The, 276, 393

Chapman, Mr. and Mrs. George, 65, 68, 342

Chechen (U.S.S.R.), 288

Cheetah, 234

Chehalis, The, 46

Chellean Man, 63, 89

Chesh Teb (Pamirs), 318

Chetco River (Cal.), 115

Chetri, Bombahadur, 5

Chiapas (Mex.), 151, 152, 153, 157, 158

Chilliwack (B.C.), 38, 39, 50, 72

Chilpa (India), 393

Chilunka (Nepal), 333

Chimpanzees, 61, 98, 121, 139, 165, 171, 186, 197, 198, 200, 202, 270, 363, 364, 449, 468, 476, 477

China, 7, 19, 41, 63, 237, 239, 241, 243, 253, 254, 256, 299, 302, 304, 306, 311, 327, 362, 369, 370, 371, 373, 386, 396, 407, 416, 447

Chinese, The, 240, 300, 301, 304, 307, 360, 369, 371, 394, 426, 442, 445

Chinese Escarpment, 325

Chinyanjas, The, 200

Chiru, 480

Choyang Valley (Nepal), 263, 270

CHRISTIAN SCIENCE MONI-TOR, THE, 309

Christie, Robert, 51

Chumbi, Kunyo, 483, 486

Franklin, Stephen, 29
Fraser River, 23, 28, 33, 38, 43,
    49, 65, 66, 79, 290
French, The, 17
Frogs, 270, 342
Frostis, Jan, 262
Fukien (China), 256
Fukuoka Daigaku Expd., 264
Fürer-Haimendorf, Prof. C. von,
    14, 264

Gabun, 182, 183, 186
Gagan Sanctuary (U.S.S.R.), 292
Galagoides demidovii, 342
Galago, 209
Galago, Demidorff's, 342
Ganges River, 261
Garwhal (India), 261
Gatti, Mrs. Attilio, 206
Gaur (*Bos gaurus*), 419
Gelada Baboon, 272
Gent, J. R. O., 8, 260
Genzoli, Andrew, 128, 130
GEOGRAPHICAL   JOURNAL,
    THE, 286
Geophagy, 160, 266
Georgia (U.S.S.R.), 288, 289, 291
Gérésun Bamburshé, 322
Germany, 386
Ghurkhas, The, 5, 257, 393, 457
Giants, 37, 61, 70, 74, 79, 105,
    111, 113, 114, 115, 121, 132,
    137, 204, 207, 249, 254, 259,
    268, 294, 325, 327, 329, 347,
    356, 359, 379, 384, 385, 386,
    387, 411, 449
Gibbons, 47, 49, 186, 214, 218,
    219, 239, 371, 469, 476, 477,
    480
Gibbon, Hoolock, 480
Gidan Mts., 312, 325
Gigantopithecus, 20, 244, 268, 326,
    353, 370, 372, 386, 476
Gilgit, 285
Gilmore, Harry, 219
Gini, Prof. Corrado, 216
Gin-Sung, 243, 244, 254, 268, 269,
    275, 325, 358, 360, 369, 371,
    412, 443
Goa (an antelope), 480
Goa (Portuguese India), 237

Goats, 265, 336, 481
Gobi Desert, 312, 318, 319, 325,
    373
Golbo (Guli-biavan; Golub-yavan;
    Bul-biyavan;   Kul-bii-aban;
    Uli-bieban;   Yavan-adam),
    459, 468, 470, 476
Gold Coast (Ghana), 204
Golden Monkey, 239
Golub-yavan (etc.), 358, 359,
    360, 370, 372, 373
Gondwanaland, 209
Coral, 480
Gorillas, 47, 49, 61, 171, 186, 200,
    201, 202, 204, 208, 270, 297,
    303, 321, 327, 336, 373, 449,
    468, 476, 477
Gosainkund Pass (Nepal), 245
Grand Altai Mts., 318
Grant's Pass, 123
Great Barrier (Eurasia), 283, 303,
    304, 306, 309, 314, 326, 353
Great Barrier (N. Am.), 25, 42,
    83
Great Basin, The, 25, 111
Great Central Lake (Vancouver
    Is.), 35
Green, John W., 50, 75, 79
Ground-Sloths, 114, 169, 172, 173,
    473
Guadalajara (Mex.), 153
Guari Sanka Range, 262
Guatemala, 149, 151, 152, 158,
    342
Guatemala City, 156
Guayaquil (Ecuador), 166
Guayazis, The, 169
Guérés, The, 190
Guereza Monkeys, 211, 271, 477
Guiana Massif, 154, 156, 166, 169,
    177, 178, 179, 180
Guinea Massif, 183, 187, 416
Guli-biavan (Gul-biavan), 291,
    312, 313, 317, 325, 327
Gunthang (Nepal), 270
Gypsies (Asiatic), 302

Hainan, 253
Hairs (of ABSMs), 143, 224, 243,
    259, 262, 263, 332, 336, 351,
    354, 369, 478, 484, 487

Hairy Ainu, The, 324, 396
Haithon, King of Armenia, 324
Haiti, 108
Hamadryad Baboon, 272
Hamitic Peoples, 193, 377
Hands (mummified), 303, 332, 333
Hantu Sakai, The, 212
Hapaloids, 477
Happy Camp (Cal.), 123, 146
Harrison Lake (B.C.), 30, 33, 38, 50, 79
Harrison Mills (B.C.), 79
Harrison River (B.C.), 70
Hatzic (B.C.), 72
Hawes, Mrs. Adelaide, 105
Helarctos, 219, 470
Henniker, Col., 277
Hershkovitz, Dr. Philip, 484
Heuvelmans, Dr. Bernard, 15, 17, 148, 172, 175, 187, 191, 204, 217, 244, 251, 267, 268, 297, 340, 346, 370, 388, 443, 469, 485
Hichens, Capt. William, 190
Hicks, Green, 40
Hill, Dr. W. C. Osman, 15, 217, 335, 469
Hillary, Sir Edmund, 21, 262, 263, 279, 309, 329, 334, 335, 342, 444, 478, 483, 486, 488
Himalayans, 343
Himalayas, 1, 7, 8, 9, 12, 15, 18, 20, 21, 23, 207, 212, 237, 238, 239, 244, 250, 254, 255, 256, 257, 258, 259, 265, 267, 275, 276, 280, 284, 285, 286, 301, 304, 307, 317, 325, 327, 337, 342, 344, 357, 358, 359, 360, 373, 374, 380, 393, 401, 406, 411, 439, 441, 444, 454, 455, 478, 484, 485
Hindu Faith, 258
Hindu-Kush Mts., 286, 289, 296
Hispanopithecus, 353
Hoatzin, 177
Hokkaido, 324, 396
Hollanders, The, 17, 217, 426
Holy Cross Mountains (B.C.), 40
Hominids, 20, 39, 60, 62, 88, 89, 141, 142, 143, 166, 171, 177,
185, 186, 202, 204, 208, 211, 212, 214, 217, 221, 228, 229, 240, 275, 327, 334, 341, 343, 348, 353, 365, 366, 370, 371, 373, 374, 378, 380, 383, 449, 466, 467, 473, 476, 477
Homo heidelbergensis, 353, 386, 476
Homo nocturnus, 399
Homo rhodesiensis, 353, 386
Homo saldhanensis, 353, 476
Homo sapiens, 353, 361, 365, 366, 378, 384, 387, 447
Homo soloensis, 353
Honduras, Bight of, 151, 163
Hong Kong, 370
Hook-Worms, 340
Hoolock Gibbon, 239, 480
Hoopa Valley (Cal.), 103, 123, 124, 133
Hope (B.C.), 39
Hoquiam, The, 46
Horba (Tibet), 323
Hottentots, The, 366
Howard-Bury, Lt. Col. C. K., 10, 431, 433, 435, 560
Howler Monkeys, 160
Hudson's Bay Company, 29
Humboldt County (Cal.), 124, 126
HUMBOLDT TIMES, THE, 104, 128, 131
Hun-guressu (Khun-goroos; Khum-chin-gorgosu), 459
Hunt, Sir John, 261
Hunt, Stanley, 79
Hüppas, The, 30, 47, 120
Huxley, Dr. Julian, 277
Hyaenas, 173
Hylobates hooloch, 239, 480
Hylobates moloch, 218
Hylopithecus, 353

I.G.Y., The 1956–59, 362, 411
Ibex, 481
Iceland, 406
Ichnology, 2, 98, 337, 346, 463
Idaho, 21, 30, 34, 104
Illinois, 92, 95

Khakhlov, Prof. V. A., 9, 19, 118, 307, 308, 313, 317
Khalkhi (Mongolia), 320
Kham (Tibet), 333
Khangai Mts., 23, 312, 318
Khani, The, 241
Khingan Mts., 23, 312, 325
Khmers, The, 419
Khumh-Dzhungh (Khumjung), 335, 483, 486, 487
Khumjung (Nepal), 335, 342, 483
Kian-su (China), 237
Kimathi, Dedan, 233
King, Mike, 34
Kirghiz, The, 291, 299, 300, 301, 314, 468
Kish-kiik, 324, 460
Klamath River (Cal.), 100, 103
Klamaths, The, 113, 119, 136, 146, 417
Knight, Hugh, 260, 441
Koenigswald, Dr. R. von, 370
Kontum (Laos), 244
Korbel (Cal.), 126
Korea, 390, 394
Kouprey (*Bos sauveli*), 239, 419, 448
Kra-Dhan, 244, 250, 454
Kraken, 279
Krasnodar (U.S.S.R.), 288
Ksy-giik, 74, 317, 321, 325, 349, 357, 360, 369, 460
Kuala Lumpur (Malaya), 231
Kubus, The, 216, 221
Kuku-khoto (Mongolia), 320
Kumaon, 393
Kung-Lu, 241, 243, 254, 268, 358, 371, 454
Kunlun Mts., 23, 286, 311, 317, 358, 373, 459
Kunming (China), 241
Kurds, The, 232
Kurriwe (Nyasaland), 200

Ladakh, 277, 457
LA GAZETA (of Santiago, Chile), 388
Lakpa Tensing, 262
Lamaist Faith, 258, 303, 457
Lamas, 262, 263, 278

Lamin-Gegen Monastery, 319
Langur Monkeys, 8, 211, 226, 240, 249, 250, 253, 271, 275, 434, 435, 477, 480
Langur, Mentawi Island, 213
Laos, 237, 394
La Plata Basin, The, 169
La Salta Prov. (Argentina), 388
Laue, Fred, 471
Leaf-Monkeys, 211, 225, 240, 271, 477
Leaf-Monkey, Banded, 226
Leakey, Dr. L. S. B., 63
Ledoux, Prof. A., 187
Lemuroids, 466, 477
Lena River, 327
Leningrad University, 309
Leontiev, Prof. V. K., 290, 291
Leopards, 205, 289, 292
Lepchas, The, 257, 265, 457
Lhakpa-La Pass, 10
Lhasa (Tibet), 4
Liard River (Canada), 42
Lichens, 266
Liddarwat (Kashmir), 262
LIFE MAGAZINE, 413, 485
Limnopithecus, 353
Limpasa Bridge (Nyasaland), 195, 199,
Linnaeus (Carl von Linné), 399
Little Khingan Mts., 325
Little People, 139, 386
Little Red Men, 93, 96, 188, 191, 192, 387
Lizards, 41
Loch Ness Monster, 97, 120
Longenot, Mt. (Kenya), 192
Loobuk Salasik (Sumatra), 217
Lorises (*Loris*), 209
Lorisoids, 209, 477
LOS ANGELES EXAMINER, THE, 122
Lotus, Flowering, 423
Louisiana, 92
Loys, Dr. François de, 169
Luli (Asiatic Gypsies), 302
Lu Ming-Yang, Lt., 323
Lung-Gompa, 276
Lutong Monkeys, 225, 226, 250, 271
Lynx, 289

Lyssodes, 244, 272, 275, 359
Lyssodes (Macaca) thibetanus, 273
Lytton (B.C.), 25, 28

Macaque Monkeys, 215, 240, 244, 272, 274, 275, 336, 466, 477
Macaques, Moor, 215
  Stump-tailed, 244, 272, 275
Mack, Oscar, 103
Mackenzie Range, 30
Mackenzie River, 42
Macroscelids, 477
Madagascar, 185, 394, 416
Mad River (Cal.), 121, 126
Magyars [Hungarians], 284, 285, 389
Maine, 417
Makah Indian Reservation, 143
Makalu (Nepal), 263, 333
Malaya, 211, 212, 214, 237, 253, 254, 256, 342, 369, 393, 395, 416
Malayan Forestry Service, 272
Malayan Security Guard, 230
Malay Peninsula, 23, 211, 213, 214, 226, 227, 229, 253, 297, 377, 396, 399
Malays, The, 32, 74, 212, 216, 219, 224, 227, 343, 360, 394, 395
Mammoth Cave (Ky.), 37
Man, 20, 58, 61, 74, 77, 88, 89, 90, 105, 141, 142, 159, 160, 185, 186, 194, 211, 212, 215, 227, 270, 275, 285, 296, 308, 336, 339, 340, 314, 355, 361, 365, 366, 368, 370, 378, 389, 390, 396, 410, 427, 435, 466
Manass [Dam] River, 315
Manchuria, 7, 304, 312, 325, 327, 328, 344, 372, 386, 471
Mandrill, 272
Mangabey Monkeys, 249, 271, 343, 477
Mangu Khans, The, 324
Manitoba, 43
Mapinguary, 174, 343, 348, 358, 370
Marapi, Mt., 214
Marco Polo, 7, 215, 225, 306

Marsden, William, 215
Marsh, Prof. O. C., 114
Marston, Dr. Alvan, 364
Masks, 317
Mastodons, 42
Matto Grosso, The, 154, 156, 166, 169, 174, 177, 386
Mau Forest (Kenya), 192
Mau-Mau, The, 233
Mauretania, 391
Maurus, 215
Maya Mts. (Brit. Honduras), 157, 163, 165
Mayas, The, 31, 152, 158, 163, 342, 357, 359
Mazaruni River (B.G.), 179
Mazatlán (Mex.), 153
McGovern, W. M., 322
Meacham, James, 93
Mediterranean, The, 365, 407
Meganthropus, 214, 353, 371, 386, 476
Megatherium, 114, 172
Meh-Teh, 74, 120, 121, 207, 248, 254, 267, 268, 271, 275, 278, 279, 294, 317, 325, 326, 327, 329, 333, 336, 337, 338, 340, 341, 346, 348, 356, 358, 359, 360, 361, 369, 370, 372, 373, 374, 385, 394, 411, 449, 459, 467, 468, 470, 473, 474, 476
Melanesians (Oceanic Negros), 377, 395, 476
Melursus, 250, 275, 334, 480
Menlung Tsu Glacier, 261, 262
Mentawi Islands, 186, 211, 213, 214
Metasequoia, 273, 371
Metoh-Kangmi, 10, 460
Mexicans, The, 343
Mexico, 42, 83, 111, 131, 147, 149, 151, 157, 158, 283, 342, 405
Mias [Orang-utans], 74, 94, 186, 213, 449, 476, 477
Mica Mountain (B.C.), 50, 76
Micronesians, 394
Middle East, 390, 391, 392
Mi-ge, 461
Mi-go, 265, 461
Mi-gu, 461

Papuans, The, 211
Paramaribo (Surinam), 415
Parasimioids, 477
Parasites (of ABSMs), 337
Parker, W. H., 286
Pasang Nyima, 262
Patagonia, 23, 156, 167, 173, 471
Peacock, Congo, 182, 195
Peissel, Michael, 279, 433
Pei Wen-Chung, Dr., 370
Pekin (China), 63, 423
Pekin Man, 63
Pennsylvania, 111
Pen-Tails, 213
Perak (Malaya), 229
Pereira, Gen., 274
Periodicticus, 209
Perkins, Marlin, 483
Persia (Iran), 23, 237, 286, 296, 297, 298, 392
Peten (Guatemala), 151, 157, 158, 165
Peter of Greece, Prince, 261, 457
Peter the First Range (Pamirs), 310
Phalut (India), 8, 260
Philippine Islands, 211, 213, 215, 253, 256, 377, 394, 395
Philology (of ABSMs), 453
Piccinelli, Corrado, 489
Pich, Audio L., 388
Pigmies, 74, 120, 137, 166, 187, 188, 190, 191, 192, 194, 206, 207, 208, 211, 213, 215, 254, 329, 343, 347, 349, 356, 367, 368, 377, 384, 385, 386, 387, 388, 395, 397, 411, 449, 470, 476
Pigmy Chimpanzee, 186
Pigmy Gorilla, 186
Pika [Whistling Hare], 337, 338, 340
Piltdown Man, 62, 63, 353, 354, 363 364, 365, 386
Pithecanthropines, 20, 211, 217, 327, 353, 362, 368, 371, 374, 395, 476
Pithecanthropus, 214, 353, 371, 476
Pithecanthropus erectus, 353
Pithecanthropus pekinensis, 353

Pithecanthropus robustus, 353, 386
Pithecoids, 477
Pleistocene, 142, 363, 386
Plesianthropus, 205, 208, 353, 476
Pliopithecus, 353
Point, William, 75
Polynesians, The, 395
Pondaungia, 353
Pongids, 20, 27, 89, 141, 186, 203, 204, 207, 219, 240, 268, 290, 341, 347, 353, 361, 371, 373, 374, 388, 466, 473, 476, 477
Porcupines, 53, 138, 336, 338, 473
Porshnev, Prof. B. F., 15, 292, 299, 300, 301, 307, 309, 446
Port Douglas (B.C.), 30
Portland (Ore.), 140
Portuguese East Africa, 191
Potaro River (B.G.), 180
Potatoes, 380
Pottos, 209
Powell River (B.C.), 53
Pranavananda, Swami, 268, 458
Presbytictis avunculus, 239
Presbytis, 225, 226
Primates, 47, 58, 61, 89, 120, 169, 178, 190, 211, 213, 239, 240, 290, 303, 321, 336, 340, 341, 342, 361, 410, 427, 447, 466, 467
Prjewalski, N. M., 306
Proboscis Monkey, 214, 271, 477
Procapra gutterosa, 480
Procapra picticaudata, 480
Proconsul, 353, 364, 476
Profelis concolor, 339, 473
Pronin, Dr. A., 19, 309
Propliopithecus, 353
Proto-Pigmies, 212, 357, 359, 360, 368, 470, 476, 477
Protosimioids, 477
Pseudovis nahura, 481
Ptilocercus, 213
Puerto Barrios (Guatemala), 163
Puget Sound, 49, 100, 103, 121
Pugets, The, 46
Pulu-Rimau (Sumatra), 221
Puma (*Profelis concolor*), 339, 473, 478
Punjab, 274
Punta Gorda (B.H.), 164

Scalps (fur caps), 279, 309, 332, 333, 335, 478, 483, 487, 488
Scatology—*see Excrement*
Schiltberger, Johann, 306
SCIENCE SERVICE, 422
Screams (made by ABSMs)—*see Calls*
Sechelt Inlet (B.C.), 57
Sedabo, 216
Sedapa, 211, 214, 215, 216, 217, 218, 219, 220, 221, 222, 224, 328, 349, 357, 360, 368, 399, 454, 469, 470, 476, 477
Seguela (Guinea), 189
Séhité, 190, 194, 357, 360, 368, 453, 476
Selenarctos, 250, 480
Semang, The, 211, 227, 395
Semitic Peoples, 383, 389
Semnopithecus, 226, 250, 253, 480
Semnopithecus entellus, 480
Semnopithecus schistaceus, 480
Senegambia, 416
Senoi, The, 212, 217, 393
Sen Tensing, 262
Serology, 336, 484
Serows, 279, 335, 478, 480, 485, 487
Shan States (Burma), 242
Sheep, 340
Sheep, Mountain, 289
Sheidim (Seirim), 379
Shensi-Babies, 339
Shensi Prov. (China), 325
Sherpas, The, 10, 212, 257, 264, 265, 267, 268, 269, 278, 341, 433
Shigatse (Nepal), 333
Shipton, Eric E., 13, 14, 261, 262, 294, 306, 433, 434, 444, 468, 474
Shiru, 166, 172, 328, 357, 359, 360, 369
Shmakov, Dr. A. A., 292, 299
Shushwap (B.C.), 36
Siak River (Sumatra), 219, 220
Siamangs, 186, 213, 214, 218, 219, 220, 226, 476, 477
Siamang, Pigmy (Lesser), 186, 213

Siamang, South Pagi Island Pigmy, 213
Siberia, 283, 304, 312, 326, 327, 372, 392, 412, 454
Sick, Helmut, 155, 173
Sierra de Chuacus (Guatemala), 151, 159
Sierra de Perijaá, 172, 180
Sierra Madre Occidental (Mex.), 151, 152
Sierra Madre Oriental (Mex.), 151, 152
Sierra Nevada, 122, 338
Sikang (China), 243, 257, 268, 269, 273, 312
Si-Kiang (China), 239
Sikkim, 2, 4, 5, 237, 260, 265, 457
Simbiti Forest (Tanganyika), 190
Simias concolor, 213
Simioids, 466, 477
Simpai (Monkey), 226
Simukov, A. D., 308
Sinai Peninsula, 379, 380, 383, 391
Sinanthropus, 371, 374
Sindai, 225
Sinkiang (China), 300, 304, 312
Sinkiangese, The, 232
Sisemite (Sisimici), 160, 162, 348, 358, 370, 476
Siskiyou Mts., 113, 116
Sivapithecus, 353
Skousen, Wendell, 159, 160
Slavs, The, 389
Slick, Thomas B. (and Expds.), 17, 263, 264, 265, 269, 338
Slossen, Edwin E., 422
Sloth-Bear, 250, 275, 334
Smell (of ABSMs)—*see Stinks*
Smithsonian Institution, 111
Smith Woodward, Prof. Sir Arthur, 363
Smythe, F. S., 261
Snaith, Stanley, 260
Snake River (Idaho), 104
Snowfields, Montane, 401
Snow-Leopard, 254, 289, 301, 334, 434, 480
Snub-nosed Monkeys, 239, 271, 434, 480

[523]

Tensing Norgay, 262
Tête Jaune Cache (B.C.), 76
Texas, 403
Thailand (Siam), 237, 238, 242,
    253, 394
Thorberg, Aage, 262
Thorold's Deer, 273
Thunderface, Chief Michael J.,
    242
Thyangboche (Nepal), 262, 335,
    486
Tibet, 3, 4, 7, 11, 20, 129, 235,
    237, 243, 251, 254, 255, 256,
    257, 258, 265, 268, 271, 273,
    274, 275, 276, 283, 308, 311,
    312, 321, 322, 324, 326, 327,
    345, 359, 360, 373, 394, 406,
    407, 442, 449, 454, 457, 458,
    475, 478, 485
Tibetan Rim, The South, 285, 301,
    304, 311, 317, 325
Tibetans, The, 257, 265, 271, 275,
    278, 283, 285, 302, 322, 324,
    333, 343, 371, 454, 456, 458
Tien-Shan Mts., 23, 286, 300, 312,
    317
Tierra del Fuego, 100, 177, 372,
    406
Tigers, 213, 242, 264, 289, 297
Tilman, H. W., 261
Tlyaratin Reserve (U.S.S.R.), 288
Toba Inlet (B.C.), 50, 53
Toirano Cave (Italy), 318, 349,
    467, 468
Tok, 242, 243, 254, 268, 275, 326,
    358, 360, 369, 371, 454, 476
Toké-Mussis, 47, 117, 119, 135
Tokyo University, 334, 486
Tombazi, A. N., 118, 260
Tonkin (Viet Min), 239
Trachypithecus, 225, 226, 250
Tracks (of ABSMs)—see Foot-
    prints
Trans-Caucasia (U.S.S.R.), 288
Trichocephalidae, 340
Trichology—see Hairs
Trichuris, 340
Trinity Mts., 113
Trollak [Trolek] Reserve (Ma-
    laya), 229

TRUE MAGAZINE, 21, 93, 100,
    143
Tschernesky, Dr. W., 347, 374
Tsin-Ling Mts., 23, 286, 311, 325,
    358, 373, 407, 471
Tumuc-Humuc Mts. (Surinam),
    413, 414, 415
Tupaioids, 477
Turfan Desert, 318
Turkey, 298, 302
Turkomen S. S. R., 286, 298
Tutkaul (Pamirs), 311
Twain, Mark, 114

Uatumã River (Brazil), 175
Ubangi-Shari, 183, 207
Udnabopithecus, 353
Uele (Congo), 192, 343
Ufiti [Fireti], 195, 198, 202, 476
Ukumar-Zupai, 388
Ulan Bator (University of), 308
Uliasutan (Mongolia), 320
Uli-Bieban, 462
ULTIMA HORA (of Rio de
    Janeiro), 388
U.S.S.R. Acad. of Sciences, 292,
    299, 307
U.S.S.R., The, 74, 237, 281, 284,
    298, 299, 302
United States (of N. Am.), The,
    85, 91, 119, 232, 283
Un. of Penna., Museum of, 160
Ural Mts., 283
Ursus, 251, 339
Ursus arctos, 251, 339, 473, 480
Ursus isabellinus, 251, 480
Ursus arctos pruinosus, 251, 334,
    480
Ursus pruinosus, 251, 480
Urubú River (Brazil), 175
Urumchi, 324
Ussure Forest (Tanganyika), 190
Usumacinta River (C. Am.), 157
Utah, 37
Uzbeks, The, 299, 300

Vancouver, 33, 49, 50, 57, 71, 79
Vancouver Island, 34, 35, 47, 49,
    50, 53, 60
Vancouver Public Museum, 70
Van Gin Shan Mts. (China), 239

Van Heerwarden, R. B., 221, 223
Vasitri, 171
Veddas, The, 395
Vegatology, 62, 87, 356, 389, 402, 407, 418, 419, 442
Venezuela, 179
Vera Cruz (Mex.), 25, 42
Vernon (B.C.), 79
Verrill, Prof. A. E., 279
Victor, Charlie, 38, 72
Victoria (B.C.), 25, 29
Victoria Memorial Mus. (Salisbury, Fed. of Nyasaland and Rhodesia), 197
Viet Min, 237
Viet Nam, 237, 394
Vitamins, 266
Voita, 313

Waddell, Major L. A., 1, 260
Walker, Capt. Joseph, 109
Wallace's Line, 215, 253, 395, 406
Wallace, Ray, 124, 130, 133
Wallace, Wilbur, 130
Washington State, 21, 30, 34, 46, 100, 146, 338, 417
Water-Civet, 182
Wavrin, The Marquis de, 167
Waw-Mooch (Frog), 342
WEEKEND MAGAZINE, 29
Weidenreich, Prof. F., 370
Wembare Plains (Tanganyika), 190
Werewolves, 228
West, The (of N. Am.), 99, 108
West Africa, 23, 165, 249, 303, 357, 375, 416
West Africans, 164, 205
Westenek, L. C., 217, 220
Westminster (B.C.), 39
West Virginia, 37
White Sea, 286
Wilkerson, Starr, 105
Willamette Valley, 140
Willow Creek (Cal.), 104, 123, 124, 128, 131, 137
Windhoek (S. W. Africa), 202, 205

Wind-Man, 19, 291
Wisent [European Bison], 289
Wisdom River (Idaho), 105
Wolofs, The, 49
Wolves, 254, 289, 434, 480
Wong Yee Moi, 229
Wood, Mrs. Ida P., 200
Wood, W. W., 262, 274
World Book Encyclopedia Co., 483
Worms, Parasitic, 337, 340, 341
Wow-Wow Gibbon, 218
Wrexham Museum, England, 37
Würzburg Fossils, The, 354
Wyss-Dunant, Dr. E., 14, 262

Xenopithecus, 353

Yablonovoi Mts., 312
Yaks, 212, 480
Yale (B.C.), 23, 29, 38, 65, 79
Yangtse-Kiang River, 41, 406, 407, 413
Yaquis, The, 152
Yavan-Adam, 462
Yazoo River, 417
Yells (made by ABSMs)—*see Calls*
Yetis, 138, 248, 264, 265, 269, 310, 332, 333, 342, 462, 478, 483, 484, 486, 487
York, Cape (Australia), 416
Yucatan, 31, 152, 342
Yukon, 21, 23, 30, 46, 327, 392
Yunnan (China), 239, 241, 257, 360, 369
Yuroks, The, 30, 47, 103, 119, 134

Zdorick, B. M., 310
Zemu Gap, 261, 262
Zeren, 480
Zerleg-khoon, 462
Zinjanthropines, 362, 476
Zinjanthropus, 89, 90, 186, 353, 386, 476
Zomba (Nyasaland), 195, 198
Zoo. Soc. of London, 217, 426
Zurich (Switzerland), 186